Design Solutions for Wireless Sensor Networks in Extreme Environments

For a complete listing of titles in the
Artech House Integrated Microsystems Series,
turn to the back of this book.

Design Solutions for Wireless Sensor Networks in Extreme Environments

Habib F. Rashvand

Ali Abedi

ARTECH HOUSE

BOSTON | LONDON

artechhouse.com

Library of Congress Cataloging-in-Publication Data
A catalog record for this book is available from the U.S. Library of Congress.

British Library Cataloguing in Publication Data
A catalogue record for this book is available from the British Library.

Cover design by John Gomes

ISBN 13: 978-1-63081-177-8

685 Canton Street
Norwood, MA 02062

10 9 8 7 6 5 4 3 2 1

We dedicate this book to our families for their support and encouragement.

Contents

Acknowledgments

We would like to thank Dr. Abolfazl Razi for his contributions to Chapters 5 and 8; Dr. James Agajo for his contribution to Chapter 9; and Dr. Juan A. Fraire for his contribution to Chapter 13. Professor Rashvand appreciates the professional consultations from Mr. John Nuttall and helps from the Advanced Communication Systems staff. We are also thankful to anonymous reviewers whose comments significantly helped in improving the quality of this book. Last but not least, we sincerely thank Artech House editors, proofreaders, and publishing staff for their help, support, and patience throughout this process.

Chapter 1

Introduction to Wireless Sensing in Harsh Environments

Sensing in harsh or extreme environments is critical for the implementation of highly efficient and reliable systems. The more information that is collected from the environment, or the system performance metrics itself, the more reliably a system can perform. Examples of such sensing applications include: vibration or chemical sensing of jet engines, temperature sensing inside a turbine, and monitoring the wind direction in a wildfire. In the past century, most sensing systems were designed using wires. The design and maintenance of wired systems is cumbersome and costly, and may even be impossible in harsh environment scenarios.

Wireless sensing on the other hand promises to deliver a low-cost and low-weight solution that presents added flexibility and new features. The elimination of the need for wiring, bundling, cabling, and adding fixtures to hold cables in place saves a significant amount in capital expenses, while reducing the need for maintenance in the future due to wires breaking or aging. Wireless sensors can be easily moved around, and new ones can be added as needed. Adding a single new wired sensor to an existing system requires design, routing, crosstalk analysis, and of course added cost.

The first step in designing wireless sensors for harsh environments is a precise and accurate requirement analysis. If this step is not completed correctly, with all aspects of the problem considered, the final design might not perform as expected or the cost might be more than necessary.

The requirement analysis starts with the input-output specifications of the sensing system itself and continues with environmental conditions analysis to ensure operations are uninterrupted. One angle is the functionality of the sensor itself and how it can detect a specific parameter, while the other angle is to consider if the material used for building the sensor can withstand the harsh environmental conditions.

This chapter presents an overview of the basics of wireless sensing and reviews the preliminary concepts required for understanding the rest of this book. We will start with an overview of sensing systems, continue with radio and network concepts review and conclude by providing a summary of the other chapters and organization of the book. A review of this chapter will help with the readability of other chapters, while helping readers to decide which chapters in the book are relevant to their applications and interests.

It is noteworthy to mention that basic knowledge of electrical engineering and communications systems and networks will be helpful for understanding this chapter and the rest of this book. Concepts such as circuit analysis, filtering, digital signal processing, and transceiver design should be reviewed prior to reading this chapter.

1.1 INTRODUCTION

Wireless sensor networks are comprised of three main parts: sensor, microprocessor, and radio. The sensor itself may include multiple blocks such as sample concentrator, front end, transducer, and interface. Some sensors may need their own power source (heating an element to absorb certain material) or operate using the power supplied through the interface wires.

The microprocessor has input channels that may be digital or analog with some having analog-to-digital converters (ADCs) of their own and some requiring external ADCs. The role of the microprocessor is to sample sensor data and send it to the radio for transmission. This unit serves as the brain of the system, and may be used for efficient power management, duty cycling to put the system to sleep, and interacting with other interfaces for monitoring or programming.

Last but not least, the radio unit includes the source coder, channel encoder, modulator, antenna, and diplexer for both transmission and reception of data. The radio unit might be the most power-hungry unit in the whole system.

In Chapter 2 all these different units are studied and their subsystems are studied.

1.2 SENSOR SYSTEM

A sensor system typically includes a sample concentrator, front end, transducer, and an interface as illustrated in Figure 1.1.

Depending on the sensitivity of the sensor, and the amount of the chemical or biological sample in the air or water that is being sampled, a concentrator may be required to bring the sample density to a level detectable by the sensor. The front end, or sensing element, is the material designed to be sensitive to specific molecules. For example, if a thin film is designed and deposited on top of a substrate so the material in the thin film can form bounds with specific molecules and change its properties, we can measure the concentration level of those molecules present in the air.

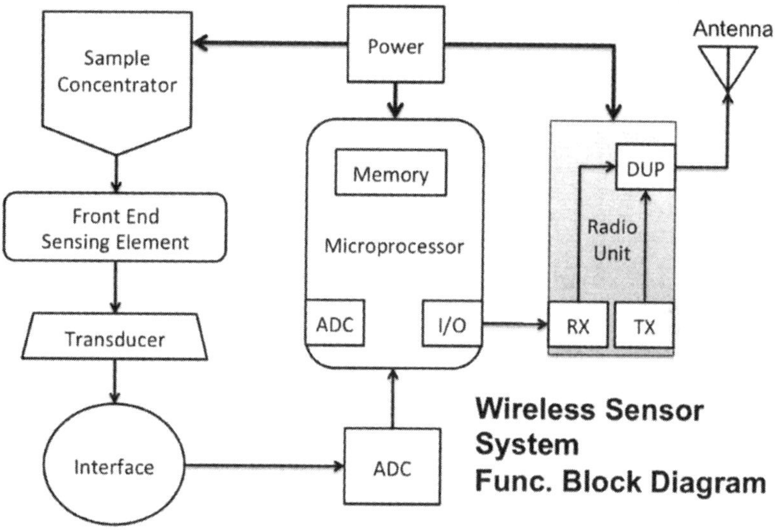

Figure 1.1 Wireless sensor system functional block diagram.

The properties changed in the thin film or substrate are then measured and changed to electric voltage or current using the transducer unit. The interface block is used to connect the transduced signal to the microprocessor of the sampling ADC. In some cases, these blocks may be combined into one circuit that does both the transducing and interfacing.

The next step is to determine how to connect the sensor to the microprocessor. If the sensor has a digital output and the microprocessor supports digital inputs, a direct connection may seem feasible. However, it is always a good idea to use buffer circuits in between to prevent accidental shorts and to protect the microprocessor.

For sensors with an analog output, if the ADC inside the microprocessor has adequate resolution and speed, the sensor can be connected there. In some cases, we may need to use external ADCs to provide higher resolutions or speeds.

1.3 MICROPROCESSOR UNIT

The microprocessor unit (also called the microcontroller) consists of internal memory, ADC, input-output (IO) interface, and programming or arithmetic operation capabilities as illustrated in Figure 1.1.

Once the data is inside the processor, either using external or internal ADC, it is sampled at predefined intervals as determined in the program. The data is then transferred to the radio.

All the other programming functions of the system, such as when to turn the sensor on or off and how often to transmit data to the radio and the overall power management of the whole system, is done using the microcontroller.

1.4 RADIO UNIT

The radio unit has a receiver (RX), transmitter (TX), and a duplexer (DUP) to use a single antenna for both RX and TX units. Both TX and RX units are comprised of a source coder/decoder, channel coder/decoder, and modulator/demodulator units as depicted in Figure 1.2.

The source coder block removes redundancy in the data and compresses it to avoid unnecessary power consumption or transmitting redundant data. Widely used algorithms such as JPEG for pictures, MPEG for video, and ZIP for data files are among many examples of efficient data compression methods. The basic principle of the data compression algorithm is to assign fewer bits to more frequent symbols rather than uniform bit assignment.

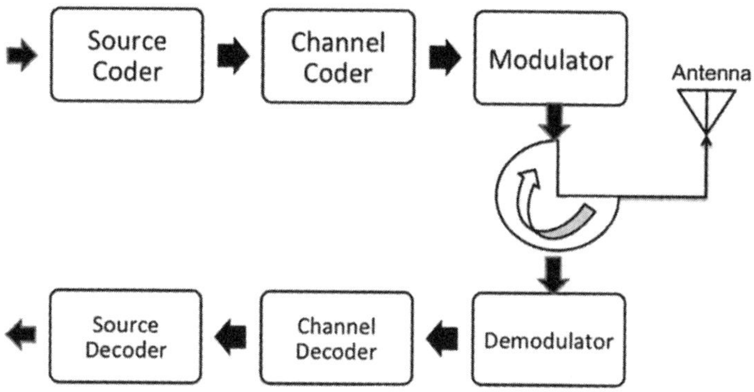

Figure 1.2 Radio unit block diagram.

Example: Consider the following five symbols representing the temperature of a sensor. What is the required number of bits if uniform bit assignment is made to digitize these symbols? What is the average number of bits if a shorter sequence of bits is assigned to the more frequent symbols? (Hint: Huffman coding may be used as an optimal data compression method.)

Table 1.1

Example

Symbol	Temperature	Probability
A	+15	0.1
B	+5	0.2
C	0	0.5
D	-5	0.15
E	-15	0.05

Solution: Using uniform binary coding, in order to present five symbols, we need 3 bits of information. Although, 3 bits can represent eight symbols (e.g., 2^3) we only use five of them.

Using Huffman coding, we need to sort the symbols from high to low probability and build the Huffman coding tree (see Figure 1.3) starting from two symbols with the lowest probabilities. The new combined symbol will have the sum of its component probabilities. The codes are then assigned based on the number of hops from symbol to root (last symbol with probability of 1) with 1 representing upper branches and 0 representing lower branches.

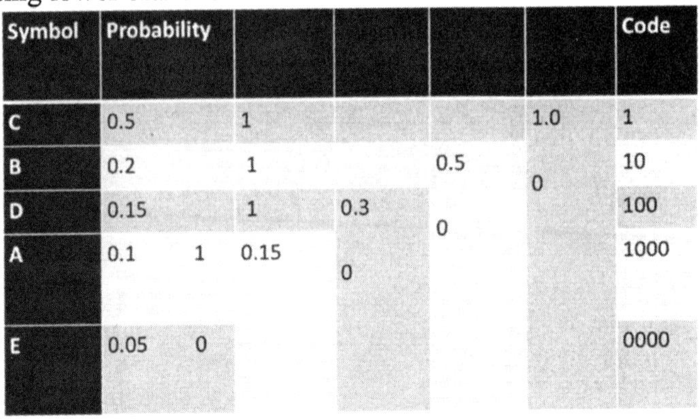

Figure 1.3 Example Huffman coding.

The average code length (see Table 1.2) or number of bits required can be calculated by taking the weighted average of code lengths as follows, which totals to 1.75 bits. This is equivalent to a 42% savings in transmission power.

Table 1.2

Example Average Code Length

Symbol	Probability	Code	Code Length	Average Code Length
C	0.5	1	1	0.5
B	0.2	10	2	0.4
D	0.15	100	3	0.45
A	0.1	1000	4	0.4
E	0.05	0000	4	0.25
			Total	**1.75**

The compressed data, as one can imagine, will be very sensitive to errors since there is no built-in redundancy to correct for flipped bits during transmission. The channel coding block will add some artificial redundancy to alleviate this problem. Of course the added redundancy is always kept to minimum, in comparison to the original built-in redundancy. A simple channel coding method is called block coding, where a sequence of bits is mapped to another sequence with a few extra bits as parity. Hamming code is one of the widely used block codes. Hamming (7,4) code maps every 4 bits of data to 7 bits of data and parity by multiplying the data vector and Hamming generator matrix.

$$G = \begin{bmatrix} 1 & 1 & 0 & 1 \\ 1 & 0 & 1 & 1 \\ 1 & 0 & 0 & 0 \\ 0 & 1 & 1 & 1 \\ 0 & 1 & 0 & 0 \\ 0 & 0 & 1 & 0 \\ 0 & 0 & 0 & 1 \end{bmatrix}$$

The resulting 7-bit codeword is then forwarded to the modulator for transmission. At the receiver side, once a new 7-bit word is received and demodulated, syndrome decoding will be used to decode the received vector and extract the transmitted 4 bits. Small amounts of error (i.e., 1 or 2 bits of error) can be corrected using this method. More powerful channel coding schemes such as low-density parity check codes or turbo codes may be used if higher error protection is desired.

The final block in the radio unit is the modulator and demodulator at the transmitter and receiver blocks. These two are connected to the same antenna using a duplexer to avoid high transmission power leaking back into the receiver (i.e., self-jamming). The role of the modulator is to translate the baseband low-frequency signal to higher frequencies for long-range communication. The demodulator, on the other hand, will translate the high-frequency signal back to baseband for detection. Some widely used modulation schemes are amplitude or frequency modulation (AM or FM). Data communication is often conducted using digital modulation as opposed to analog methods (AM or FM). Digital modulation methods such as phase shift keying (PSK) or quadrature amplitude modulation (QAM) are widely used in a variety of data communication scenarios.

1.5 HARSH ENVIRONMENTS

In this section, the main implications of harsh environments on the three main components of a wireless sensor system are studied. Starting with the sensor device itself, the main challenge is its material properties and its capability to withstand very high or very low temperatures. If the material is not capable of handling those harsh environments, certain types of coating, or even packing, may help in some cases. A harsh environment is not necessarily limited to extreme temperature fluctuations; chemical gas vapors, or operations in liquids may also pose deteriorating effects on the sensor device. Some material options and design considerations are discussed later in this book.

Battery and power source placement remains a challenge, even if the sensor is designed for a harsh environment; there are not many choices for batteries that can still work effectively in very hot or cold temperatures. Some solutions to this problem are presented later in the book using passive sensing technology.

The radio unit faces a different set of challenges in electromagnetically harsh (i.e., highly reflective mediums) environments. In other words, the challenges of temperature or corrosive gas are not the only issues to be considered. Reflections and scattering of radio signals and higher noise levels are additional issues to be considered in harsh environments, inside reflective enclosures, and in industrial settings. More discussions are included later in the book to answer these questions.

The microprocessor is perhaps the most sensitive part of the whole system and needs to be cooled or heated (depending on the environment) to maintain its performance. In some applications, the sensors may be placed in different locations, and wired to the microprocessor, which has to be further away from heat source or shielded from gas or liquids. Hence, a hybrid wired/wireless method may be the only solution in some cases.

1.6 BOOK ORGANIZATION

The rest of this book is organized as follows. Chapter 2 covers throughput and interference trade-off for wireless sensor networks operating in harsh and challenging environments. Boundary conditions and limiting factors that restrict environmentally sensitive applications using conventional interference management methods in such settings are analyzed. Wireless sensing requirement analysis in challenging environments from energy, spectrum management, and maintenance point of views are presented. Practical approaches to interference management are also presented.

Chapter 3 focuses on sensor resiliency and longevity by presenting recent developments in material science, sensor systems, and networking science and engineering that contribute to the longevity of sensor systems in harsh environments. System reliability and hazard rate are defined and several examples with solutions are provided for more clarity.

Design considerations and major challenges for the practical deployment of wireless sensing systems in extreme environments such as space and under water are studied in Chapter 4. In space applications, radiation tolerance is another major issue, while in underwater situations waterproofing is a major challenge. Hardware and electronics deployed in harsh industrial environments need to be designed under a completely different set of boundary conditions as compared to common electronic devices that are used in our day-to-day activities on earth. All these issues are presented with examples to reinforce the concepts.

Material science perspective is presented in Chapter 5 to enable the implementation and use of passive wireless sensors in harsh environments. Passive sensor modeling and design from material science and signal processing perspectives are presented in this chapter.

Chapter 6 discusses the principles, technical properties, and potentials of beamforming as a natural property of the wireless sensor network (WSN). In doing this it examines various technical aspects of the system/method including arrayed antennas, beam patterns, and the importance of directivity. Looking for the energy and high propagation problems of WSN, it then leads us to the greatly untapped potential of beamforming, which is unique to wireless sensor networks. The fundamental property of WSN beamforming stems from the fact that if nodes of the network are able to collaborate as a cluster or any other discipline to share their propagation signaled synchronized together, then they can dynamically change their overall propagation pattern to direct very high energy (lobe) towards a particular direction (e.g., required base station) while avoiding sending virtually zero energy (null) to nearby friendly nodes. In high traffic WSN networks, combining the beamforming features with multi input-multi output (MIMO) technology can provide a new high performance platform for a new superior large-scale WSN networking paradigm.

Chapter 7 presents overlay technology. Stemmed from traditional network management, overlay as a concept is maturing steadily while going along its development moving through strength to strength while being distracted by the surge of internet and its P2P service provision. In order to clarify the landscape we define four overlaying techniques: (1) traditional supervisory overlay monitoring (SOM), (2) lightweight mission overlay management (MOM), (3) service network overlay (SNO), and finally (4) network management overlay the stages of development. Although MOM and SNO have been quite popular with internet services we recommend the final stage of development network management overlay (NMO) to be deployed through a new algorithm called dynamic network management (DNM). This dynamic overlaying technique features new properties to enhance new range of practical sensor rich applications suitable than ever the extreme environments, including dynamic early warning systems (DEWS).

Chapter 8 continues the discussion with the focus on a specific variant of wireless sensor networks, where data sources are located in hard to reach areas and therefore additional challenges and issues arise. This covers an important scenario of field monitoring under extreme environmental conditions. The concepts of joint source and channel coding and distributed coding of correlated sensor data are presented and complemented with detailed explanations and examples.

Chapter 9 presents a new investigation of a wide range of technological advancements and achievements for using smart sensors in the form of lightweight UWSN applications. The potential for new opportunities are huge, but unfortunately hampered by commonly available low bandwidths acoustic propagation technology. Examining far end models acoustic signals, which can be enhanced enabling new applications through remote sensing, networking, and use of underwater autonomous vehicles (UAV). They, in conjunction with new gateways, buoys, and other offshore platforms make scientific and geological explorations safe and enjoyable. UWSN also provides new industrial opportunities for new, vast, and available oil and gas resources located on the seabed and ocean floors, where using smart sensor technologies make many new operations both feasible and safe.

Chapter 10 presents an intensive investigation of using WSN to enable new applications in two different dimensions of (1) safeguarding ourselves from the crust related disasters (e.g., earthquakes) and (2) empowering underground sensing for industrial and agricultural productivities. The first involves extensive use of seismic sensors (geophones) and building global scale databases for understanding the UG, the earth and the Pangaea theory (GSTM). The second part is for improving the safety aspect of mining using WUSN technologies, and then enhancing agricultural industries using WSN to find better propagation methods including electromagnetic radio, wideband seismic, and magnetic induction. Then upon a promising medium range ad hoc wireless underground sensing (WUSX) future innovations in agile and fast reacting smart sensors can make better use of all available propagation technologies.

Chapter 11 discusses the possibility of wide range of innovative production processes, products, and service provision through the WSN in harsh environments such as production and transportation. We take the analysis through various aspects of development and design from the innovation point of view so that the needs and requirements have been analyzed and discussed prior to setting up the projects. Taking the production process through recent developments of I^2oT and IWSN we examine various extreme environments unique design and developments methods to enable many upcoming industrial WSN paradigms.

Space applications of wireless sensor networks are studied in Chapter 12. A comprehensive view of wireless sensor systems applications in space explorations based on the NASA technology roadmap is presented. Various technology areas to guide technology development for space explorations are studied, and opportunities for wireless sensor systems are identified.

Chapter 13 presents a complementary device related compilation of samples for sensors, devices, and systems to complement other Chapters in the book. In doing that we propose a new coding method for classification of the items related and for the extreme environments. Finally, we propose a modular system design for integrating successful developments as well as a requirement for a new family of processors to enable low cost and low energy WSN services for the extreme environments.

Chapter 2

Networking Wireless Sensors in Challenging Conditions Throughput and Interference Trade-offs

This chapter presents recent developments in throughput and interference trade-offs for wireless sensor networks operating in harsh and challenging environments. Boundary conditions and limiting factors that restrict environmentally sensitive applications using conventional interference management methods in such settings are analyzed.

Several real world examples from space-related extreme environments are presented. Wireless sensing requirement analysis in challenging environments from the energy, spectrum management, and maintenance point of view are presented.

Finally, throughout this chapter we examine some existing and up-coming practical approaches on how to *manage interference* and to present practical solutions to design efficient networks with acceptable *throughput performance* working under specific harsh conditions.

This chapter does not cover electronic hardware issues, materials science behind designing sensors, and batteries operating in harsh environments. Refer to Chapters 3, 4, and 5 for detailed discussions of these related topics.

2.1 INTRODUCTION

Wireless sensor networks promise high spatial resolution when it comes to sampling measureable parameters of interest working under different conditions being in one or a mixture of physical, chemical, or biological environments, much like advanced ADC techniques enabling high temporal resolution outputs.

Often, understanding the changes in a scalar field and determining if an event occurred is of high importance to a variety of applications [1]. Typical examples include tracking wildfires in forests, monitoring oil spills in the ocean, tracking biological or chemical gases spread over a region, predicting storm landings close to population centers, and monitoring space habitats.

In all these examples a large number of samples from as many locations as possible as fast as needed are required to map the field of interest. Looking at the wildfires example, temperature is the parameter of interest describing a scalar field. Predicting how the fire will spread and its future direction is critical in planning evacuations wisely, not to overwhelm roads with unnecessary traffic (see Figure 2.1).

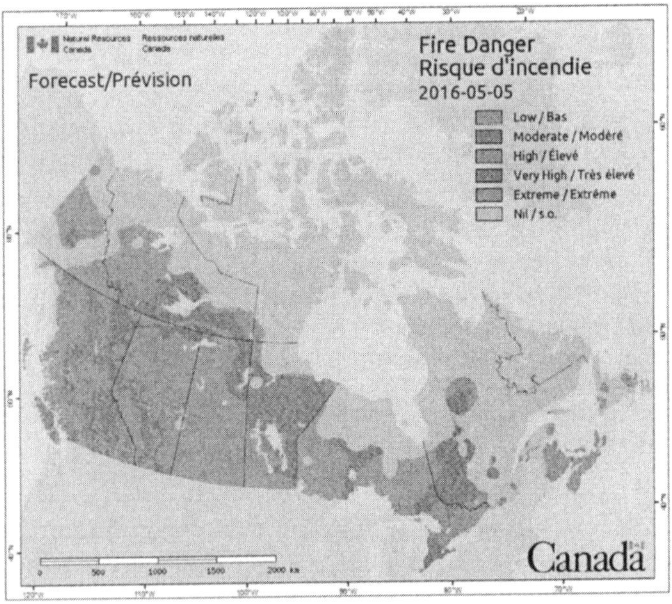

Figure 2.1 Canadian Wildland Fire Information System forecasted fire danger map for 2016-05-05. Reproduced with the permission of the Canadian Forest Service, Natural Resources Canada, 2018.

Figure 2.2 Researchers at UM WiSe-Net Lab testing sensors inside the inflatable lunar habitat.

Another example is monitoring homes for energy efficiency or even space habitats for life support. Figure 2.2 shows the inside view of NASA's inflatable lunar habitat, a 42-ft-diameter concentric torus that is built by NASA and instrumented by the University of Maine's WiSe-Net laboratory researchers with 124 passive and 48 active sensors. These sensors monitor the structural integrity of the habitat. They localize impact, find leaks, and track temperature and humidity.

It is clear that making useful decisions based on data in all these examples requires a large number of samples. Increasing spatial and temporal resolution is equivalent to increasing data throughput in the network, particularly when data becomes closer to the fusion center (in centralized algorithms) or to the cluster heads (in distributed algorithms) [2–4]. The question is how much data is enough and when we should stop increasing the resolution or sampling rate. An obvious by-product of high throughput in a wireless network is increased interference, which is going to be a critical limiting factor determining the upper bound on the throughput.

Interference management in infrastructure-based wireless networks has been well studied and many elegant solutions such as interference alignment have been proposed [5, 6]. However, noting the limited energy, spectrum, and computational capabilities of wireless sensor networks in harsh and challenging environments, more research and development in interference management is required.

This chapter presents recent developments in throughput and interference trade-offs for wireless sensor networks operating in harsh and challenging environments, which pose yet another set of boundary conditions and limiting factors on what can be done in such settings [7, 8].

The rest of this chapter is organized as follows. Section 2.2 presents the boundary conditions imposed on a wireless sensor network while operating in harsh and challenging environments. Throughput requirements and trade-offs with interference, which relate to simple receiver design are discussed in Section 2.3. Interference in wireless sensor networks and management strategies in challenging and extreme environments are presented in Section 2.4. The chapter concludes in Section 2.5.

2.2 WIRELESS SENSING IN CHALLENGING ENVIRONMENTS

The boundary *conditions* imposed on a wireless sensor network in harsh and challenging environments determine the path forward on developing effective throughput and interference management methods. In this section, these limitations and their effects on relevant typical methodologies are discussed. The requirements are categorized into three major topics: energy, spectrum, and maintenance. Some design solutions, appropriate to extreme environments, are provided to serve as guidelines for practicing engineers.

2.2.1 Energy Requirements

The first and foremost challenging requirement for harsh environments is access to an energy source in order to operate an electronic circuit. Battery operation in very high or very low temperatures may be problematic to say the least. Frequent battery replacement is also costly and often cumbersome. The solution therefore is either running wires from the power source (in controlled environments) to sensors in harsh environments or developing batteries that can withstand harsh environments. Depending on how we define harsh environments, recent developments in battery design provide solutions that can withstand temperatures from -40 to 85 degrees Celsius (C) (Tadiran Co., NY, USA) or even up to 125 degrees C (SAFT America Inc., NC, USA).

In addition to energy requirements and battery failures, other electronic components of the system are prone to failure working in harsh and extreme environments. Recent development in high-temperature circuitry is another interesting solution for specific temperature ranges [9]. For example, electronic circuits are available that can operate in temperatures as high as 400 degrees C (NASA GRC, OH, USA). However, as temperature requirements of applications get wider, available solutions get narrower. Running wires to sensors is another solution that has been used in furnaces and ovens based on thermocouple principles for years. Transmitting power wirelessly to the sensors in harsh environment is in fact a new solution and under research and development in recent years.

Some examples of wireless power transfer include the recent development of RF power harvesting chips (Powercast Co., PA, USA) for longer range applications or the use of near-field magnetic coupling (NFC Standard) in shorter range applications when accessing inside a chamber is not possible with destructive protrusion.

Another example is the battery-less solution based on remote powering of sensors in [10]. As seen in Figure 2.3, a passive or battery-less sensor can be read using a field-programmable gate array (FPGA) based *interrogator* acting as both transmitter and receiver. The interrogator consists of a synthesizer, FPGA, amplifier, RF switch, and filters. The FPGA board creates a spread spectrum signal from the synthesizer that is matched to a specific sensor code. This signal is then amplified and transmitted through the RF switch to the sensor. The sensor reflects a modified signal back that contains changes, which can be quantified based on temperature change. At this point the RF switch creates the receiving path towards the filter, amplifier, and signal processing unit that can also exist in the FPGA or perform offline using a MATLAB embedded emulating program. This is an example of passive sensors that are capable of addressing energy needs at the sensor level without having a battery present.

Other examples of the energy requirements for a wireless sensor include wireless power transfer using highly focused microwave beam as utilized in the space solar power research community; RF energy harvesting from existing RF signals in the environment; and optical power transfer using high-intensity sun-like light sources and solar arrays. Each of these methods is extensively studied in the literature and has its own benefits and shortcomings.

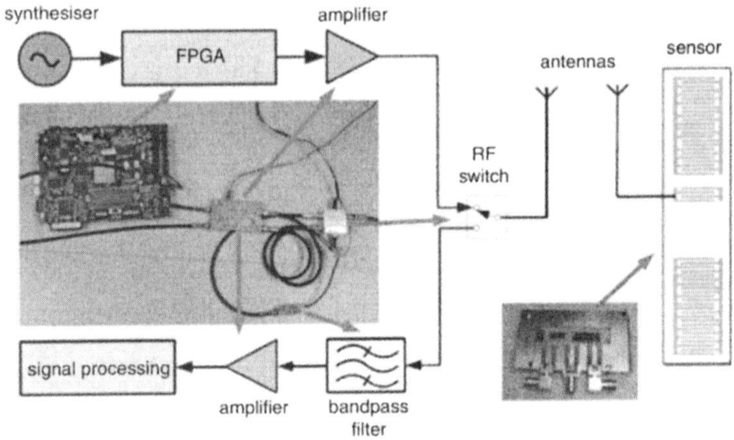

Figure 2.3 Block diagram of a battery-less sensor system developed at UMaine [13]. Reproduced by permission of the Institution of Engineering & Technology.

2.2.2 Spectrum Management

Applications requiring high spatial resolution such as structural monitoring impose a large demand on spectrum availability in the sensor network. In harsh environments such as hard to reach areas or confined metallic spaces, spread spectrum techniques are often used to combat interference and multipath reflections. This puts even a larger demand on broadband signal requirements. Compressing the transmitted data or using spectrum efficient methods may address this issue to some extent, but at some point all these conventional methods may fall short, calling for a paradigm changing approach to spectrum sharing. Several recent techniques such as node cooperation, dynamic spectrum access, cooperative spectrum sensing, and the use of channel state information (CSI) are summarized in [11].

Node cooperation or relaying other nodes' data is one way to efficiently use spectrum. For instance, if a direct link from a source to destination is experiencing low signal to noise ratios or in the case of multipath channels being in a deep fade, there is no point for the source to waste precious spectrum and power and try to transmit packets that are destined to be lost or received too noisy to be reconstructed. This action also poses interference on other neighboring nodes. Instead, the source can use CSI to predict such cases and use a relay that is closer and hence use less power, affect fewer neighboring nodes, and increase its chance of successful transmission.

Noting that the spectrum is not used all the time and that all the users are not going to transmit at all times, identifying spectrum holes and sharing spectrum significantly increases the overall spectrum use efficiency of a network.

Although this approach seems simple to explain and optimize for a simple three-node network, a large-scale network with multiple nodes and various traffic priorities may render a complicated optimization problem when it comes to solving the resource (power and spectrum) allocation challenge. Game-theoretical approaches such as the one in [12] offer a tractable solution to analyze such systems. Assuming different users with different priorities, as depicted in Figure 2.4, lower priority users (S_2) may act as relays for higher priority users (S_1) during a portion of each time slot to increase their throughput and open up more opportunities in future time slots for low priority users to send their data. Ensuring that a dedicated portion of time slot is allocated to high priority users guarantees their cooperation in this game-theoretical resource allocation scheme.

The main advantage of cooperative relaying is extending the coverage of a network by relaying information from nodes in hard to reach areas as needed. As discussed before, relaying high priority data when the main channel experiences low quality is another notable advantage of relaying. Cognitive radio networks, on the other hand, promote a more efficient use of available resources such as spectrum through well-planned sharing mechanisms.

This combined cooperative and cognitive radio approach may be the key for future implementation of cognitive networks. Noting that complete interference alleviation is unavoidable, high priority users (or in cognitive radio terminology, primary users) will never risk their licensed spectrum without any tangible gain.

Cooperative relaying is the price that low priority users (or secondary users) pay to gain access to the licensed spectrum during a predetermined but dynamic portion of each time slots.

One example of using cooperative relaying is in emergency management situations during disaster relief operations. Imagine a first responder in the middle of a wooded area without any cellular coverage and miles away from closest base station. A vehicle with more capable wireless node, as compared to handheld devices used by the responder, can be used as relay when parked in between the disaster area and base station.

In such a scenario, it is often possible that several different agencies are responding to the event including firefighters and medics.

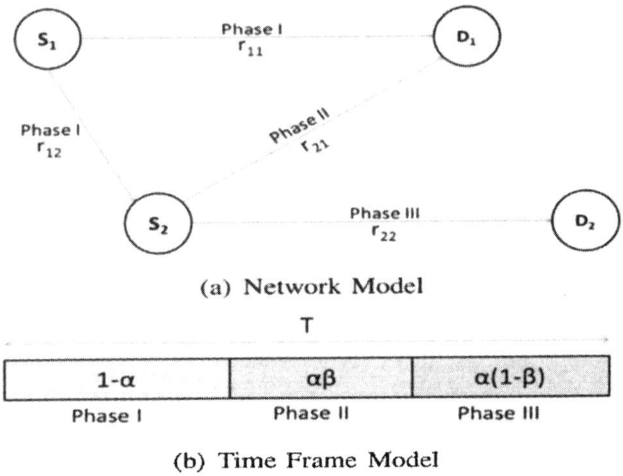

Figure 2.4 Resource allocation with cognitive cooperation [12].

At some point the vehicle serving as a relay will be overwhelmed with data that might include lifesaving videos or information critical for mission success. At this point the relay needs to look at data stream priority and decide when to relay which data stream to ensure that critical data is not left in a queue behind less critical data. A game theoretical modeling of this problem and related simulation results ensures that design parameters of the relay are chosen appropriately to avoid data loss or extensive delays.

2.2.3 Maintenance

Another set of boundary conditions imposed on sensors in hard to reach areas is the maintenance requirement. Minimal or no maintenance may be the condition in several applications were the sensor is embedded in a structure permanently and extraction/repair is either not possible or extremely costly to perform. Materials that can withstand high and low temperatures and that are rugged enough to operate in harsh environments become desirable from the maintenance point of view. Some low maintenance realization of wireless sensors on piezoelectric substrates such as Lithium-Niobates [13] operate in room temperature, while other materials such as Langasite can be used for high temperature applications if coated properly [14]. More information on this issue from material science perspective is provided in Chapter 3.

One practical example from space applications includes the structural monitoring of inflatable space habitats. Passive sensors that are placed on the interior wall of the habitat can be used for shape monitoring if each one is localized properly. Various methods such as the received signal-strength, angle of arrival or time of arrival can be used individually or combined for a high-resolution localization of all sensors. Incorporating material properties, dictating the curvature based on air pressure or strain measurements can yield to development of a three-dimensional dynamic model of the structure (Figure 2.5).

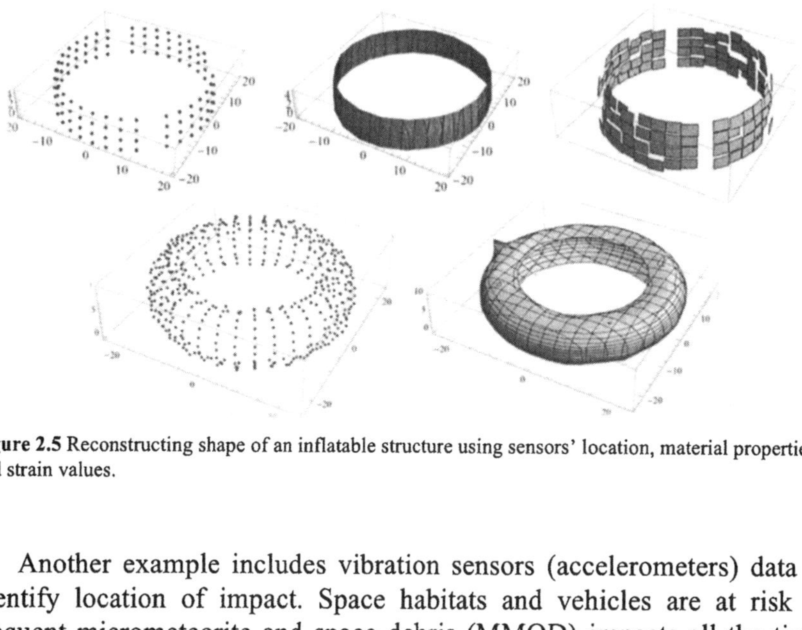

Figure 2.5 Reconstructing shape of an inflatable structure using sensors' location, material properties, and strain values.

Another example includes vibration sensors (accelerometers) data to identify location of impact. Space habitats and vehicles are at risk of frequent micrometeorite and space debris (MMOD) impacts all the time. Having a wireless sensor network to monitor the severity of impact and report the location back is of utmost importance for crew safety and mission success. Vibration sensors often operate based on a change in capacitance of a microelectromechanical (MEMS) cantilever beam. A similar MEMS concept may be used to scavenge energy (similar to RF energy harvesting) in harsh environments.

The three main categories of energy, spectrum management, and maintenance impose a unique set of conditions that will guide design and implementations of wireless sensor networks in harsh and challenging environments. Section 2.3 provides an overview of the throughput requirements and their effect on interference.

2.3 THROUGHPUT REQUIREMENTS

Imagine an accident causing harmful gases to spread over a highly populated city, such as London (see Figure 2.6) with an over 600-mi^2 area [15]. Noting the 300-ft typical communication range of an IEEE 802.15.4 sensor node, it takes over 180,000 wireless nodes to cover the whole area. For a typical low data rate node with 250 Kbps, the fusion center in a centralized algorithm makes it necessary to process data at a massive of rate 45 Gbps.

Figure 2.6 Map of London from Google-map overlaid with sensors [15].

Even a distributed cluster-based network needs to deal with several hundreds of megabits-per-second data rates. This might seem like an over-exaggerating example, but it helps put the problem into the correct perspective. As described in Section 2.2, limited capabilities of sensor nodes in challenging environments call for lowering the throughput to maintain interference below a manageable threshold. This ensures simple receiver designs to remain practical in simple node implementations.

2.3.1 Throughput Efficiency

Looking at throughput, passive sensors are more efficient as compared to active sensors due to the large control overhead required in active sensors. A passive sensor can be read with a waveform transmitted from an interrogator with no need for digital transport overhead such as packetizing and bundling as required in active sensing. Active sensors in this text are defined as traditional sensor/radio/microprocessor triplets, while passive sensors are just simple transponders that reflect a distorted version of the reader signal based on sensed parameters. This is the first level of efficiency that is noteworthy.

2.3.2 Throughput Enhancements

Exchanging data between two sensor nodes with no direct link through another node requires at least four time slots per packet. As illustrated in Figure 2.7, the first time slot is used by node A to send its packet to the relaying node that will forward that data to node B in the second time slot. The same process is followed by node B to send its packet to node A.

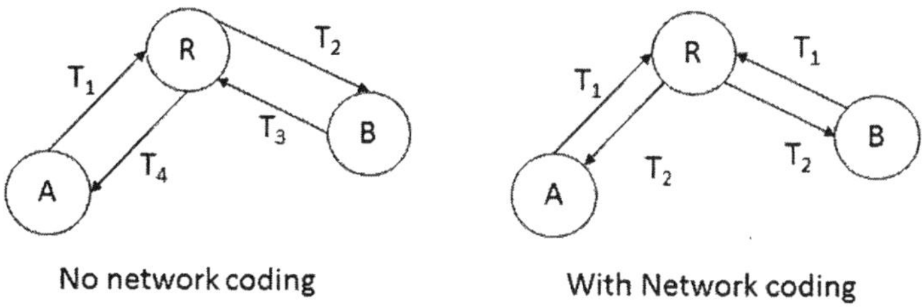

Figure 2.7 Routing over two way relay channels with (right) and without (left) network coding.

Simple linear coding methods such as network coding may be used to enhance throughput by a factor of two, when both nodes send their data at the same time and the relay node broadcasts the superimposed signal from A and B back to them. In this case, each node can extract the other packet using a simple mathematical XOR operation. Network coding can be performed at the network layer or physical layer. A simple XOR operation at the network layer can provide notable efficiency as compared to traditional two-way relay channels without network coding. However, significant throughput enhancement can be achieved if physical layer network coding is used.

Physical layer network coding takes advantage of the broadcast nature of wireless signals and utilizes interference to its advantage. When two users simultaneously transmit their signals to a relay, if nothing is done at the relay to avoid interference, those two electromagnetic waves are superimposed on top of each other over air due to the laws of physics. The relay then receives one signal. While it is not possible for the relay to decompose that signal into two users' signals, it is possible for it to just broadcast that signal back to both users. This way each user can find its counterpart's signal after some signal processing (a bit more complicated than simple XOR). The relay can perform some simple operations such as amplification, de-noising, or even decoding to further increase the performance.

2.3.3 Latency Reduction

Latency in getting packets from one point to the other in a network is often a limiting factor that directly impacts the end-to-end throughput. A delay optimized packetization policy is proposed in [16] that adjusts the length of each packet based on channel error rate in order to minimize latency. It is shown in Figure 2.8 that minimum delay can be obtained for various BER values, if appropriate packetization time is used.

Latency is often a more important parameter than throughput in real time and mission-critical applications such as dynamic flight control, voice or video applications delivered over low data rate networks, and cyber physical systems. As will be discussed in Section 2.4, adding more sensors with various latencies will increase the overall network latency.

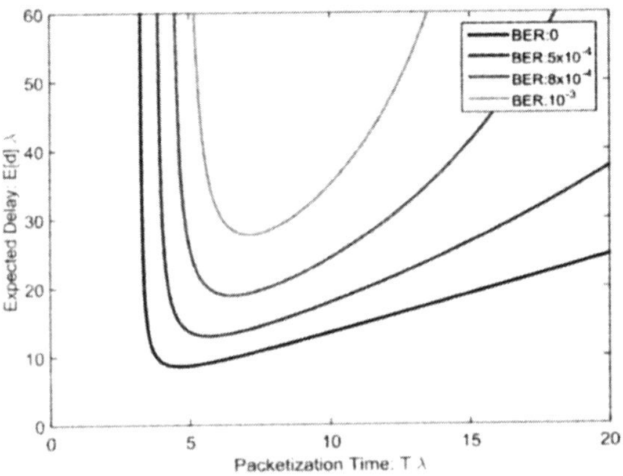

Figure 2.8 Optimum value of expected delay versus packetization time for various BER values [16].

2.4 INTERFERENCE IN SENSOR NETWORKS

A single pair of transmitter and receiver devices may suffer from just a single interfering node if not designed properly. Adding multiple new sensor nodes to a well-designed network increases the interference to the point that it will eventually break the network. It is important to understand the nature of the interference and determine effective ways to manage it in the networks deployed in harsh environments.

In this section, with a focus on harsh and challenging environments, the interference problem is studied for passive sensors that are appropriate for use in such environments. Various levels of interference mitigation in time, frequency, and code are studied next.

2.4.1 Intersymbol Interference (Time)

When multiple sensors are impinged at the same time, their responses will be superimposed on top of each other making single sensor data hard to extract. This is similar to intersymbol interference in communication systems. In order to avoid this situation, different fixed delay values can be implemented in each sensor to make sure their responses arrive at the receiver at different times. Time domain separation can give us flexibility in terms of the number of sensors, while it poses challenges on the sensor size.

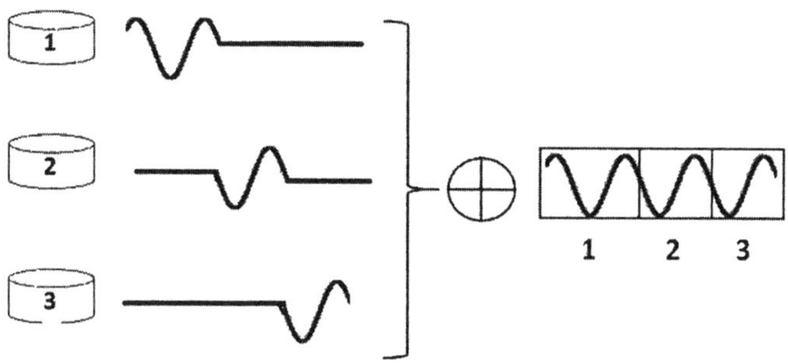

Figure 2.9 Three-sensor-network with three time-separated responses.

Note that adding delay inside the sensor structure translates into longer sensors. This limiting factor makes one think about other methods to separate sensor responses.

Figure 2.9 illustrates the time division concept in a network with three sensors. Each sensor responds at different time slots in a consecutive manner and the aggregate signal is received at the receiver. Sensors can be of a different kind or may have longer or shorter response signals, but as long as they do not overlap in time, the data extraction at the receiver is as simple as partitioning the received signal.

This deterministic time slot allocation will be performed at the design level, and each node is built with their specific delay value built in the hardware. In active systems, where user nodes can reconfigure their radio, more effective time-sharing methods may be used.

One example of such a network is ZigBee, which is based on the IEEE 802.15.4 standard. Carrier sense multiple access with collision avoidance (CSMA/CA) is used for sharing the medium among many sensors. This randomized time-sharing approach (in contrast with the deterministic predefined method in passive networks) is capable of handling more nodes or users in a network. The process involves sensing the spectrum to avoid transmission if other carriers are discovered. In that case, a random back-off timer is set to keep nodes from transmitting at the same time, hence limiting the interference and packet loss.

The benefit of this method is supporting more low data rate nodes in a network, but the shortfall is its lack of support for passive sensors.

2.4.2 Cross Talk (Frequency)

Similar to different radio or TV stations operating on different channels, sensors can be designed to operate on various central frequencies. A simple filter bank or tunable filter at the receiver side can easily separate sensor responses received at different frequency bands. The main limitation of this method is scarcity of spectrum and limited bandwidth availability. If combined with time domain separation, these two methods can increase the number of sensors that can be used in proximity of each other without too much unmanageable interference. However, new methods are still required to increase the number of sensors.

Similar to sensor response separation in time as illustrated in Figure 2.9, separation in frequency can yield an aggregate signal that can extract each sensor response using a filter tuned to that specific band. This key design criterion ensures some guard band is placed in between each two bands to allow for a practical filter design.

This is a well-known design criterion in conventional communication systems, but it is more critical for systems operating in harsh and challenging environments. The filter design in such environments has more limitations in terms of available and appropriate components and much more rigorous spurious emissions and out of band interference requirements.

Dynamic frequency allocation that is used in active wireless systems presents a more efficient use of spectrum by allowing more sensor nodes to transmit at the same time over an almost similar frequency band. One method of avoiding collision is called frequency hopping. For the ZigBee example presented in Section 2.4.1, each user can hop among a fixed number of frequencies with speed of up to 1,800 channels per second. As long as no two users are on the same channel at the same time, collision will not occur. Unfortunately, this method is not applicable to passive sensor networks.

2.4.3 Cross-Correlation (Code)

Spread spectrum techniques, which led to the development of CDMA systems, assign different spreading code to each user. Passive sensors can also be designed with different ID codes that are used for spreading the output response.

As long as the peak to side-lobe ratio (PSLR) of the cross-correlation between two different sensor responses is minimized, yet another degree of freedom to separate sensor responses is realized. The same parameter, PLSR is also used to design the orthogonal codes for each sensor and ensure autocorrelation of those codes has maximum possible PLSR.

As an example, ZigBee networks use one of the 16 PN-sequences that are 32 chips long to spread the signal and provide a coded operation. These quasi-orthogonal codes provide another degree of separation among various users' signals. Each 4 bits of incoming data will select one of those 16 sequences. In passive sensors a spreading code is already built in for each sensor node and is used like an ID for each user.

Time, frequency, and code domains altogether can provide a reasonable separation among a large number of sensors in harsh and challenging environments with limited spectrum and energy.

Even with all these methods of separation, due to manufacturing imperfections and design error margins, a small amount of interference may still occur. Sensors placed at different relative distances to the reader may respond with different amplitudes due to signal attenuation and multipath fading effects. These challenges call for new interference mitigation methods in order to ensure the sensor responses are properly extracted.

Figure 2.10 Single-sensor response (left) compared to multi-sensor response (right).

2.4.4 Interference Reduction and Avoidance

In this section, two methods for interference reduction are presented. The first method is active interference cancellation [8] and the second one is passive interference avoidance [17].

Active interference cancellation in a multisensor network may be conducted in different ways. For a passive sensor, a more appropriate alternative for challenging environments is a filter bank matched to all sensor nodes in the network work in parallel to detect every single sensor response (left side of Figure 2.10) from the superimposed received signal (right side of Figure 2.10). The filter responses are then sorted from the strongest signal with respect to peak to side lobe ratio to the weakest signal. The next step is to remove the portion of the received signal that is related to the strongest signal and repeat this procedure for the remaining sensor responses until the weakest signal is detected. This process is summarized in Figure 2.11.

We note that the theoretical ideal response for each sensor is known and can be used to improve signal quality in each sensor response before removal from the received aggregate signal.

For performance analysis purposes, average interference caused by one unknown sensor at each time slot can be calculated using the following equation:

$$f_1 = M(C_i) + \frac{1}{n}\sum_{j=1, j\neq i}^{n} N(C_j) \tag{2.1}$$

where, M(.) and N(.) represent autocorrelation response of desired sensor and cross-correlation of the interfering sensors, respectively, n is the number of sensors in the network (including the ith sensor under interrogation). Extending this equation to include average interference caused by all *n-1* sensors in the network yields

$$f_m = M(C_i) + \frac{1}{n^m}\sum_{j_1}^{n} \cdots \sum_{j_m}^{n} N(C_{j_1}) + \cdots + N(C_{j_m}) \tag{2.2}$$

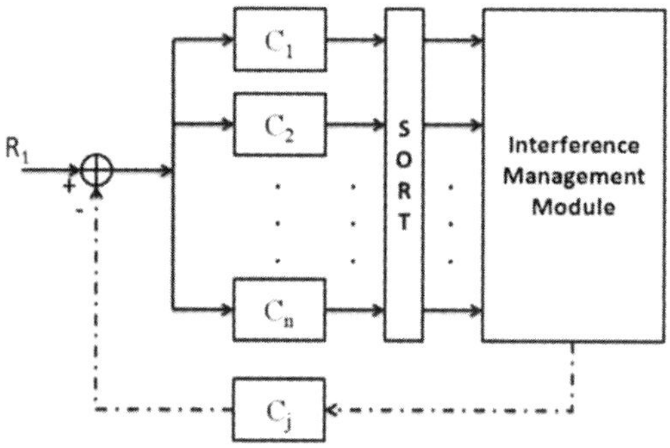

Figure 2.11 Iterative interference mitigation [8].

which is a very complex metric to calculate for large *n*. A reduced complexity equation is proposed in [8] as follows that is significantly less complicated:

$$f_n = f_{n-1} + \frac{1}{n}\sum_{j=1, j\neq i}^{n} N(C_j) \tag{2.3}$$

2.5 CONCLUDING REMARKS

The problem of interference in wireless networks operating in challenging and harsh environments is studied in this chapter. There is no single method that is superior to all others. For different applications and networking scenarios, one or a combination of various interference management or avoidance methods may be used to achieve the desired throughput performance.

The trade-off between throughput and interference is a long-standing problem in communication networks. It is getting even more challenging as applications with high throughput demand emerge, and the number of users or required spatial accuracy increases.

In harsh and challenging environments, access to energy source is often the main hurdle. Maintenance is also the second most important obstacle that calls for developing rugged maintenance free systems. When the number of communicating devices increases, the interference is also increased and spectrum efficiency becomes critical for system reliability. It is important to quantify the interference at least from a statistical point of view (i.e., average interference) to provide a somewhat accurate guideline to design engineers.

A large network with multiple nodes may pose a very complicated interference calculation problem. A reduced complexity method based on a simple recursive equation is proposed in this chapter can address this issue. Even after separating signals from different users in time, frequency, and code, it is still necessary to mitigate the interference from other users. This interference is due to multiple reasons including imperfections in manufacturing filters and amplifiers to name a few. A novel patent pending interference mitigation at the source proposed in this chapter can ease the processing requirements on the aggregate signals at the receiver. The idea is based on hard limiting the interfering signals at the source, and not allowing them to aggregate and pile up to an unmanageable level.

References

[1] Farah, C., F. Schwaner, A. Abedi, and M. Worboys, "Distributed homology algorithm to detect topological events via wireless sensor networks," in *IET Wireless Sensor Systems*, Vol. 1, No. 3, pp. 151–160, September 2011.

[2] Akyildiz, I. F., W. Su, Y. Sankarasubramaniam, and E. Cayirci, "A survey on sensor networks," in *IEEE Communications Magazine*, Vol. 40, No. 8, pp. 102–114, August 2002.

[3] Heinzelman, W. R., A. Chandrakasan, and H. Balakrishnan, "Energy-efficient communication protocol for wireless microsensor networks," *Proceedings of the 33rd Annual Hawaii International Conference*, System Sciences, 2000, vol.2, pp. 10.

[4] Younis, O., and S. Fahmy, "HEED: a hybrid, energy-efficient, distributed clustering approach for ad hoc sensor networks," in *IEEE Transactions on Mobile Computing*, Vol. 3, No. 4, pp. 366–379, October-December, 2004.

[5] Cadambe, V. R., and S. A. Jafar, "Interference Alignment and Degrees of Freedom of the K -User Interference Channel," in *IEEE Transactions on Information Theory*, Vol. 54, No. 8, pp. 3425–3441, August 2008.

[6] Maddah-Ali, M. A., A. S. Motahari, and A. K. Khandani, "Communication Over MIMO X Channels: Interference Alignment, Decomposition, and Performance Analysis," in *IEEE Transactions on Information Theory*, Vol. 54, No. 8, pp. 3457–3470, August 2008.

[7] Razi, A., and A. Abedi, "Interference reduction in Wireless Passive Sensor Networks using directional antennas," IEEE Fly by Wireless Workshop (FBW), 2011 4th Annual, Montreal, QC, 2011, pp. 1–4.

[8] Abedi, A., and K. Zych, "Iterative interference management in coded passive wireless sensors," SENSORS, 2013 IEEE, Baltimore, MD, 2013, pp. 1–4.

[9] Ponchak, G. E., et al., "High temperature, wireless seismometer sensor for Venus," Wireless Sensors and Sensor Networks (WiSNet), 2012 IEEE Topical Conference on, Santa Clara, CA, 2012, pp. 9–12.

[10] Abedi, A., "Battery free wireless sensor networks: Theory and applications," Computing, Networking and Communications (ICNC), 2014 International Conference on, Honolulu, HI, 2014, pp. 287–291.

[11] Kompella, S., et al., "Special issue on cognitive networking," in *Journal of Communications and Networks*, Vol. 16, No. 2, pp. 101–109, April 2014.

[12] Afghah, F., et al., "A reputation-based Stackelberg game approach for spectrum sharing with cognitive cooperation," 52nd IEEE Conference on Decision and Control, Firenze, 2013, pp. 3287–3292.

[13] Dudzik, E., A. Abedi, D. Hummels, and M. P. Da Cunha, "Wireless multiple access surface acoustic wave coded sensor system," in *Electronics Letters*, Vol. 44, No. 12, pp. 775–776, June 5 2008.

[14] Pereira da Cunha, M., et al., "Wireless acoustic wave sensors and systems for harsh environment applications," Wireless Sensors and Sensor Networks (WiSNet), 2011 IEEE Topical Conference on, Phoenix, AZ, 2011, pp. 41–44.

[15] Google map, Google Inc.

[16] Razi, A., F. Afghah, and A. Abedi, "Channel-Adaptive Packetization Policy for Minimal Latency and Maximal Energy Efficiency," in *IEEE Transactions on Wireless Communications*, Vol. 15, No. 3, pp. 2407–2420, March 2016.

[17] Abedi, A., Systems and methods for interference mitigation in passive wireless sensors, US Patent 9,607,187, Issued: March 28, 2017.

Chapter 3

Resilient Sensor Systems: Design for Longevity

Sensor network lifetime is determined by a variety of parameters (see Figure 3.1) during the design and operation phases, and is influenced by environmental conditions. Sensor deployment in harsh environments requires special attention to these parameters to extend the sensing system's lifetime. This chapter presents recent developments in material science, sensor systems, and networking science and engineering that contribute to the longevity of sensor systems in harsh environments.

3.1 INTRODUCTION

Design parameters such as power requirements for data collection and transmission, and adaptability to the environment are the more important considerations to start with when designing a wireless sensing system for extreme environments. Sensor material is also another important aspect of the design consideration. It is obvious that if only one of the main components (i.e., sensor, processor, or radio) fails, the whole system will become inoperable.

Looking at the operation cycle from a network point of view, the number of sensors, networking protocols, and placement criteria are another set of critical parameters that can be optimized to enhance network longevity.

Ultimately, the environmental conditions set the boundary conditions for design and operations and directly influence the overall system performance and cost.

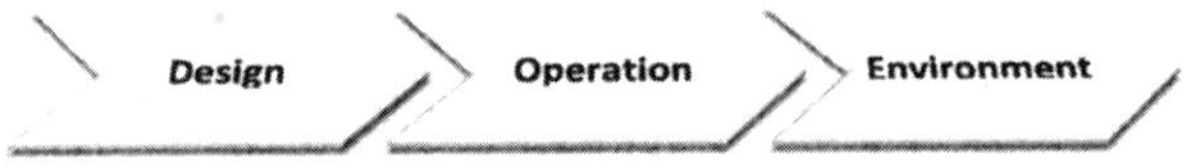

Figure 3.1 Longevity depends on design, operation, and environmental parameters.

3.2 DEFINITION OF LONGEVITY

Definition of longevity starts with each individual component. One widely used model to predict the expected lifetime of an electronic component is based on exponential probability distribution.

Example: Experimental tests show that before 2,000 hours, half of the sensors under the same operating conditions fail. What is the expected lifetime of one sensor?

Solution: In this example, an exponential probability distribution may be used to model the lifetime of each sensor. For this model, the probability density function is

$$f_L(t) = \propto e^{-\propto t}, t > 0$$

where the unknown parameter, $\propto$ can be estimated using the results of experimental tests. Noting that half of sensors will fail before 2,000 hours, one can write

$$P(L < 2000) = \int_0^{2000} f_L(t)dt = 1 - e^{-2000\propto} = 0.5$$

Therefore, $\propto$ can be found as follows

$$\propto = -\frac{\ln 0.5}{2000} = 3.47\times 10^{-4}$$

Now, the sensor expected lifetime can calculated using the following property of exponential probability:

$$E[L] = \frac{1}{\propto} = 2885 \; hours \; and \; 23 \; mins$$

Most electronics systems including sensor systems include many components each with a different lifetime. Some of these components are connected in series, while others are placed in parallel, or switch in or out of the circuit. For each of these architectures, the overall lifetime can be calculated as follows.

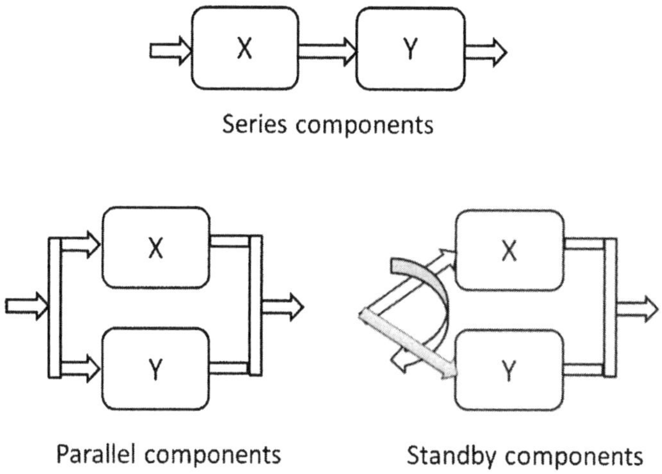

Figure 3.2 Three different component (or system) interconnections.

For series interconnection of two components with lifetimes represented by random variables X and Y, the overall lifetime is calculated as min{X,Y} (see Figure 3.2). In other words, the component with shorter life will dominate the overall lifetime. An example of such a system is a cascaded filter and amplifier. Feedback control systems also fall into this category.

Parallel interconnected components create a more reliable system with an overall lifetime of max{X,Y}. As long as one of the components is in good working order, the overall system is considered live. An example of such architecture is found in the multilink redundant network connections in a packet network.

Standby interconnection is the ultimate design for reliability when one component is working and the other one is on standby. An automatic switch will put the standby system online as soon as the first system fails. The overall lifetime in this case is $X+Y$. An example of such a system includes redundant communication systems on board marine vessels.

All of the above components' interconnections can be extended to systems of systems or extended to multiple components. System reliability is formally defined as probability of lifetime being more than a specific threshold and represented in terms of probability distribution function as follows:

$$R(t) = P(X > t) = 1 - P(X \leq t) = 1 - F_X(t)$$

Expected lifetime can be calculated using system reliability function, *R(t)* as follows.

$$E[\mathrm{X}] = \int_0^\infty R(t)\, dt$$

Longevity of a sensor system that has been operating for a while can be estimated more accurately using the concept of conditional reliability or hazard-rate. For example, if a system or component has been working at time *t*, the probability that it fails before time x is

$$F_X(x|X > t) = \frac{P(x \leq X, X > t)}{P(X > t)} = \frac{F_X(x) - F_X(t)}{1 - F_X(t)}$$

Taking derivatives from both sides results in the definition of hazard-rate function, $H_X(t)$, as follows

$$f_X(x|X > t) = \frac{f_X(x)}{1 - F_X(t)} = \frac{-R'(t)}{R(t)} = H_X(t)$$

This preliminary discussion is the foundation for the stochastic modeling of longevity and includes network longevity in the following sections. Although proper design can significantly impact the longevity parameters and probability functions that govern the lifetime, it is noteworthy to remember that components', systems', and networks' lifetime are all random variables and we can only talk about their average or expected values. This analysis also helps in determining lifetime and reliability limits of a complex system comprised of several different components with different lifetimes and reliabilities. Identifying critical components in a system is an important step to eliminate bottlenecks in performance and reliability.

3.3 SENSOR SYSTEM DESIGN

3.3.1 Sensor Materials

Sensors' substrate material not only needs to be designed to withstand harsh environments, but it also needs to operate during fast changes in the environmental conditions. One technique that can improve tolerance to change and improve longevity is casing or coating, which is described in this section.

Sensors are often created using multiple layers of different materials with different expansion coefficients. This process may create a variety of sensors with electrical responses suitable for integrating into wireless or wired systems. However, due to differences in material types and how they react to the environmental changes such as extreme temperatures, adhesion and connectivity between these various layers may be compromised. In order to address this challenge, wireless sensors are required to be either housed in a temperature controlled box or environment (which defeats the whole purpose of their operation in extreme environments), or coated with materials that are tolerant of such extreme changes in the temperature.

One example of such protective coating is a 50 nm Alumina layer that protects the sensor for operations between 300 and 750°C [1]. Another example highlights the challenges caused by the environmental gases and not the temperature. Most gas sensors are built using SnO2, which is not stable in high temperatures in the presence of hydrogen. To prevent dissolving this gas sensing later, a coating of micro-porous ceramic is proposed in [2] to allow gas sensing, while preventing the hydrogen effect on the sensing layer.

3.3.2 Sensor Systems

The main topic in sensor systems design as relates to longevity is the issue of power requirements. Power requirements dictate the battery size, which in turn limits the choice of battery type and therefore its operational conditions. In addition to the ultralow power microelectronics design techniques that have been developed in recent years, intelligent signal processing, duty cycling, power allocation and management techniques in a network setting can significantly maximize the longevity of sensing systems.

Basic signal processing techniques such as data compression and change detection alleviate the need for unnecessary data communication and wasting limited on board power. Advanced signal processing methods can extend the lifetime even further by taking the correlation among sensors in close proximity into the account.

Example: A simple run-length coding applied to a string of characters such as 'aaaaaaaaffffrrrraaaaaaaa' compresses the string to '8a4f4r8a'. Assuming that each character transmission requires 5 mA over a 1 ms time slot and that the on board battery has 480 mAh capacity, how long will the battery last if such strings are transmitted continuously with or without run-length compression?

Solution: The uncoded string includes 24 characters; therefore it requires 120 mA to transmit. A 480 mAh battery will last for 4 hours. If the string is compressed, only 8 characters will need to be transmitted which requires 40 mA. In this case the battery will last for 12 hours.

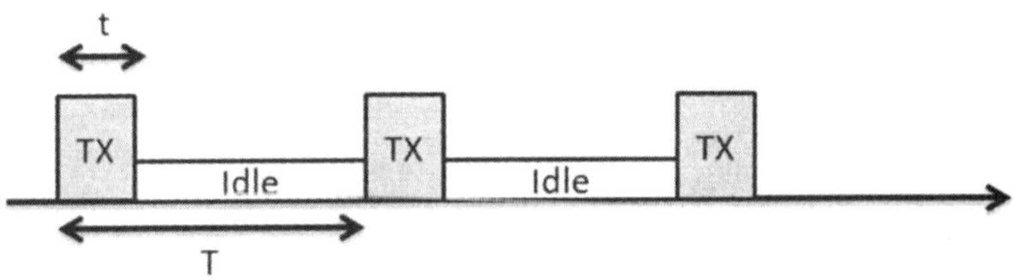

Figure 3.3 Duty cycle is defined as ration of t divided by T. Sensor transmits during the TX period and stays idle (to save energy) during idle time slots.

Duty cycling (see Figure 3.3) can significantly reduce the power requirements and extend the network lifetime. For example, a temperature sensor output may not change that fast to require 1 sample per second; therefore, maybe a duty cycle of 1 sample per minute or even 10 minutes would be more than enough. It is worth noting that the rate of change in the measured parameter determines the optimum duty cycle, which can significantly save the on board battery power. Here is an example to illustrate how much savings can be obtained using duty cycling.

Example: Assuming measuring and transmitting one temperature value requires 10 mA over a 1 sec time slot, how much extra life time is achieved in a sensing system with a 18,000 mAh battery, when switching the duty cycle from 1 sample/sec to 1 sample/hour?

Solution: A higher sampling rate of 1 sample/sec requires 3,600 measurements and transmissions during one hour for a total of 36,000 mA. An 18,000 mAh battery will only last for 30 mins in this scenario.

However, the slower sampling rate of 1 sample/hour only requires one transmission during the same time for a total of 10 mA. Therefore, the same battery will last for 1,800 hours, or 75 days.

Power allocation to sensors with lower power requirements and lower selectivity (LPLS acting as triggers), while keeping sensors with higher power requirements and higher selectivity (HPHS) off until triggered, can significantly improve longevity. Once a change discovered by a low power sensor in a specific geographic area, high power sensors with more selectivity and sensitivity can be turned on in the area of interest.

This technique can be further exploited among multiple nodes or radios to minimize overall power requirements, while keeping the interference low. Figure 3.4 illustrates a sensor network with three types of sensors (HPHS, HPLS, and LPLS) that start at trigger stage, moving to measurement stage, and staying in operation stage, until all sensors' estimates go below a predetermined threshold. In addition to power saving, this architecture may be used for determining the topological shape of events.

Example: Consider a hybrid sensor network with 900 low power sensors that only require 1 mA to operate and 100 high power sensors that require 100 mA to operate. Assuming that low power sensors are on at all times (100% duty cycle), but high power sensors are on only 10% of the time, what is the overall power consumption of such network in 1 hour. Compare your answer to a heterogeneous network with 1,000 high power sensors that are on at all times.

Solution: The overall power consumption during one hour includes two parts, low power sensors, and high power sensors. The former requires 900 mAh, while the latter requires 1,000 mAh to operate at a 10% duty cycle. Therefore, the total will be 1,900 mAh. For the heterogeneous scenario, 1,000 high power sensors will require 100,000 mAh for one hour of operations, which is 52 times more.

(a) Trigger stage.

(b) Measurement stage.

(c) Operation stage.

Figure 3.4 High/low power high/low selectivity (HPHS/LPLS), heterogeneous sensors.

3.3.3 Network Science and Engineering

Network life then depends on a variety of parameters including the components' lifetime, the systems' lifetime, and of course managing the resources in the network. The first two topics (components' and systems' lifetime) were discussed in the previous sections. In this section, efficient management of network resources is studied. The focus will be on spectrum and power management at the physical layer as well as efficient routing at the network layer.

3.3.3.1 Physical Layer Resource Allocation

Spectrum and power are the most scarce resources in a remotely deployed, battery powered, wireless network. The problem of resource allocation can be studied from different perspectives. One can pursue a solution to find optimum resource allocation to achieve higher throughput, to minimize interference, or to increase network lifetime.

When lifetime is considered as the main objective, both circuit power and radio power need to be considered and minimized. Several sensor selection and power allocation algorithms have been proposed in [3], where significant power savings are achieved (2 dB for correlated and 10 dB for uncorrelated noise). It is important to note that resource allocation should not be optimized to the point that quality of service (QoS) is compromised. Low quality data or unreliable radio links are not useful in decision making after all. An optimal resource allocation scheme with QoS provisioning is presented in [4]. In this work, downlink energy assignment and uplink power control are jointly optimized. The main objective in [4] is to maximize the aggregate effective capacity.

In scenarios with wireless energy transfer, the boundary conditions of the power allocation problem are different. A multi-antenna power station can be used to transfer energy to sensor nodes in an optimal manner, in the sense that energy beamforming and time allocation are jointly designed [5]. In this work, the sum throughput of sensor nodes is maximized. A game theoretic approach is used to provide incentives if two networks of sensor nodes coexist.

Using game theoretical channel and power allocation in wireless sensor nodes can help with network longevity. An iterative power and channel allocation method for sparsely distributed wireless sensor nodes based on game theory is proposed in [6]. The performance of concatenated power and channel allocation and interlaced methods are compared showing significant power savings of around 55%.

The power allocation problem in an asymmetric wireless sensor network is considered in [7]. In this work, the mutual information between the source and decoded data after transmission is maximized under the sum power constraint. A scheduling algorithm proposed in this work singles out the sensor nodes that are dominating the bit error rate performance. A similar work uses ratedistortion as a metric (rather than mutual information) proposed in [8], where fading is also considered.

Several power allocation methods are compared in [9] in a systematic manner. These methods are called optimal, extended optimal, uniform, extended uniform, and single-node selection. The extended versions of optimal and uniform methods consider the possibility of faulty sensor nodes as well. Thinking about the possibility of node failure in the context of network lifetime brings up the concept of robustness, which may present different definitions for network lifetime.

In an ideal case, we would like to have maximum convergence and therefore, may define network lifetime as the time that the first node fails. Another more relaxed definition might be when a few nodes (p%) of the nodes fail, but we can still interpolate data at locations with failed sensors. This definition may work for cases where a continuous field is being measured, where differential equations can be used to model the field and therefore interpolate data in the missing nodes area.

In summary, the power and channel allocation problem in various application scenarios maybe a difficult and in some cases intractable problem. Depending on whether a centralized or a decentralized detection method is used in the network, the parameter being measured, and how much accuracy or spatio-temporal resolution is needed, may change our approach to address the resource allocation issue. Nature-inspired methods such as Cat Swarm and Cuckoo search have been proposed in [10] to solve the power allocation problem in decentralized sensor networks. Biologically inspired methods might be an interesting approach for these situations.

3.3.3.2 Efficient Routing

Similar to the physical layer, the routing problem can be solved with different objectives in mind. The minimum number of hops, higher throughput, and lower latency are a few parameters of interest when a routing method is designed. In order to optimize for the desired benefits listed above, we need to consider the costs associated with each hop and node. This cost may be determined based on traffic, node speed, hop length, and several other factors. In this section, we focus on increasing network lifetime and designing routing methods that can support the longevity objective.

A typical routing problem is defined as follows. Let us assume that a sensor network with wireless connectivity is collecting data and transmitting it to a fusion or decision center. In this scenario, the routing problem is to find the optimum path with the lowest delay, highest throughput, or lowest power. It would be ideal to jointly optimize all of the above, but that may not be possible in most cases. Figure 3.5 depicts a network of seven wireless sensors with a centralized fusion center. Each node has two layers: the physical and network layers. The physical layer includes the wireless radio and measures the channel conditions from an electromagnetics point of view by providing an estimate of bit or symbol error rates, while the network layer is responsible for routing the packets and measures packet error rate and latency.

The connectivity graph can be modeled using the following matrix, **A**.

$$\boldsymbol{A} = \begin{bmatrix} a_{11} & \cdots & a_{17} \\ \vdots & \ddots & \vdots \\ a_{71} & \cdots & a_{77} \end{bmatrix}$$

where a_{ij} represents signal to noise ratio (SNR) or link quality between nodes i and j. For most cases, this matrix is symmetrical (i.e., $a_{ij} = a_{ji}$). For the special case of Figure 3.5, the value of $a_{ij} = 0$ where there is no link between the nodes i and j (e.g., $a_{34} = a_{35} = a_{37} = 0$).

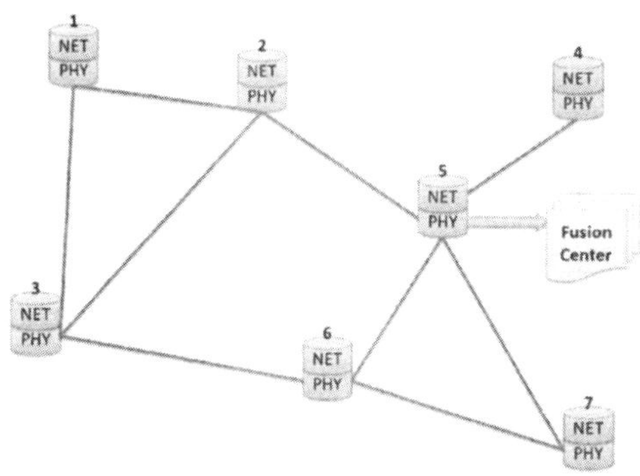

Figure 3.5 Routing diagram for a network of wireless sensors with a central fusion center.

This matrix with SNR values, can be used by the physical layer to determine the most power-efficient or spectrum-efficient way to transmit bits or symbols (modulated bits) between two nodes. A similar matrix with bit error rate (BER) replacing SNR values can be created to help the network layer solve the routing challenges.

The network layer needs to assign a cost to each path and BER might be a good parameter to start with. The lower the BER the higher the chances that a packet (which might include several bits) can be successfully transmitted on the first attempt. If a small number of bits in a packet is corrupted, forward error correction code implemented in the packet can repair the incorrect bits, but if the number of errors are more than correction capabilities of a particular code, that packet will be discarded and a request for retransmission is sent to the transmitting node. This will waste significant power and add large delays to the network, both working against our main objective of network longevity.

Example: Consider the network depicted in Figure 3.5 and the following connectivity matrix with SNR values presented in decibels.

$$A = \begin{vmatrix} - & 10 & 20 & 0 & 0 & 0 & 0 \\ 10 & - & 10 & 0 & 20 & 0 & 0 \\ 20 & 10 & - & 0 & 0 & 25 & 0 \\ 0 & 0 & 0 & - & 10 & 0 & 0 \\ 0 & 20 & 0 & 10 & - & 10 & 15 \\ 0 & 0 & 25 & 0 & 10 & - & 10 \\ 0 & 0 & 0 & 0 & 15 & 10 & - \end{vmatrix}$$

What is the best route (with highest SNR) from node 3 to the fusion center?

Solution: Examining the graph there are two routes from node 3 to 5, or the fusion center. One is through node 2, and the other is through node 6. The aggregate SNRs going through nodes 2 or 6 are as follows:

$$SNR_{325} = a_{32} + a_{25} = 10 + 20 = 30\ dB$$

$$SNR_{365} = a_{36} + a_{65} = 25 + 10 = 35\ dB$$

It is clear that the path through node 6, denoted by SNR_{365} has higher SNR value and can provide lower BER if used.

Adding the processing cost at each node may change the solution presented in this example. There has been a large body of literature devoted to routing protocols, some with energy awareness and some with coding awareness as their main focus. In most of these cases, one challenge that arises in centralized networks with a single fusion center is the asymmetric power consumption of the nodes adjacent to the fusion center. These nodes, also called immediate neighbors, should use their own power to not only transmit their own data but also other packets from the rest of the network. Load balancing strategies may be used to distribute the relaying responsibilities among a larger group of nodes and distribute the bottleneck as much as possible. The network longevity heavily depends on bottleneck nodes like these.

In the previous example, nodes 1, 3, 4, and 7 may last longer as compared to nodes 2, 6, and 5. Node 5, being the connection to the fusion center, may be the first node that runs out of power. The challenge with node 5 is obvious and can be alleviated by adding more on board energy, but the solution for nodes 2 and 6 might not be as easy to handle. This problem becomes more critical when the network scales up.

A low-energy adaptive clustering hierarchy (LEACH) is proposed in [11] that evenly distributes the energy load in the network. This method utilizes data fusion information when make routing decisions. This method has been shown to double the network lifetime in many cases. Since the introduction of this method, there have been numerous extensions to this work with a variety of objectives and refinements that fit different application scenarios.

3.4 ENVIRONMENTAL ADAPTABILITY

The concept of environmental adaptability can be studied from two different perspectives. We will first focus on the physical and chemical characteristics of the environment that may adversely affect our sensor systems. These include extremely high or low temperatures, the presence of corrosive gases or chemicals, and strategies to alleviate or manage these situations. The second part of this section includes discussions of communication channel parameters (focusing on the radio rather than sensor) that may affect network longevity. Propagation challenges in inclement weather or harsh environments are the main focus of these discussions.

3.4.1 Physical and Chemical Parameters

Temperature variations from extremely low, cryogenic temperatures to very high (above 1,000°C) pose certain challenges on sensor, electronics, and radio design. If these challenges are not properly addressed they may cause significant reduction in network longevity. To study these challenges, let us group them based on application time requirements. Applications requiring long term monitoring may need a different design approach as compared to short-term event-based deployments.

One example of long-term monitoring using wireless sensors in harsh environments is engine health-related applications. Whether chemical gases composition or strain on engine components are monitored or temperature is measured, this long term monitoring application can lead to significant savings in conditions-based maintenance rather than traditional schedule-based maintenance [12]. The main challenge in conditions-based maintenance is ensuring that the rate of false alarms is minimized. In harsh environments such as inside a jet engine with a high temperature and corrosion constraints, sensor and radio material should withstand high temperatures for prolonged periods of time. Passive wireless sensors may offer a good solution in such scenarios, with radios and other electronics put outside the engine's hot zone. A simple insulated packaging may work for short term monitoring, but not for this case.

Short-term event-based monitoring network is deployed when an incident such as oil spill in the ocean is reported. Packaging techniques with robust material capable of withstanding harsh, salty, and wet oceans and in some cases cold water may work for short periods of time. In cold environments heaters may be used and in hot sunny places UV light protection may be required. The main point is that in these cases short-term solutions with longevity factors relative to the operation time span may satisfy the lifetime needs.

Ionizing radiation, a characteristic of harsh environments, will reduce the wireless sensing systems lifetime, causing damage to the components. One example of such short-term applications is in space as related to radiation protection. Items used in space for short periods of time may be built using conventional electronic components, while long-term deep space missions may require radiation-hardened electronics (see Figure 3.6) and spacial casing, adding significantly to the mission cost [13].

Figure 3.6 Some examples of radiation-tolerant electronics by API technologies.

In addition to packaging and choice of materials, another approach to overcome harsh environment challenges involves indirect estimation methods. For instance, if the parameter under measurement is studied well it might be possible to identify other parameters outside of the harsh environments that change with a known relationship to the original parameter. For example, temperature changes inside a metallic enclosure may cause the metal's exterior surface to emit IR rays, which can be measured easily using an IR sensor outside the harsh environment. Another example of such a measurement method is spectral analysis of light coming from other planets to determine the material composition of their atmosphere.

3.4.2 Communication Channel Parameters

The communication channel quality may be compromised or degraded due to a variety of reasons. Environmental obstacles such as buildings, hills, or propagation media changes such as rainy or snowy days are among a few of the events that change the channel parameters. Multipath fading, scattering, and shadowing effect add another degree of complexity to communication channel parameter estimation, but they are crucial in optimizing physical and network layer operations to increase network lifetime. Last, but not least, adding mobility to the nodes will introduce another challenge, called the Doppler effect, where the operation frequency is slightly changed.

Multipath fading (see Figure 3.7) is caused by the reception of multiple copies of the same signal at the receiver side. These copies may have traveled through different paths with various lengths, causing them to be in phase or out of phase. When in phase, rays will add up constructively creating a stronger signal than the original single ray free space ray, while out of phase rays may cancel each other creating a very poor low power signal. Scattering is caused when a radio signal passes close to sharp edges of buildings or other obstacles. This effect is notable when the wavelength is small as compared to obstacle dimensions (i.e., in high frequencies) [14]. The shadowing effect happens when a radio wave is partially or totally blocked by obstacles. Figure 3.8 depicts one example of shadowing and scattering in urban networks. Mobile sensor networks pose another challenge: the Doppler effect. This change in the frequency due to the relative velocity of transmitter and receiver units in mobile monitoring networks may be easily addressed in conventional environments, but in harsh environments its effect may become more pronounced.

Dealing with all of these challenges may be understood in well-behaved controlled environments, but those techniques may not necessarily work in harsh environments. The question is how to make sensing systems resilient to these changes in the environment.

Power consumption limitation, performance degradation, and limited choice of materials and components that can operate in harsh environments hinder the application of well understood coding and information theory techniques as well as electronics design methods to address the propagation media challenges in harsh environments. Simple cooling techniques that often go unnoticed in transmitter systems may not work in heated environments where energy dissipation happens at a much slower rate than at room temperature. Adaptive packet length determination based on transmitter internal temperature may be a good solution to avoid overheating.

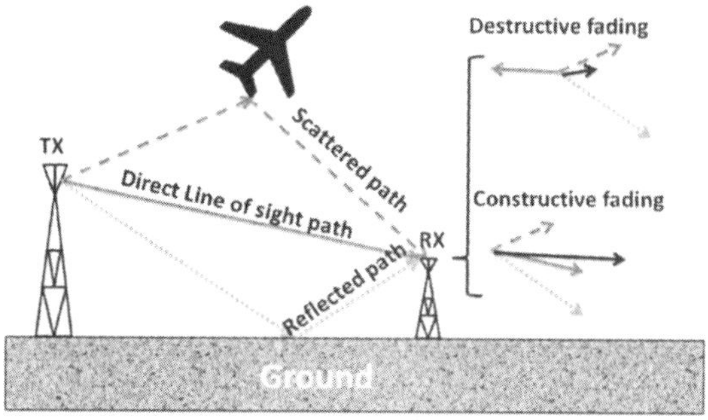

Figure 3.7 Fading phenomena.

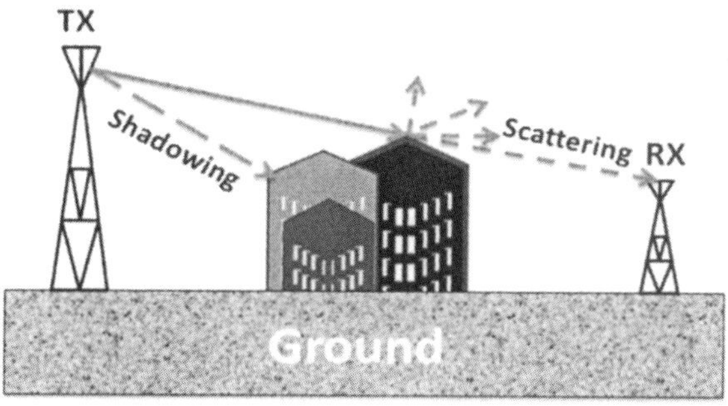

Figure 3.8 Scattering and shadowing effects.

The price to pay is higher delays as packets get smaller and contain less data. An optimal packetization scheme that minimizes end-to-end delay in a wireless network by regulating packet size is presented in [15]. Figure 3.9 shows that a minimum expected delay maybe achieved if the packetization time is selected appropriately. A similar approach may be used to regulate packet size in harsh environments with minimizing the transmitter load as the main objective. The higher the packet size, the higher the throughput, and the lower the delay, while increasing transmitter temperature. These two contradicting arguments (higher temperature lower delay versus lower temperature higher delay) can be used to formulate an optimization problem and find the best possible packet length for each scenario.

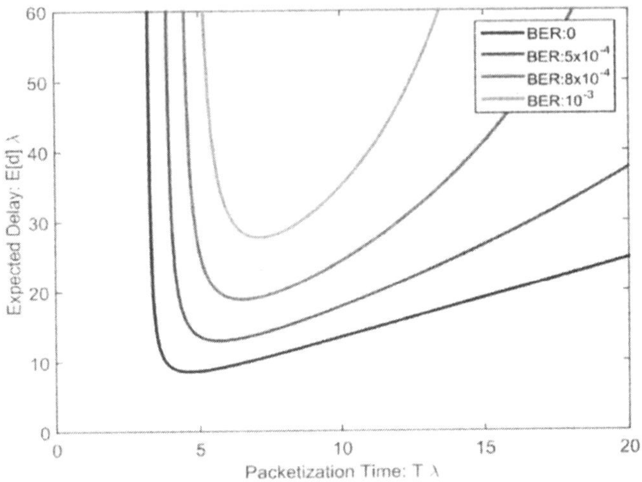

Figure 3.9 Expected delay versus packetization time [15].

Trade-offs between power and performance, lower throughput or higher latency may be a promising direction when designing such systems. Redundancy and efficient use of available degrees of freedom is another tool that should be exploited in such scenarios.

3.5 CONCLUDING REMARKS

The problem of longevity is studied in this chapter. Sensor lifetime and reliability are defined based on the stochastic modeling of sensor failure times and experimental parameter estimation techniques. Interconnecting different systems in series, parallel, and standby are studied and the overall lifetime and reliability of such complex systems are presented using min, max, and sum functions.

The concept of conditional lifetime, which assumes that a system has been in operation for a specific period of time, is also presented. Several examples are provided to add more clarity to all of the above concepts. The chapter continues with sensor system design with discussions on materials used for sensor design as well as power management and duty cycling to render higher longevity of the network. We follow a device, system, network progression to address the longevity in harsh environments at each stage of the process of data collection, preparation, and transmission.

The chapter concludes by shining light on environmental adaptability from two perspectives of physical/chemical parameters as well as propagation mediums. The challenges in these scenarios are highlighted and some directions to address those issues are presented.

This chapter is complemented by a rich selection of highly cited references in archival journals and conference proceedings that offer insight and new solutions for improving network longevity in general and, if adopted, for harsh environments.

References

[1] Maskay, A., A. Ayes, R. J. Lad, and M. Pereira da Cunha, "Stability of Pt/Al2O3-based electrode langasite SAW sensors with Al2O3 capping layer and yttria-stabilized zirconia sensing layer," *2017 IEEE International Ultrasonics Symposium (IUS)*, Washington, DC, 2017, pp. 1–4.

[2] Prasad, R.M., et al., "Sensing in harsh conditions: How to protect SnO2 sensing layer," *2010 IEEE Sensors*, Kona, HI, 2010, pp. 755–758.

[3] Niyazi, L.B., et al., "Energy-Aware Sensor Networks via Sensor Selection and Power Allocation," *2017 IEEE 86th Vehicular Technology Conference (VTC-Fall)*, Toronto, ON, 2017, pp. 1–6.

[4] Gao, Y., W. Cheng, H. Zhang, and Z. Li, "Optimal Resource Allocation with Heterogeneous QoS Provisioning for Wireless Powered Sensor Networks," GLOBECOM 2017 - 2017 IEEE Global Communications Conference, Singapore, 2017, pp. 1–6.

[5] Chu, Z., et al., "Wireless Powered Sensor Networks for Internet of Things: Maximum Throughput and Optimal Power Allocation," in *IEEE Internet of Things Journal*, Vol.5, No. 1, pp. 310–321, Feb. 2018.

[6] Strauss, R., and A. Abedi, "Game Theoretic Power Allocation in Sparsely Distributed Clusters of Wireless Sensors (GPAS)," IWCMC'09, Leipzig, Germany, June 2009, pp.1454–1458.

[7] Jiang, W., X. He, and T. Matsumoto, "Power Allocation in an Asymmetric Wireless Sensor Network," in *IEEE Communications Letters*, Vol. 21, No. 2, pp. 378–381, February 2017.

[8] Behroozi, H., F. Alajaji, and T. Linder, "On the optimal performance in asymmetric Gaussian wireless sensor networks with fading," *IEEE Trans. Signal Process*, Vol. 58, No. 4, pp. 2436–2441, Apr. 2010.

[9] Alirezaei, G., O. Taghizadeh, and R. Mathar, "Comparing Several Power Allocation Strategies for Sensor Networks," WSA 2016; 20th International ITG Workshop on Smart Antennas, Munich, Germany, 2016, pp. 1–7.

[10] Tsiflikiotis, A., and S. K. Goudos, "Optimal power allocation in wireless sensor networks using emerging nature-inspired algorithms," *2016 5th International*

Conference on Modern Circuits and Systems Technologies (MOCAST), Thessaloniki, 2016, pp. 1-4.

[11] Heinzelman, W. R., A. Chandrakasan, and H. Balakrishnan, "Energy-efficient communication protocol for wireless microsensor networks," *Proceedings of the 33rd Annual Hawaii International Conference on System Sciences*, 2000, pp. 10, Vol.2.

[12] Ellis, B. A., and A. Byron. "Condition based maintenance." *The Jethro Project* 10 (2008): pp. 1–5.

[13] Rockett, L., et al., "Radiation hardened FPGA technology for space applications," *Aerospace Conference, 2007 IEEE*. IEEE, 2007.

[14] Ghassemi, A., A. Abedi, and F. Ghassemi, *Propagation Engineering in Radio Link Design*, 553 pages, Springer, January 2013, ISBN: 978-1-4614-5313-0.

[15] Razi, A., F. Afghah, and A. Abedi, (2015). Channel-Adaptive Packetization Policy for Minimal Latency and Maximal Energy Efficiency, *IEEE Transactions on Wireless Communications*, 15. 10.1109/TWC.2015.2503750.

Chapter 4

Sensing System Challenges in Extreme Environments
Electronics and Hardware Issues

This chapter studies design considerations and major challenges for practical deployment of wireless sensing systems in extreme environments such as those in space, and under water. A common thread in these environments is large temperature variations from extremely low cryogenic temperatures to very high temperatures. In space applications, radiation tolerance is another major issue, while in underwater situations waterproofing is a major challenge. Hardware and electronics deployed in harsh industrial environments need to be designed under a completely different set of boundary conditions as compared to common electronic devices that are used in our day to day activities on earth.

In addition to tolerating the harsh conditions, sensing systems need to maintain an acceptable degree of reliability when measuring specific physical, chemical, or biological parameters and transmitting or storing the measured parameters for further processing. Several techniques, including direct, indirect, and proximity-based sensing methods, are presented in this chapter. Several practical examples are presented throughout the chapter to clarify the presented concepts.

4.1 INTRODUCTION

Harsh or extreme environments may have different definitions in different types of applications. Extreme temperature variations from very cold cryogenic levels inside liquid-filled fuel tanks to very hot levels inside jet engines present one perspective on how to define harsh environments. Industrial applications where many metallic or reflective surfaces or objects create scattered or blocked radio waves is another definition for extreme environments.

Chemicals such as corrosive gas or liquid-filled environments present yet another definition for harsh environments. In general, any environment where special design considerations for satisfactory operation of electronic circuits are required may be classified as harsh environment. It is obvious that this chapter cannot address all the possible extreme scenarios, but a recent comprehensive review of existing techniques for designing robust electronics systems and methods to relax some of the extreme environment conditions is presented.

Harsh environments pose several challenges when it comes to deploying sensors and electronic circuits. Extremely high temperatures require special material design to withstand the heat, while still operating with an acceptable accuracy and performance. Low temperatures (as low as cryogenic levels) are also detrimental to optimum operation of electronic circuits including sensors and radios.

Electromagnetically harsh environments with multiple reflective surfaces and obstacles create signal scattering and fading, adding another layer of complexity in design of wireless sensing systems. Wireless sensing has been proved to be advantageous in harsh environments as compared to wired sensing for a variety of reasons. Running wires or replacing batteries may be extremely difficult if not impossible in harsh environments.

Wireless sensing eliminates the need for running wires and if designed properly alleviates the need for battery replacement. Figure 4.1 depicts an example of temperature sensing inside a fuel tank, where running wires is not possible. Wireless magnetic coupling in the near field is one possible solution depicted in Figure 4.1.

One of the most versatile and widely used electronics components are the various types of transistors. InAlN/GaN transistors for high temperature operations (see Figure 4.2) are reported in [1]. In this design, critical parameters such as drain current, threshold voltage, and trans-conductance remain the same from room temperature all the way up to 600°C.

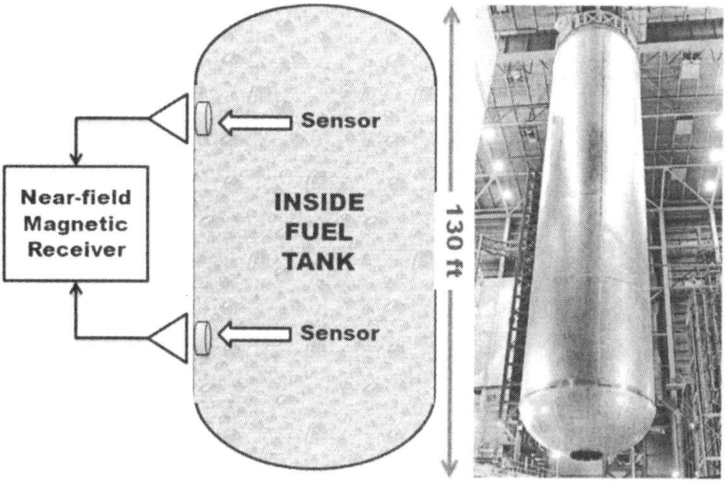

Figure 4.1 Wireless sensors measuring temperature inside space launch system fuel tank, right-side picture (courtesy of NASA).

Designing electronics components and packaging for high temperature, low temperature, and everything in between should consider temperature cycling effects and aging of the components (see Figure 4.3). If components materials are not tolerant of frequent temperature cycling, they may expand and shrink beyond their structural strength resulting in cracks and loss of operation.

Packaging material have been tested by cycling temperatures between 196 and 245°C as in [2]. Germanium Gold solder material has been shown to be resistant to these temperature fluctuations even compared to lead based solders.

Thermal conductivity and dielectric breakdown were optimized for Bismaleimide Resin (BMI) – Boron Nitride Nanoparticle (BNNP) composites in [3] using instruments based on ASTM standards.

Silver based bonding methods have been proposed to replace lead based electronics, but due to their high cost, other alternatives may be required. Zinc has acceptable thermal and electric properties making it a good choice to replace high cost silver. The downside of using Zinc is its high bonding temperature of 430°C. Sn-plated Zn solder can reduce this bonding temperature to 250–350°C [4]. High temperature environments combined with high pressure and vibration as well as chemical and corrosive material call for a new generation of electronics components.

Design and development of reliable packaging systems for these environments are presented in [5]. Ceramic and polymer materials have been tested in [6] to evaluate their properties as packaging and understand their limitations. Popular CMOS technology for 300°C and above have been developed in [7] for mixed signal applications using silicon carbide.

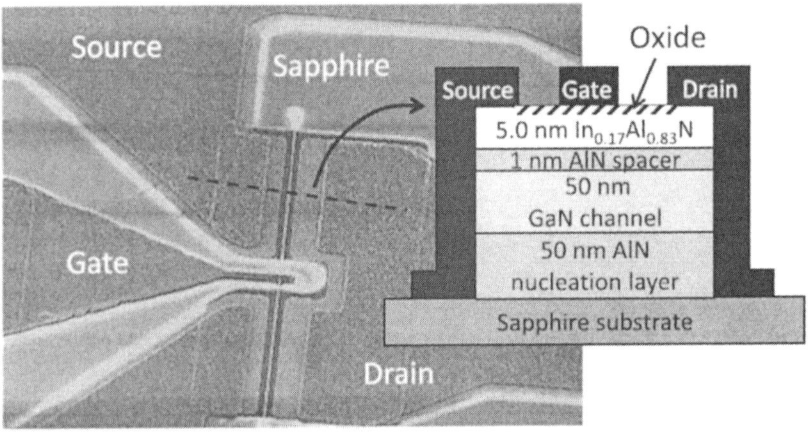

Figure 4.2 Schematic diagram of high temperature transistor built in [1].

One of the weakest points in high temperature electronics systems is the soldered points. Lead based solders can tolerate high temperatures over 200°C; however, with lead being a contaminant, is regulated by environmental protection agencies and should be limited. A new approach replacing lead by Bismuth to create high temperature lead-free solder alloys is presented in [8]. The proposed alloy contains Cu and Sb to improve poor thermal plasticity of Bismuth for temperature ranges from -55 to 200°C.

Summarizing the aforementioned approaches, it is clear that before design commences, specific application requirements including temperature fluctuations and desired operation times need to be considered. Depending on the project budget, lower cost options such as Zinc as compared to Silver may be used to achieve the desired performance while keeping the project cost within the limits. Table 4.1 summarizes various solder types with their temperature tolerances.

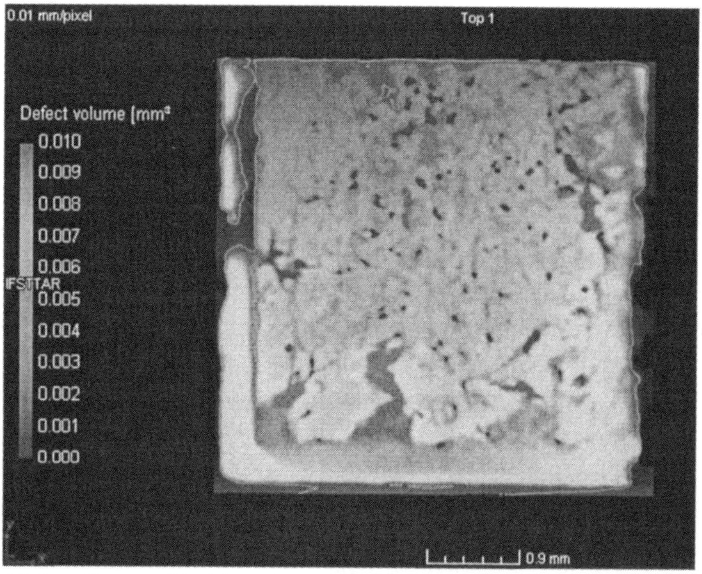

Figure 4.3 Aging effect on solder joints after temperature cycling [2].

Table 4.1

Various Solder Types and Their Temperature Tolerance

Solder Type	Germanium Gold	Sn-plated Zn	Cu and Sb added to Bismuth Alloy
Temperature Tolerance	196 to 245°C	250 to 350°C	-55 to 200°C

4.2 SENSING HARDWARE CHALLENGES

4.2.1 Radiation in Space

In addition to physical and chemical conditions discussed later in this chapter, the space environment poses yet another serious challenge in dealing with ionizing radiation. Radiation tolerance may be classified in three different categories depending on the radiation caused by various particles: particles trapped in the Earth's magnetic field, particles shot into space during solar flares, and galactic cosmic rays.

The last item, galactic cosmic rays, probably the mostly used term, caused by high-energy protons and heavy ions from outside the solar system that can create ionizing radiation. Low energy radiation (also called nonionizing) may not have enough energy to remove ions from the surface of electronic materials, while ionizing radiation may cause serious damage to the electronic components if not radiation hardened properly. It is important to note that the total radiation over a prolonged period of time may cause noticeable damage, while short exposure to radiation in time-limited applications may not be as destructive.

Radiation hardness or tolerance may be improved using different methods. Adding a N_2O annealing step after the interpoly oxidation has been shown to enhance the robustness of a flash memory in [9]. Another approach is described in [10], where carbon nanotubes are used for CMOS technology development with radiation tolerance.

A three-layer structure including a thin oxide, a doped polysilicon, and a thick oxide layer is proposed in [11] to enhance radiation tolerance for CMOS applications. Shallow trench isolation is coined radiation hardened by design in [12]. It has been shown that isolated transistors are hard to at least 500 krd in worst-case conditions. The sample chip in this work was built using 0.25-μm CMOS technology.

4.2.2 Underwater Sensors

Underwater sensors need to be designed to estimate a parameter while tolerating random movements of water that may change the measured value quickly. Also, waterproofing the connections to the transmitter box adds yet another level of complexity to types of devices that can be used in underwater applications.

One example of such a system for underwater sensing is the wireless water quality sensor system built at UMaine's WiSe-Net laboratory funded by the National Science Foundation (NSF) in 2015. This system is comprised of four submerged sensors to measure pH, temperature, conductivity, and pressure.

The collected data is stored on an onboard secure digital (SD) card and wirelessly transmitted to next node or relay as part of a larger mesh network. The submerged sensors were fully sealed and connected through waterproof cables and connectors to a box set higher than water level alongside small streams of water.

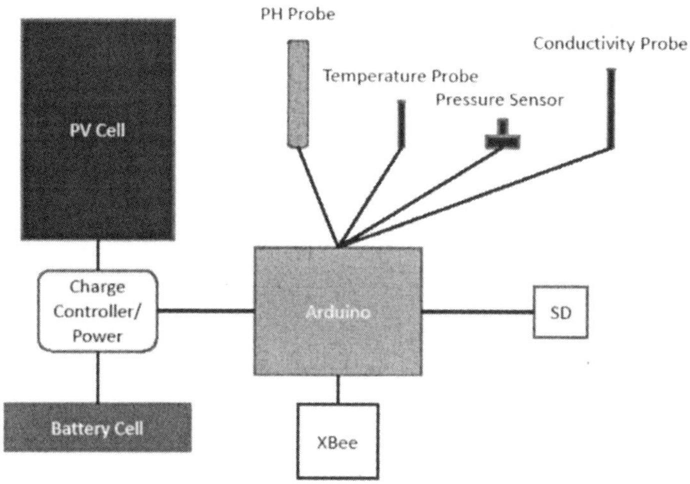

Figure 4.4 Wireless water quality sensor systems built at UMaine's WiSe-Net laboratory.

The design challenges faced in this application included solar cell and antenna placement, which requires a line of sight to the next receiver as well as a view of the sky for charging. Getting too close to the water body, may compromise the electronic circuits when flooding happens, but it provides more efficient charging and transmission (due to lower cable loss).

Getting too far from the water makes the system more reliable, while introducing loss due to the longer cables needed for the radio antenna and solar array. Figure 4.4 illustrates a common design trade-off between reliability and efficiency. Pushing the efficiency limits may not necessarily render a reliable system and vice versa.

4.2.3 Industrial Applications

Measuring various parameters using wireless sensors in industrial applications requires operations in the presence of chemicals, corrosive gases or liquids at high or low temperatures. The choice of material to make sensors resilient to chemicals or encapsulate them to minimize exposure while still maintaining contact with the environment are two different approaches commonly used in industrial settings. An approach based on two closed ring resonators inside a dielectric ceramic matrix is proposed in [13] to enable sensing in corrosive harsh environments. Different ring structures are depicted in Figure 4.5.

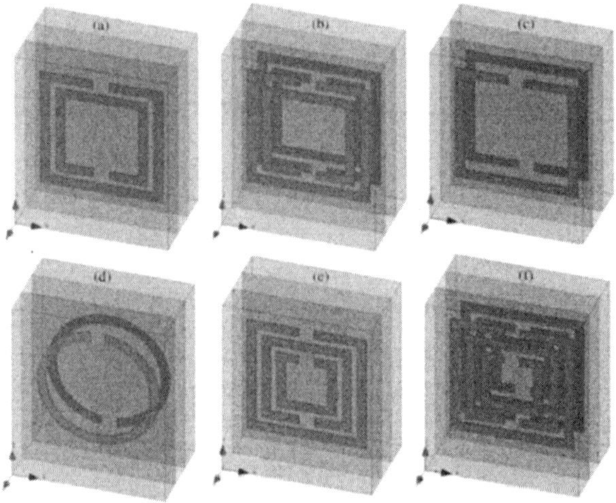

Figure 4.5 Closed ring resonator alternative structures [13].

4.3 COMMUNICATION CHALLENGES

4.3.1 Space

Sensors' communication range in space and easy access to their locations in space vehicles are two major issues considered in this section. A typical low power sensor node such as the IEEE 802.15.4-based ZigBee radios can run on batteries for prolonged period of times yet provide an acceptable range of several hundred feet. Although the node life can be prolonged with signal processing (data compression), duty cycling (sleep/stand by modes), and power allocation at some point the battery still needs to be recharged or replaced.

Passive or battery free sensors such as surface acoustic wave devices provide a shorter range of few feet, but eliminate the need to replace batteries. The trade-off between range and battery requirements can be further explored by designing a hybrid active-passive network to provide both range and accessibility for battery replacement. A sample-architecture of such a hybrid network is illustrated in Figure 4.6.

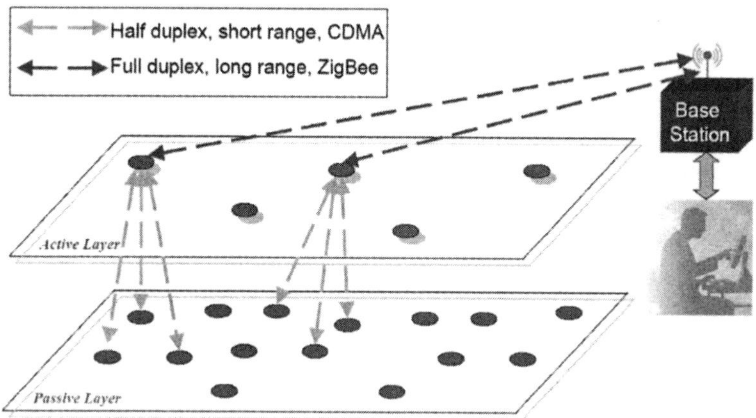

Figure 4.6 Hybrid wireless sensor network with active and passive sensors.

Another example is the use of energy harvesting technology to store enough energy harvested from radio frequency waves to be sufficient for transmitting one packet of data, then wait, store, and transmit when more energy is available. This approach can be refined even further if transmission scheduling is designed to limit transmission when stored power is low or data channel is in poor signal to noise condition to avoid wasting energy.

4.3.2 Underwater

Underwater wireless systems may operate at radio frequency (RF), acoustic, or even optical wavelengths. High attenuation of RF signals and optical communication in water do not allow for long range operations. Acoustic signals on the other hand have been a more widely used mode of operation due to their smaller loss.

For all of these modes of communication, due to random movements of water, alignment of directional antennas is a major challenge. Also, waterproofing the transmitter and receiver boxes add yet another level of complexity to types of devices that can be used in underwater applications.

As seen in Figure 4.7, the penetration depth of radio waves in sea water, fresh water, or ice water are significantly different at different frequencies. At lower frequencies where wireless sensors are going to be close to the water surface, RF may still be a good choice; however, as we get deeper into the water, acoustic waves become more preferable.

For example, radio communications between submarines are usually carried out at the 3–30 KHz band also known as the VLF (very low frequency) band. This means that the antennas will have to be very large, since the wavelengths at these frequencies are very large. Another challenge in designing such systems is the significant loss of signals coming from water to air, in the order of 100 dB for distances less than 20m.

The incoming and outgoing waves will have a phase shift of 90 degrees according to electromagnetic boundary conditions. This means a vertically polarized wave under water will travel horizontally when it enters the air on top of the water and vice versa.

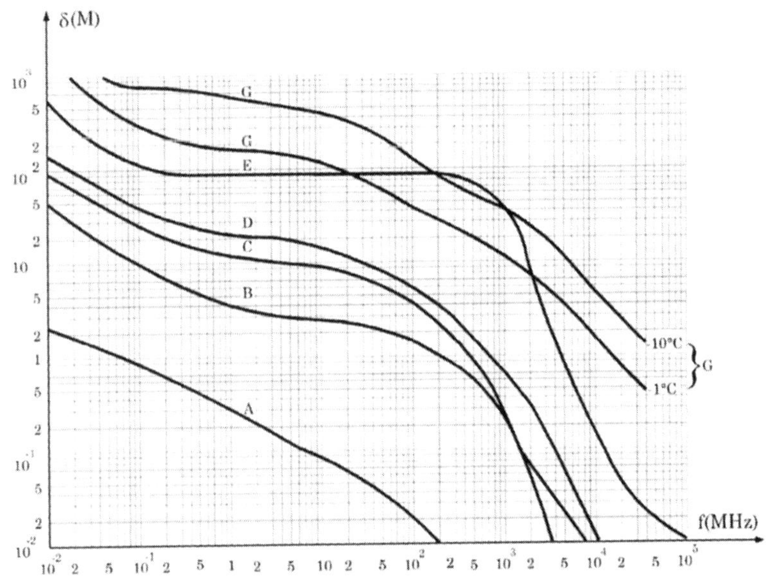

Figure 4.7 Radio frequency wave penetration depth in A: sea water, B: wet soil, C: fresh water, D: dry soil, E: very dry soil, and G: ice based on ITU-R recommendations.

With all these challenges and losses, the transmitter under water should be pumping out 100 KW or more to get a tiny small signal with at least 20 dB signal to noise ratio out. This may be acceptable for radio communication with submarines but certainly not for power limited wireless sensor networks.

A hybrid acoustic and RF architecture (Figure 4.8) with acoustic waves working under water and RF waves above the surface may be the most efficient method to create an underwater sensor network. A dual mode gateway that can communicate in both acoustics and RF modes will link these two sensor networks. For reliability purposes, several of these gateways need to be deployed to create a mesh network.

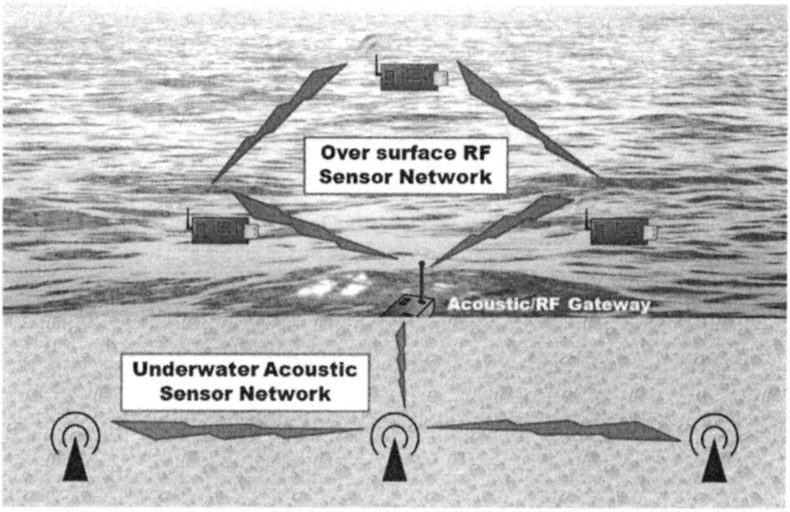

Figure 4.8 Underwater wireless sensor network with acoustic and RF links.

4.3.3 Industrial Applications

Electromagnetically harsh environments with multiple reflective surfaces create a notable amount of interference due to signal scattering and fading. Obstacles in most industrial settings create significant signal attenuation, calling for development of new approaches when designing wireless sensing systems.

Most notable challenges in industrial settings include shadowing, metallic cavities, and interference. Shadowing prevents direct line of sight signal to travel from transmitter to receiver, causing significant degradation in signal quality. A simple solution to address this challenge is relay deployment (see Figure 4.9). Adding a new node to act as relay may increase the cost (noting the scale and number of nodes required to cover a large factory floor); however, if the existing sensor nodes are repurposed to act as relays and still perform their primary function, there will not be any hardware cost increase. The price to pay is using more complicated software implementing the cooperation algorithm. Once challenge that the relay node may face at each time slot is to weather transmit its own packet, relay the TX packet to RX, or stay silent and save power. Game theoretical approaches have been proposed to address this problem.

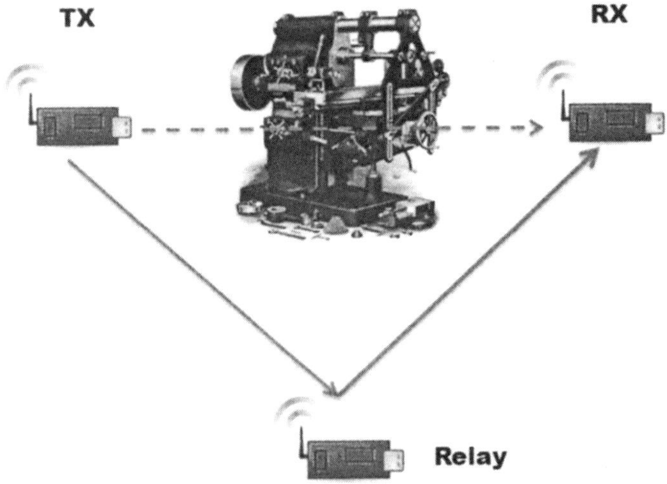

Figure 4.9 Relay deployment to address obstacles causing shadowing.

Two Markovian stationary game methodologies are proposed in [14] to model cooperative relaying. In this model, a two-player game with a finite number of states and a finite number of possible actions for each player is proposed. Both cases of noncooperative and cooperative games are studied. In the noncooperative game model, each player selfishly maximizes his/her own stationary utility function to reach the best response Nash equilibrium strategies. While in cooperative game, players collaborate with each other to jointly maximize the total utility of the game. Different relay decision methodologies may be considered.

In the first methodology, the relay node is assumed to be reliable in the sense that it transmits the received packets from the source node in the next time slot, even if it has its own packets in the transmit queue. This model, called the reliable relaying method, is more efficient in terms of channel use. In the second methodology, another relaying approach is proposed where the relay node can decide to transmit either its own packet or the received packet from the source node, when both are in the queue. The advantage of this methodology, which is called the flexible relaying method, is that a more flexible strategy is utilized to forward the packets in the relay node.

A best response Nash equilibrium strategy can be obtained using the graph depicted in Figure 4.10. It can be seen that the probability of sending a packet from source to destination decreases as the energy cost between source and destination is increased. A reverse trend for communication between source and relay is observed. For a fair solution where both probabilities are equal, the crossover point energy cost may be used as a base line.

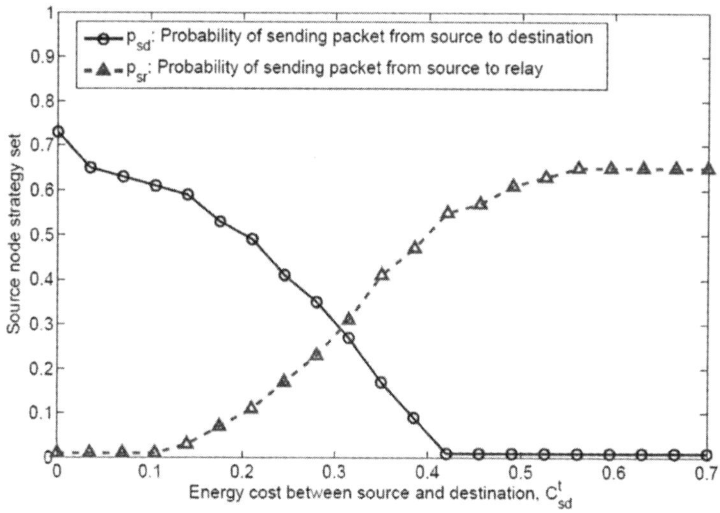

Figure 4.10 Relay deployment to address obstacles causing shadowing.

Another challenge in industrial settings is the hard to reach areas. In some cases, antenna placement and height adjustment might solve the problem as depicted in Figure 4.11.

Finally, the most challenging issue is the interference from devices operating in the same frequency band. The most common source of interference is microwave ovens used for warming food. These are high power radio frequency devices that operate in the 2.4 GHz unlicensed band as do most widely used wireless standards such as WiFi and Bluetooth. The NIST recommendation to address this challenge is depicted in Figure 4.12, with shielding the room containing the microwave oven.

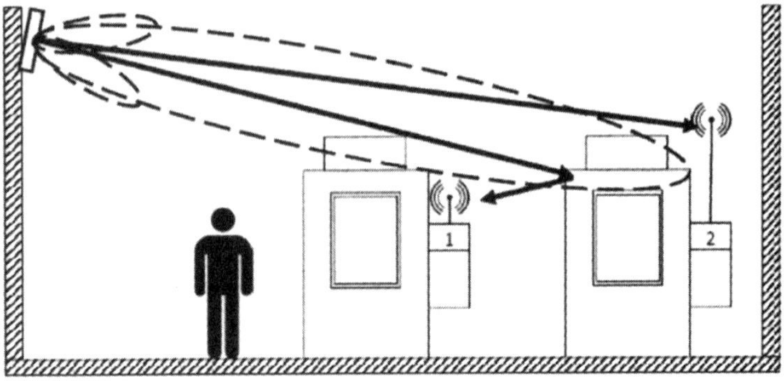

Figure 4.11 Addressing the hard to reach areas using antenna placement [15].

Figure 4.12 Shielding the room containing the microwave oven [15].

In some cases, the interference may come from other radios operating in the same area. In addition to trivial approaches of time division or frequency division multiple access schemes, proper antenna placement, and change of polarization, more effective measures may have to be taken to minimize or eliminate interference.

If the interference is strong, it is recommended that it be decoded along with the actual signal of interest. This may limit the rate of main signal and may be complicated; however, methods of multiuser detection can be used to make it practical. In the other extreme case of very weak interference, we can treat it as noise and utilize well known information theoretic approaches to extract the signal. In a more commonplace case, where the interference has a similar power level as the main signal, orthogonalization methods may be used to manage the interference. An interesting approach called interference alignment, introduced in [16], restricts all interference at every receiver to approximately half of the received signal space, leaving the other half interference-free for the desired signal. This can be accomplished by considering propagation delays on each path in the time domain as well.

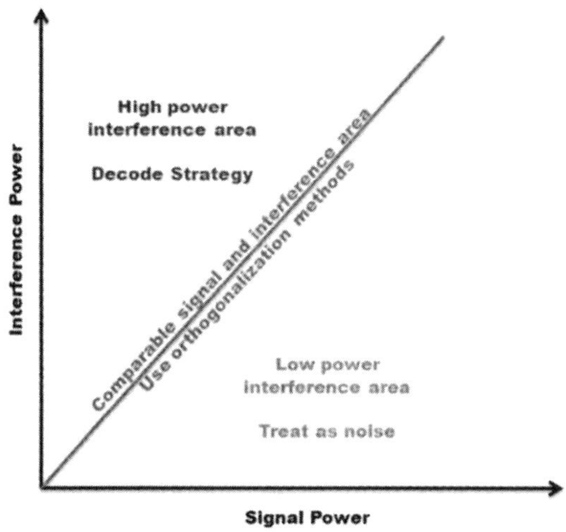

Figure 4.13 Interference management strategy.

4.4 PARAMETER ESTIMATION

4.4.1 Direct Method

Direct placement of sensors in the location of interest is the first order approach to the estimation of desired parameters. In this scenario, assuming the under measurement parameter is denoted by X, and estimation error is denoted as ε, we have

$$\hat{X} = X + \varepsilon \tag{4.1}$$

where $\hat{X}$ is our measured value of X. In order to ensure the measured value is as close to the true value as possible, the minimum mean square error method may be used. This method may be implemented using multiple measurements and linearly combining them with optimized coefficients or using a nonlinear function. The linear method is as follows.

$$\text{Min } E\left[|\hat{X} - (aX + b)|^2\right] \tag{4.2}$$

where parameters a, b can be found by setting derivatives of the E[.] equal to zero. This yields

$$a_{opt} = \frac{C_{\hat{x}x}}{\sigma_{\hat{x}}^2} \tag{4.3}$$

where the optimum value of a_{opt} is obtained by dividing correlation of $\hat{x}, x$ by variance of x. The optimum value of parameter b can be found as follows

$$\mathrm{b}_{opt} = \mu_{\hat{x}} - a\mu_x \tag{4.4}$$

This method results in the minimum possible error with the error orthogonal to the data. This approach can be applied to any estimation problem assuming that the relationship between the data coming from measurements and the parameter under estimation is linearly defined.

In cases where the sensors cannot be placed in the location in which estimation is required, other indirect or proximity based methods may be used. Examples of such scenarios include hot areas, or hard to reach locations.

4.4.2 Indirect Method

Infrared (IR) or laser based temperature measurement methods have been widely used to indirectly measure temperature without having to put a sensor in an extremely hot area. Other indirect measurement examples include a laser signal that is modulated by vibrations on a window glass to measure the seismic wave on a high-rise building. All these examples can be modeled using a hidden Markov model (HMM) where a random experiment is conducted and observations are reported, while the state of the experiment is kept hidden. Several methods such as the expectation propagation or forward backward method have been used in the literature to find the hidden states and provide a measure of the desired parameter. These methods may fail to work when relationships between parameters, states, and observations are not linear. Also, in most solutions, Gaussian noise is assumed, which may not be the case in all practical applications. Another approach, which is called the CEO problem, discussed next.

4.4.3 Proximity Based Technique

The CEO problem is defined by making an estimation based on multiple unreliable observations [17]. In the case of a wireless sensor network where nodes are in proximity of the desired location, all sensors provide noisy or unreliable observation of the desired parameter in their proximity. Since all nodes are close to the location of interest, they all have correlated data streams that can be jointly coded and transmitted efficiently, while providing the best possible estimate of the desired parameter. A distributed coding model for proximity based sensing (see Figure 4.14) is presented and analyzed in [18].

In this work, the power allocation for sensor nodes is optimized and shown to minimize bit error rate if the appropriate power ratio is used. Figure 4.15 shows the bit error rate versus power ratio (between the inner and outer channels) for various SNR. Each graph has a minimum point where the least error is achieved.

4.5 CONCLUDING REMARKS

In this chapter, sensing system challenges in extreme environments including space, underwater, and industrial settings are studied from an electronics and hardware point of view.

Various semiconductor technologies that can withstand high temperatures or corrosive environments are studied. Electronics hardware challenges in space, underwater, and industrial settings including radiation, extreme temperatures, and corrosive gases are reviewed.

Communications challenges in these environments are considered next. Methods combining passive and active sensors in a hybrid network for space to achieve efficiency are presented. For underwater applications a similar hybrid approach combining acoustic and RF sensors is considered. The main challenges presented by communication underwater or between water and air, are hashed out in detail from the electromagnetic propagation point of view.

In industrial settings, the effect of shadowing and interference from commonly used devices such as microwave ovens and WiFi access points are studied, and recommendations to address these issues are presented.

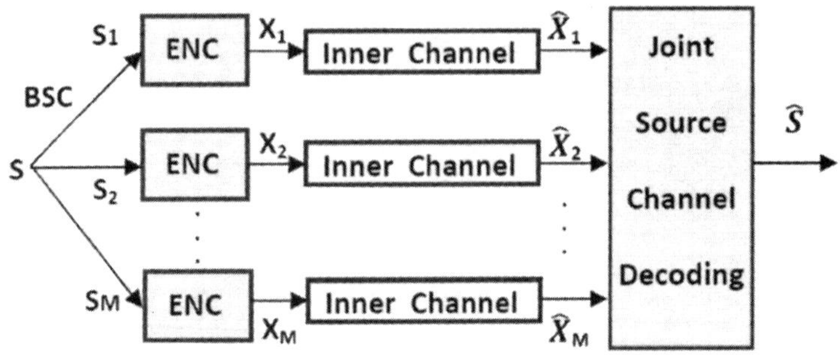

Figure 4.14 Distributed coding for proximity based sensing [18].

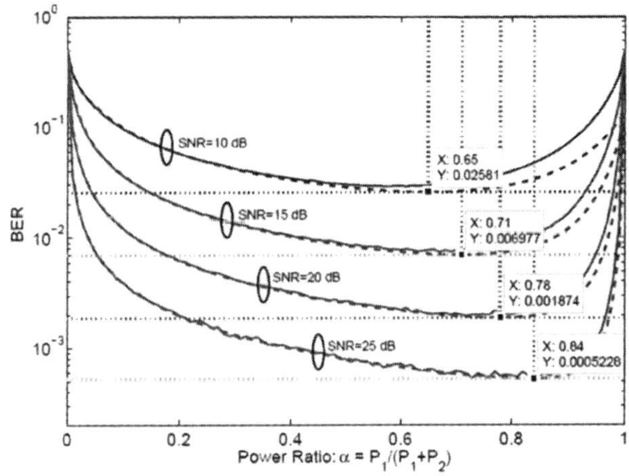

Figure 4.15 Minimizing bit error rate using optimum power allocation [18].

Finally, the chapter concludes by reviewing estimation theory techniques for direct and indirect estimation. A new method for proximity based estimation combined with distributed coding for energy efficient communication is introduced and studied. The bottom line in communication in hard to each or extreme environments is the need to focus on energy efficiency and to strive to design a system that takes advantage of novel algorithms to transmit data with lowest possible power, which means longer lifetime, range, and reliability.

References

[1] Herfurth, P., et al., "Ultrathin Body InAlN/GaN HEMTs for High-Temperature (600 C) Electronics," in *IEEE Electron Device Letters*, Vol. 34, No. 4, pp. 496–498, April 2013.

[2] Ibrahim, A., and Z. Khatir, "Power cycling ageing tests at 200°C of SiC assemblies for high temperature electronics," *2013 15th European Conference on Power Electronics and Applications (EPE)*, Lille, France, 2013, pp. 1–10.

[3] Coker, Z., H. Diaz, N. D'Souza, and T-Y. Choi, "Boron Nitride Nanoparticles-based thermal adhesives for thermal management of high-temperature electronics," *Fourteenth Intersociety Conference on Thermal and Thermomechanical Phenomena in Electronic Systems (ITherm)*, Orlando, FL, 2014, pp. 421–425.

[4] Park, S. W., et al., "Partial transient liquid phase bonding for high-temperature power electronics using Sn/Zn/Sn sandwich structure solder," CIPS 2014; *8th International Conference on Integrated Power Electronics Systems*, Nuremberg, Germany, 2014, pp. 1–6.

[5] Riches, S., and C. Johnston, "Electronics design, assembly and reliability for high temperature applications," *2015 IEEE International Symposium on Circuits and Systems (ISCAS)*, Lisbon, 2015, pp. 1158–1161.

[6] Shaddock, D., and L. Yin, "High temperature electronics packaging: An overview of substrates for high temperature," *2015 IEEE International Symposium on Circuits and Systems (ISCAS)*, Lisbon, 2015, pp. 1166–1169.

[7] Rahman, A., et al., "A family of CMOS analog and mixed signal circuits in SiC for high temperature electronics," *2015 IEEE Aerospace Conference*, Big Sky, MT, 2015, pp. 1–10.

[8] Cho, J., et al., "Exploring Bismuth as a New Pb-Free Alternative for High Temperature Electronics," *2016 IEEE 66th Electronic Components and Technology Conference (ECTC)*, Las Vegas, NV, 2016, pp. 432–438.

[9] Tiao-Yuan Huang et al., "Improving radiation hardness of EEPROM/flash cell by N2O annealing," in *IEEE Electron Device Letters*, Vol. 19, No. 7, pp. 256–258, July 1998.

[10] Brock, D. K., et al., "Carbon Nanotube Memories and Fabrics in a Radiation Hard Semiconductor Foundry," *2005 IEEE Aerospace Conference*, Big Sky, MT, 2005, pp.1–9.

[11] Manchanda, L., et al., "A high-performance directly insertable self-aligned ultra-rad-hard and enhanced isolation field-oxide technology for gigahertz silicon NMOS/CMOS VLSI," in *IEEE Transactions on Electron Devices*, Vol. 36, No. 4, pp. 651–658, April 1989.

[12] Lacoe, R., et al., "Application of hardness-by-design methodology to radiation-tolerant ASIC technologies," in *IEEE Transactions on Nuclear Science*, Vol. 47, No. 6, pp. 2334–2341, December 2000.

[13] Kairm, H., et al., "Concept and Model of a Metamaterial-Based Passive Wireless Temperature Sensor for Harsh Environment Applications," in *IEEE Sensors Journal*, Vol. 15, No. 3, pp. 1445–1452, March 2015.

[14] Afghah, F., "Design and Analysis of Cooperative Communications Networks Using Game Theory," PhD Dissertation, University of Maine, Orono, August 2013.

[15] Candell, R., et al., "Guide to Industrial Wireless Systems Deployments", NIST Advanced Manufacturing Series, 300–304, April 2018.

[16] Cadambe, V. r., and S. A. Jafar, "Interference Alignment and Degrees of Freedom of the K-User Interference Channel," in *IEEE Transactions on Information Theory*, Vol. 54, No. 8, pp. 3425–3441, August 2008.

[17] Berger, T., Z. Zhang, and H. Viswanathan, "The CEO problem [multiterminal source coding]," *IEEE Trans. Inf. Theory*, Vol. 42, No. 3, May 1996.

[18] Razi, A., F. Afghah, A. Abedi, ``Power Optimized DSTBC Assisted DMF Relaying in Wireless Sensor Nets with Redundant Super Nodes," *IEEE Trans. on Wireless Com*, Vol. 12, No. 2, pp. 636–645, 2013.

Chapter 5

Sensor Design for Harsh Environments: Material Science Perspective

This chapter presents recent developments in material science that enable the implementation and use of wireless sensors in harsh environments. The common property among passive sensors is variation of an electric property such as voltage or current with the measured parameter. As long as the relation between the changed electric current or voltage and the measurand is known, a fairly reasonable estimate can be obtained. Active sensors with radios, microprocessors, and batteries may not be easily placed in harsh environments such as high temperature areas. Passive sensors can solve this issue since they do not require onboard processing and batteries. In this chapter, we study passive sensors modeling and design from material science and signal processing perspectives.

5.1 INTRODUCTION

Passive sensors are sensors that do not include battery-powered electric components. As such they are rigid devices that can operate in extremely harsh environments. For instance, passive sensors can be used in temperatures as high as 1,000 degrees Celsius. Further, they require minimal maintenance during their long lifetimes. Based on these facts, passive sensors are the only solution for some applications (e.g., jet engine temperature monitoring) and the most economical solutions for some other applications. The operation of passive sensors, regardless of their technological variations, is based on receiving an interrogation signal from a remote interrogator and reflecting it back to the interrogator for further processing.

The response signal may include controlled distortions and morphological features that convey information about a parameter of interest such as stress, pressure, temperature, and chemical particles. Figure 5.1 depicts two examples of passive sensors. The surface acoustic wave sensor (left) built at the University of Maine (2008) operating at 103

MHz in the FM band was used for measuring temperature. RFID based sensing using Alien Technology tags (right) operating in the 900 MHz UHF band and University of Maine localization software were used in the shape monitoring of inflatable space habitats (2011).

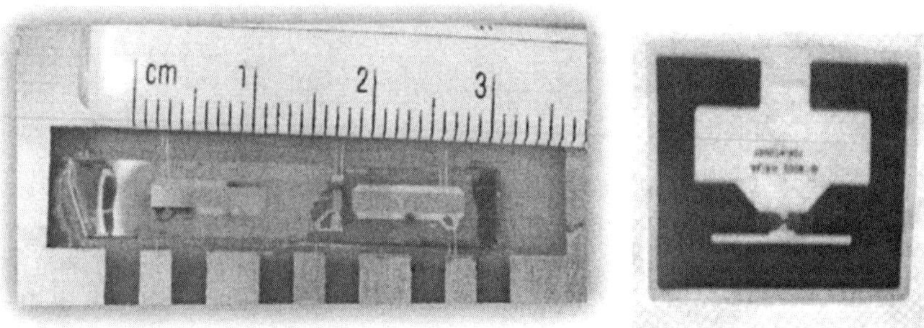

Figure 5.1 Some examples of passive sensors: surface acoustic wave sensor (left), RFID based sensor (right).

An important challenge that arises in using passive sensors is the concept of interference; since these devices tend to respond in parallel, hence using conventional multiple access methods is challenging. Interference is an important phenomenon that arises in wireless communication systems due to an undesired mix of intended signals with other signals. Interference can substantially deteriorate the received signal quality and corrupt the payload data. The interfering signals produced by other users in the system or by users of other systems. Interference can also be due to the malfunction of electric components when they produce signals that violates subtle multiple access criteria. Interference is a well-studied concept in the wireless community.

Comprehensive information about interference sources and interference management techniques can be found in communication system textbooks as well as review papers including [1, 2, 3, 4]. This chapter studies the design of a measurement system composed of wireless passive sensors. Its main focus is on interrogation system design and interference mitigation techniques to improve measurement fidelity. Interested users are referred to [5, 6] for further details.

5.2 SENSOR RESPONSE SEPARATION

Multiple sensors' responses may interfere with each other if not separated properly. In this section, sensor response separation in time, frequency, and code domains are studied, and several examples are provided for further clarity.

5.2.1 Time Domain

The first approach to separate sensor responses is time domain separation (see Figure 5.2). In this method, each sensor is designed in such a way that its responses are received at nonoverlapping times at the interrogator, even though all sensors receive the interrogating signal at the same time. The main advantage of this method is its simplicity at the sensor side as well as at the interrogator side. In order to understand the limitations of this approach, the following example is presented.

Example: Consider a sensor network with 50 nodes each responding to the interrogating signal after a predetermined internal delay of j ms for sensor j. Assuming that distances between sensors and the interrogator is negligible compared to the speed of light (or RF waves), what is the upper bound on the sampling frequency?

Solution: After the first interrogating signal is transmitted, the first sensor will respond with a 1 ms delay, the second sensor will respond with 2 ms delay, and so forth. The last sensor will respond after 50 ms. Therefore, in between each two interrogating signals there is a 50 ms delay. This renders the maximum frequency to be less that 1/50 ms or 20 Hz. The previous example illustrated one of the shortcomings of sensors that are designed using this method, with time-separated responses. The sampling rate limitation may not be as important when estimating a slow changing parameter such as temperature, but it is a limiting factor when fast changing parameters such as structural modes of vibration are estimated. The next method of separation will address this limitation.

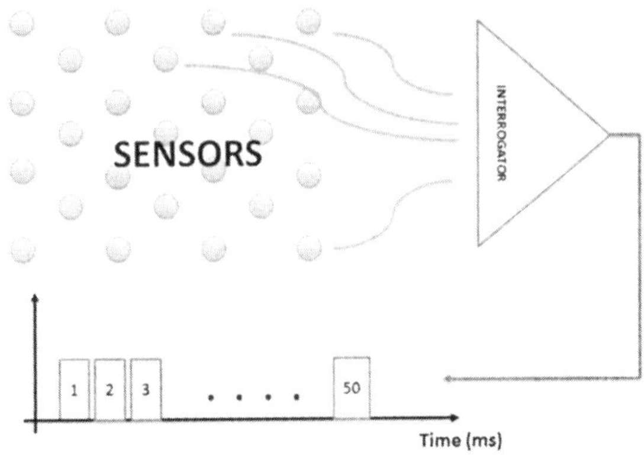

Figure 5.2 Time-separated sensor responses.

5.2.2 Frequency Domain

In addition to the time domain separation discussed in Section 5.2, sensors' responses may be separated from each other in frequency domain (see Figure 5.3). For example, dividing a predetermined bandwidth to several subchannels, enables all sensors in a network to respond at the same time. This method can practically sample fast-changing parameters as fast as the interrogator can handle them. Let us study another example to understand the limitation of this approach.

Example: A 100 node sensor network is operating on a licensed band from 895–896 MHz. Assuming that all sensors have subchannels with similar bandwidths allocated to them, what is the main limitation of this sensor network when it comes to sampling?

Solution: The total available bandwidth of 1 MHz is divided among 100 sensors, providing each 10 KHz bandwidth. This subchannel allows for communicating a signal with 5 KHz bandwidth sampled at the Nyquist rate. Phone voice quality is one example of such a signal measured in distributed sensor network for surveillance and monitoring. This example shows that the main limitation comes from the original total bandwidth allocation. For instance, if a video network were to operate with 6 MHz channels per sensor, a 600 MHz bandwidth is required. A hybrid solution that allows sensors to operate on various bands and respond at various times may allow for using even more sensors without the need for large amounts of bandwidth. A third degree of freedom is explored in Section 5.2.3.

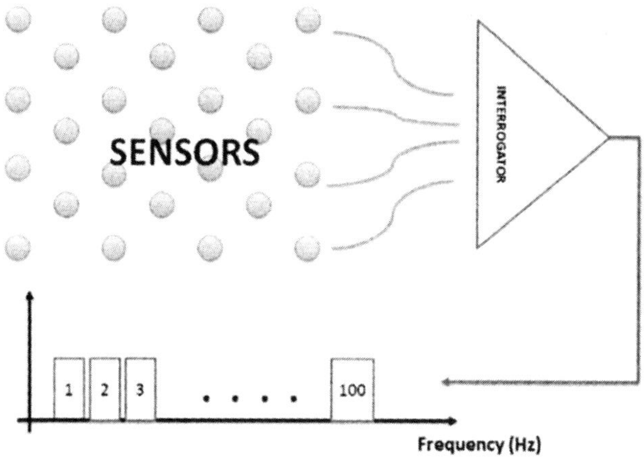

Figure 5.3 Frequency-separated sensor responses.

5.2.3 Code Domain

Code domain separation can be used by itself or in conjunction with other methods such as time and frequency and can add another degree of freedom to design phase. This method, also called code division multiple access (CDMA), allows the sensors to respond at the same time on the same frequency band. The sensors are designed in such a way that each response is orthogonal to the others, allowing for matched filter detection at the interrogator. Several families of orthogonal codes are available for such sensor designs. One point to consider is that all these digital codes will be implemented in sensors with an analog response, which renders the signals to be quasi-orthogonal.

Therefore, the optimization of signals' correlations should take sensor response into account when selecting good codes. There are various techniques to design codes for direct sequence spread spectrum (DSSS) systems. A desired property of these codes is a low pairwise cross-correlation to minimize the undesired interference effects [7]. A popular variant of these codes is the Gold code, which is easy to implement for conventional transceivers using simple components such as shift register and basic arithmetic operations. An implementation of differential reflective delay line (DR-DL) sensors based on gold codes using surface acoustic wave (SAW) devices is presented in [8].

Example: The following codes of length 8 chips are used to provide multiple-access in a DSSS system. Binary phase shift keying (BPSK) is used for modulation.

$$C_1 = [00001111], C_2 = [11110001], C_3 = [00110011]$$

The respective signals are depicted in Figure 5.4. Which pair of the codes provides a better separation property?

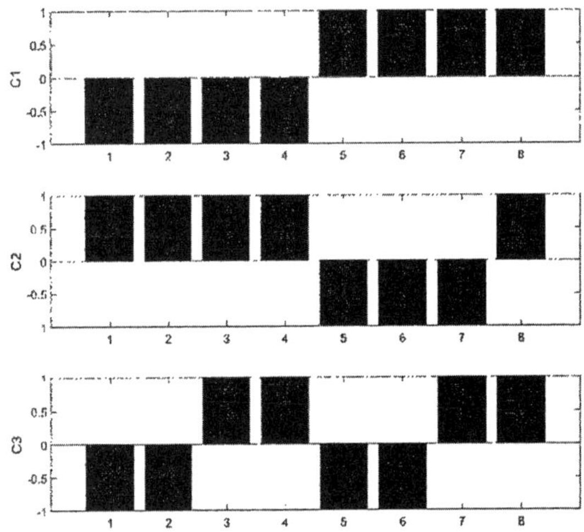

Figure 5.4 Code-separated sensor responses.

Solution: We first obtain the BPSK modulated signals by mapping (0→-1, 1→+1). Note that, this mapping is arbitrary and (0→+1, 1→-1) can be used as well. The resulting modulated codes are:

$$B_1 = [-1\ -1\ -1\ -1\ +1\ +1\ +1\ +1]$$

$$B_2 = [+1\ +1\ +1\ +1\ -1\ -1\ -1\ +1]$$

$$B_3 = [-1\ -1\ +1\ +1\ -1\ -1\ +1\ +1]$$

Now we calculate cross-correlation among the codes as follows:

$$\text{Cross}(B_i, B_j) = \frac{1}{n}\left|\sum_{k=1}^{n} B_{ik}B_{jk}\right|$$

where B_{ik} is the kth chip of code B_i and $n = 8$ is the length of each code. We obtain:

$$\text{Cross}(B_1, B_2) = \frac{|-6|}{8} = \frac{3}{4}$$

$$\text{Cross}(B_1, B_3) = \frac{|0|}{8} = 0$$

$$\text{Cross}(B_2, B_3) = \frac{|2|}{8} = \frac{1}{4}$$

The results suggest that (B_1, B_3) is the best choice. In fact, they are fully orthogonal codes as part of the wellknown Walsh codes widely used in CDMA systems.

5.3 PASSIVE SENSOR MODELING

The essence of passive sensor operation is based on inducing controlled changes in the response signal, which depends on a physical parameter under measurement. In other words, the response signal conveys information about the desired physical parameters. Hence, response signals convey useful information about an environmental or a physical parameter that can be extracted by advanced signal processing methods. Wireless passive sensor networks are typically comprised of multiple passive sensors, one or multiple interrogators, a receiver, and a data fusion center.

The interrogator, receiver, and processing units are typically combined into one computer-based system called the interrogation system. The interrogation module operates similar to a wireless transceiver by propagating a radio wave to the surrounding passive sensors. The electromagnetic waves are received by the passive sensors and reflected back after some distortions, which depend on the physical parameter under measurement. The interrogator receives the reflected signals and extracts measurement information based on the changes on the received signals.

There are different types of sensors to realize a mapping from a physical parameter such as temperature, stress, and light into signal distortions. Here, we focus on a specific class of passive sensors built using SAW technology. The majority of these devices operate based on the fact that the shape of sensors changes with a parameter under measurement. For instance variation of temperature causes contraction or expansion of the device.

SAW sensors are built in different types in order to incorporate the physical changes into the response signals. The idea of resonation-based sensors is to map the measurand to the resonance frequency. The interrogator typically transmits a broadband signal, but the reflected signal includes a peak at a resonance frequency, which is related to a specific parameter under measurements [9]. For instance, SAW resonators are proposed to implement vapor sensors, where the resonance frequency is a function of temperature and humidity [10].

A more advanced version of these devices has been developed in the University of Central Florida by using orthogonal frequencies, where different reflectors have different resonance frequencies and a combination of these reflectors can be used to realize multiple access for systems with more than one sensor [11].

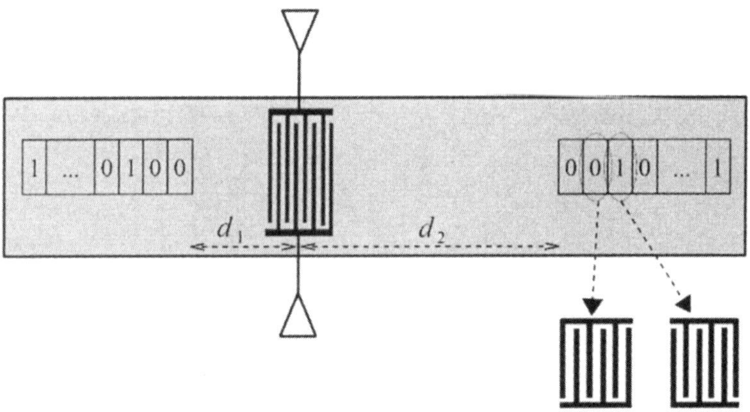

Figure 5.5 A block diagram of a differential reflective delay line based on SAW technology.

Here, we turn our attention to a new variant of SAW sensors called DR-DL devices (see Figure 5.5). The operation of these devices is based on techniques borrowing from spread spectrum systems by producing CHIP sequences using a special structure called reflectors. These devices are built at the University of Maine; and the idea is to use orthogonal binary codes to enable multiple access systems similar to DSSS systems in wireless networks.

Each sensor is composed of an antenna, an inter-digital transducer (IDT), and two series of reflectors with mirrored patterns, all printed on a temperature-sensitive substrate such as Lithium Niobate Y Z - LiNb03. The electromagnetic wave is received by the antenna and is converted to an acoustic wave by the IDT. The received electromagnetic wave induces an electric voltage difference between the upper and lower bars of the IDT, which energizes the free electrons on the substrate with a crystal structure. The vibrating electric charges generate acoustic waves that can easily propagate through the device surface. These acoustic waves hit the two sets of reflectors in both sides of the devices and reflected back towards the IDT in middle. The reflected acoustic waves are converted back to electromagnetic signals and propagated back by the antenna. Since the distance of two reflective fingers in both sides of the device from the IDT is different, the received signals do not overlap.

Let us note that a key property of surface acoustic waves is their relatively lower propagation velocity, which is 100,000 times slower than that of an electromagnetic wave. As such, relatively small distances are translated into measureable propagation delays. Another key property of the utilized piezo-electric material is that its length changes with temperature almost linearly over a wide temperature range. If the temperature rises, the relative distance between the reflectors increases and as such the arrival time difference between the two copies of the received signal increases. Therefore, using a simple mapping table, the temperature can be measured. These devices are rigid with a wide range of operations that can go up to 1,000 degrees Celsius and hence are of particular interest in extremely harsh environment, where the conventional battery-operated sensors fail.

Example: A SAW-based passive sensor of type DR-DL is deployed to measure temperature inside a jet engine. During a test phase, the interrogator sends a single pulse and inspects the reflected signal. The received signal includes two pulse sequences whose relative distance changes with temperature. Table 5.1 shows the results of training phase:

Table 5.1

The relation between the temperature and the relative distance between the two pulses in the response signal of a passive sensor of type DR-DL

Temperature (Celsius)	T1=100	T2=200	T3=?
Time between pulses	t1=5 μsec	t2=5.5 μsec	t3=7.5 μsec

Find the temperature for the third measurement (T3). If the velocity of the acoustic wave is 3×103 m/sec, find the device expansion coefficient versus 1 degree of temperature change.

Solution: We assume a linear operation. Therefore, the rate of device expansion versus temperature is: (t2−t1)/(T2−T1) = (5.5×10−6−5×10−6)/(200−100)= 5×10−9 sec/Celsius. T3 can be easily found as T3=T1+ (t3−t1)/ 5×10−9=100+2.5×10-6 /5×10-9=100+500=600. Note that 1 degree of temperature variation translates into 5×10-9 sec, which is equivalent to a physical expansion of d= 5×10−9 × 3×103 =15μm.

In DR-DL sensors, multiple-access is enabled using reflectors with unique patterns. These unique patterns are also considered device IDs and can be used for identification purpose. The reflectors have two polarizations to represent digits +1 and −1 as depicted in Figure 5.5. In BPSK modulation, the signals for digits +1 and −1 are simply flipped versions of the same sine wave.

Therefore, we can have multiple sensors operating in parallel with negligible interference. Gold codes can be used to realize a good separation property. The reflectors have mirrored patterns, and as such the reflected signal includes two copies of the same bitstream with different delays. Figure 5.6 demonstrates a sample impulse response of a DR-DL sensor with pattern 11001, where each reflector includes three fingers.

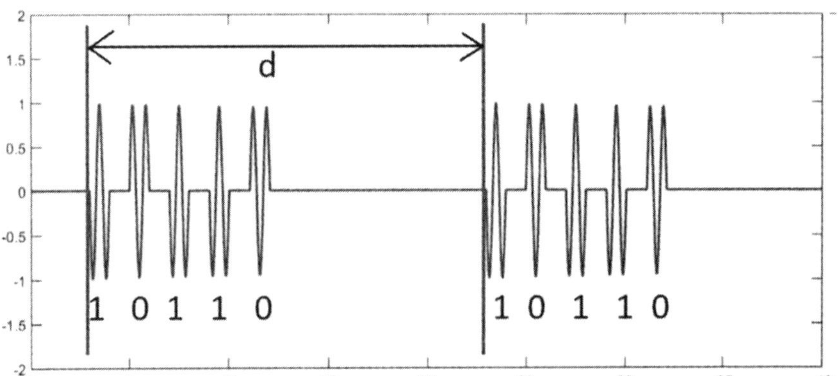

Figure 5.6 A sample impulse response of a SAW-based passive sensor of type DR-DL with pattern: 10110. The impulse response includes two BPSK-modulated bitstreams of pattern 10110 with different delays. The distance between the two patterns (d) is proportional to the length of the sensors and conveys information about the current temperature, stress, or pressure; and hence can be used to measure these parameters.

Finally, we note that analytical modeling of SAW devices is based on considering the first order terms and using simplified assumptions such as the full absorption of leaking signals at the endpoints of the device. However, these assumptions do not hold in practice. Further, fabrication variations impact the device operation and should be taken into account. Therefore, finding a reliable analytic model for a passive sensor can be extremely challenging. One shortcut for this problem is using computer-based software solutions [12, 13].

5.4 INTERROGATOR DESIGN

In Section 5.3, fundamentals of SAW sensor operation are provided. In DR-DL devices, similar to other types of SAW devices, an electromagnetic wave is radiated by the interrogator. One of the drawbacks of passive sensors is their limited transmission power, since the reflected signal is significantly weaker than the interrogation signal. The main reason for this issue is that passive sensors are not equipped with any sort of amplification modules in contrast to active sensors. This is an important drawback of these devices, as it restricts their operation distances to as low as a few meters.

In order to enhance this property, instead of using a single pulse by interrogator, we can use a sequence of pulses with a pattern matching the time reversal of the target device's reflector pattern. The only cost paid is a slightly longer interrogation phase, which is acceptable in systems with slow-varying measurands. In this approach, the response signal is the convolution of the input signal and the impulse response with the reversed patterns. Consequently, it includes two large pulses. However, if the interrogation signal pattern does not match the time-reversal of the undesired device pattern, the response signal is more like a random noise. Therefore, this method reduces interference among devices operating at the same frequency.

Figure 5.7 illustrates this concept by presenting the response signal of a target device with pattern 01001010, as well as that of an interfering device with pattern 11100111. The interrogation signal pattern is 01010010, which is time reversal of the target sensor's pattern. The target sensor exhibits a response signal with two large peaks whose distances can be used to measure a parameter of interest. Figure 5.8 represents an actual response signal obtained using the SAW sensors built at the University of Maine [8].

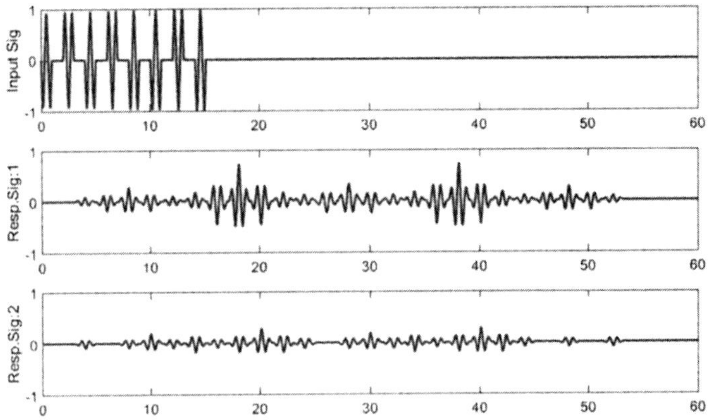

Figure 5.7 An illustrative example of the operation of DR-DL sensors. The interrogation signal pattern is 01010010. The target sensor (sensor 1) whose pattern 01001010 is the time reversal of the interrogation signal presents a response signal with two strong peaks. The distance between the peaks is proportional to the temperature and can be used to read the temperature. The second sensor with a different pattern of 11100111 exhibits a noise-like response signal and hence does not significantly interfere with the target sensor.

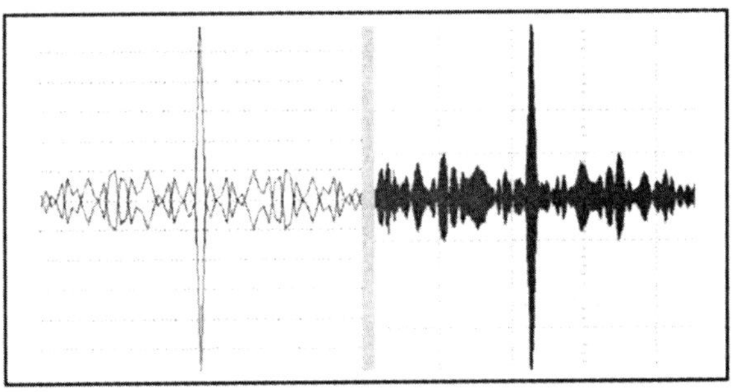

Figure 5.8 Theoretically driven response signal vs. a measured response signal for DR-DL sensor. (Source: [8]).

Practical implementation of interrogators can be done based on various platforms. One popular option is using FPGA boards due to their flexibility, robustness, high performance signal processing, and wireless connectivity [8, 14]. However, specific radio lines and antennas should be designed for better performance.

5.5 INTERFERENCE MITIGATION

In order to obtain accurate sensor measurements, it is preferable to have interference-free communication from the sensors to the reader using one of the aforementioned multiple access methods. However, complete elimination of interference is not feasible. Firstly, interfering signals may be related to a system other than our designed system, and hence are out of our control. Secondly, the above transmission systems rely on idealistic assumptions that do not often hold in practice. For instance, in time domain multiple access methods, full separation among different sensors' pulses is required, as elaborated in Section 5.2.1. To realize a full separation between pulses, rectangular time-domain pulses should be used. However, a rectangular pulse includes infinite frequency spectrum and hence is not practical, and we usually settle using pulses with interfering tails, which cause inter symbol interference (ISI). Further, full synchronization is hard to achieve in practical systems. Other multiple access methods are also prone to interference due to imperfect electric components and antennas, fabrication artifacts, Doppler effects due to the mobility of sensors, loss of frequency synchronization, and nonzero cross-correlation among codes in CDMA. Therefore, dealing with interference in sensor networks is unavoidable. Two major approaches include interference avoidance and interference management.

Techniques, in the first approach used by transmitter are to minimize the level of interference. The most recent one is interference alignment by employing precodes in the transmitters to align all interfering signals into one direction to be easily filtered out by the receiver [15]. However, these methods cannot be directly implemented on passive sensors and they require more advanced signal processing in the interrogator side, which can be considered as a potential future research direction. In the second approach, advanced signal processing is used in the receiver to eliminate interfering signals after occurrence. Advanced methods such as parallel interference cancelation (PIC), serial interference cancelation (SIC), matched filtering, and channel equalization are used in order to separate the desired signals from the interfering signals [16]. This is more crucial in passive sensors, since due to the lack of flexibility in the sensor sides, utilizing interference mitigation techniques is extremely challenging. Therefore, the level of interference is higher than usual; this calls for a more intelligent reader to polish the received signal.

Ideas recently have been proposed to use directional antennas and beamforming techniques to reduce the interference level in passive sensor networks [17]. In this work, the authors proposed using a cyclic array antenna with 16 antennas, where only four adjacent antennas are active at each time point. The selection of active antennas is performed electronically by controlling the feeding current, which enables steering the propagation direction towards the target sensor location. The 3rd order Chebyshev polynomial $T(x) = 4x^3 - 3x$ is used to adjust feeding current amplitudes to minimize the side lobe level with respect to the given main lobe width. This work proposes a two-step interference reduction algorithm, where in the exploration phase, the existing sensors are recognized and in the measurement phase, the target sensors are read by respective interrogation signals. This flexible approach enables the use of a universal interrogation system with any set of sensors with arbitrary patterns. Figure 5.9 presents a block diagram of this interrogator along with the simulated antenna pattern using 4nec2 software.

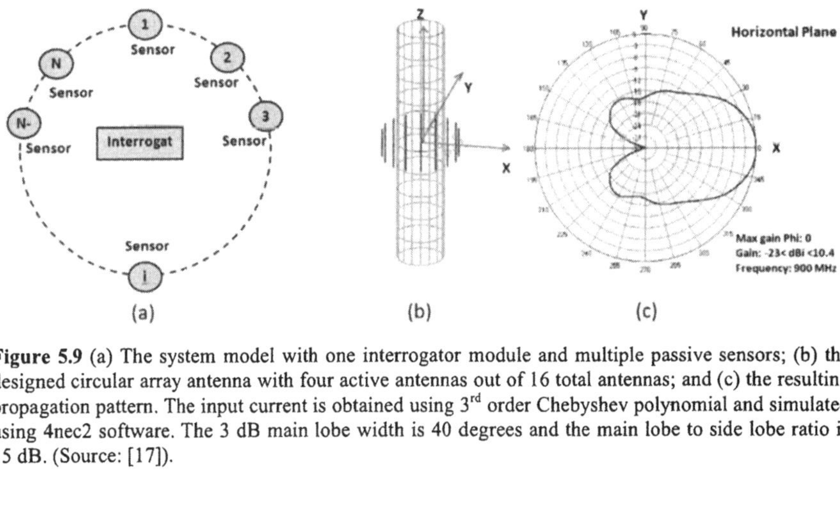

Figure 5.9 (a) The system model with one interrogator module and multiple passive sensors; (b) the designed circular array antenna with four active antennas out of 16 total antennas; and (c) the resulting propagation pattern. The input current is obtained using 3rd order Chebyshev polynomial and simulated using 4nec2 software. The 3 dB main lobe width is 40 degrees and the main lobe to side lobe ratio is 15 dB. (Source: [17]).

5.6 DESIGN CHALLENGES

The use of passive sensors for field monitoring and data aggregation in harsh environments is still in its beginning stages. Sensor fabrication, sensing, and communications algorithms are being studied by several research groups in academia and industry to further leverage this technology in real world applications such as space exploration [17], remote sensing [18], wearable biosensors [19], RFID-based navigation [20], underwater applications [21] , structural health monitoring [22], and even passive WiFi [23]. There are several challenges that limits the applicability of these devices including short communication ranges (in term of a few meters), interference management, more efficient transducer design, realistic modeling of devices including nonlinear terms; and fabrication cost. This section addresses the two important aspects of communication ranges and interference management through using directional antenna arrays and advanced signal processing methods. Various models including conceptual models [24], differential equation-based models [25], models based on the coupling of modes theory [26], mathematical models [27], and recently simulation-based models [28] are proposed in the literature. However, the need for a unified model for different types of passive sensors as well as computer-based simulation environments, similar to antenna simulators, is essential to improve sensing and interrogator design by considering more realistic models for the utilized sensors. To reduce implementation costs, the idea of using commercial software-defined radio and embedded systems [29], low-cost printable sensors [30], and hierarchical networks of passive sensors [31] is under investigation.

5.7 CONCLUDING REMARKS

This chapter presents recent developments in passive sensor design with a special focus on SAW devices. We review multiple access (MA) techniques developed for passive sensors to realize a scalable passive wireless sensor network, where multiple passive sensors read by a single interrogator module in a cluster. This approach enables a new generation of applications and improves the performance of current single-sensor systems. A recently developed passive sensor, the so-called differential reflective delay line based on SAW technology, is elaborated in full detail as an illustrative example. In these devices, chip sequences with high autocorrelation to cross-correlation values based on Gold codes are used to eliminate the interference effect. However, this technology is still in the beginning stages, and there are many research challenges yet to be solved. Current analytical models are too basic, and more accurate models can be developed by considering second term effects and relaxing simplistic assumptions such as linear operations and loss-less acoustic wave propagation through the device surface. Also, the most current interrogation systems use transmission techniques that are primarily designed for wireless networks with active nodes. However, a more elegant interrogation signal design can further improve the performance of these systems. Interference mitigation using new chip design, directional antennas, and beamforming represent another potential research direction.

References

[1] Rappaport, T.S., *Wireless Communications: Principles and Practice*, New Jersey: Prentice Hall PTR, 1996.

[2] Martin, H, and R. K. Ganti, "Interference in large wireless networks," *Foundations and Trends in Networking,* Vol. 3, No. 2, pp. 127–248, 2009.

[3] Su, I. F., W., Y. Sankarasubramaniam, and Cayirci, "Wireless sensor networks: a survey," *Computer Networks,* Vol. 38, No. 4, pp. 393–422, 2002.

[4] Farid, S., Y. Zia, A. Farhad, and F. B. Hussain, "Homogeneous interference mitigation techniques for wireless body area network under coexistence: A survey," in *IEEE Asia Pacific Conference on Wireless and Mobile (APWiMob)*, 2016.

[5] Farah, C., F. Schwaner, A. Abedi, and M. Worboys, "Distributed homology algorithm to detect topological events via wireless sensor networks," *IET Wireless Sensor Systems,* Vol. 1, No. 3, pp.151–160, 2011.

[6] Kompella, S., et al., "Special issue on cognitive networking," *Journal of Communications and Networks,* Vol. 16, No. 2, pp. 101–109, 2014.

[7] Hara, S., and R. Prasad, "Overview of multicarrier CDMA," *IEEE communications Magazine,* Vol. 35, No. 12, pp. 126–133, 1997.

[8] Dudzik, E., A. Abedi, D. Hummels, and M. d. Cunha, "Orthogonal code design for passive wireless sensors," in *Biennial Symposium on Communications*, 2008.

[9] Reindl, L., et al., "SAW devices as wireless passive sensors," in *In Ultrasonics Symposium*, 1996.

[10] Grate, J. W., and M. Klusty, "Surface acoustic wave vapor sensors based on resonator devices," *Analytical Chemistry,* Vol. 63, No. 17, pp. 1719–1727, 1991.

[11] Puccio, D., et al., "Orthogonal frequency coding for SAW tagging and sensors," *IEEE transactions on Ultrasonics, Ferroelectrics, and Frequency Control,* Vol. 53, No. 2, pp. 377–384, 2006.

[12] Wilson, W. C., and G. M. Atkinson, "Mixed modeling of a SAW delay line using VHDL," in *In Behavioral Modeling and Simulation Workshop*, 2006.

[13] Scheiblhofer, S., S. Schuster, and A. Stelzer, "Modeling and performance analysis of SAW reader systems for delay-line sensors," *IEEE Transactions on Ultrasonics, Ferroelectrics, and Frequency Control,* Vol. 56, No. 10, pp. 2292–2303, 2009.

[14] Xuan, W., et al., "A film bulk acoustic resonator oscillator based humidity sensor with graphene oxide as the sensitive layer," *Journal of Micromechanics and Microengineering,* Vol. 27, No. 5, pp. 1–8, 2017.

[15] El Ayach, O., S. W. Peters, and R. W. Heath, "The practical challenges of interference alignment," *IEEE Wireless Communications,* Vol. 20, No. 1, pp. 35–42, 2013.

[16] Stavroulakis, P., "Interference suppression techniques: a twenty-year survey," *International Journal of Satellite Communications and Networking,* Vol. 21, No. 1, pp. 1–12, 2003.

[17] Razi, A., and A. Abedi, "Interference reduction in wireless passive sensor networks using directional antennas," in *4th Annual Caneus Fly by Wireless Workshop (FBW)*, Montreal, 2011.

[18] Jones, M., A. Lucas, and K. C. M. J.S. Kimball, "Satellite passive microwave remote sensing for monitoring global land surface phenology," *Remote Sensing of Environment,* Vol. 115, No. 4, pp.1102–1114, 2011.

[19] Rocha-Gaso, M., C. March-Iborra, A. Montoya-Baides, and A. Arnau-Vives, "Surface generated acoustic wave biosensors for the detection of pathogens: A review.," *Sensors,* Vol. 9, No. 7, pp.5740–5769, 2009.

[20] Yang, L., J. Cao, W. Zhu, and S. Tang, "Accurate and efficient object tracking based on passive RFID," *IEEE Transactions on Mobile Computing,* Vol. 14, No. 11, pp.2188–2200, 2015.

[21] Etter, P., *Underwater Acoustic Modeling and Simulation*, CRC Press, Boca Raton, FL, 2018.

[22] Zhang, J, G. Tian, A. Marindra, A. Sunny, and A. Zhao, "A review of passive RFID tag antenna-based sensors and systems for structural health monitoring applications, *Sensors*, Vol. 17, No. 2, p. 265, 2017.

[23] Kellogg, B., V. Talla, S. Gollakota, and J. R. Smith, "Passive Wi-Fi: Bringing Low Power to Wi-Fi Transmissions," *NSDI,* Vol. 16, 2016.

[24] Scheiblhofer, S., S. Schuster, and A. Stelzer, "Modeling and performance analysis of SAW reader systems for delay-line sensors," *IEEE transactions on Ultrasonics, Ferroelectrics, and Frequency Control,* Vol. 56, No. 10, pp. 2292–2303, 2009.

[25] Luo, W., Q. Fu, J. Wang, Y. Wang, and D. Zhou, "Theoretical analysis of wireless passive impedance-loaded SAW sensors," *IEEE Sensors Journal,* Vol. 9, No. 12, pp.1778–1783, 2009.

[26] Genji, T., and J. Kondoh, "Analysis of passive surface acoustic wave sensors using coupling of modes theory," *Japanese Journal of Applied Physics,* Vol. 53, No. 7, 2014.

[27] Hashimoto, K. Y., "Surface acoustic wave devices in telecommunications: modelling and simulation," *Springer Science & Business Media,* 2013.

[28] Borrero, G., et al.,"Design and fabrication of SAW pressure, temperature and impedance sensors using novel multiphysics simulation models," *Elsevier Sensors and Actuators A: Physical,* Vol. 203, No. 1, pp. 204–214, 2013.

[29] Humphries, J., D. Armstrong, A. Weeks, and D. Malocha, "Standalone SAW sensor interrogator using an embedded computer and software defined radio," in *IEEE*

International Conference on Wireless for Space and Extreme Environments (WiSEE), 2015.

[30] Gao, C., Y. Li, and X. Zhang, "LiveTag: Sensing Human-Object Interaction through Passive Chipless WiFi Tags." 15th {USENIX} Symposium on Networked Systems Design and Implementation (NSDI 18).," in *USENIX Symposium on Networked Systems Design and Implementation*, 2018.

[31] Razi, A., F. Afghah, and A. Abedi, "Hierarchical network development of wireless passive sensors," in *IEEE/Caneus Fly by Wireless Workshop (FBW)*, Orono, 2010.

Chapter 6

Wireless Beamforming for Extreme Environments

This chapter presents a generalized design view of adopting beamforming (BF) in wireless sensors systems for eight interrelated technological aspects: (a) array antennas for directivity, (b) propagation efficiency, (c) distributed beamforming, (d) improved energy efficiency, (e) interference control, (f) effective ad hoc networking, (g) effective smart sensing at the device level, and (h) advanced integrated MIMO-BF configuration. These aspects should help the development of WSN design in extreme environments through enabling direct and narrowbeam signal propagation.

6.1 INTRODUCTION

Despite some controversy concerning the invention of radiotelegraphy some believe Marconi not only invented the radio but also practiced rudimentary beamforming, in his experiments for receiving weak low frequency radio signals using multiple antennas [1, 2].

Before discussing the beamforming aspects of *wireless smart sensing* we need to understand that there are significant beamforming differences between *common wireless communication systems* and *wirelessly distributed sensing systems*. This may not sound significant to many designers but we believe that this trivial misunderstanding, caused by common practices in wireless disciplines, is retarding the progress of the WSN. Poor utilization of beamforming and its directivity capabilities have slowed down progress. In other words, directivity in propagation beams plays a vital role in practical aspects of the sensing process as we try to apply classic cellular wireless communication techniques. We strongly believe that we are loading our WSN systems with too many unwanted overhead signals and excessive processes that designers should avoid.

Furthermore, in examining the importance of beamforming by looking at its practical aspects, using these techniques we can tweak the progressive capabilities of wireless propagation techniques for WSN applications. We examine these aspects under three operational management factors: (1) efficiency, (2) certainty, and (3) trust.

The efficiency factor comes in when we use wireless technology without beamforming. Here due to a lack of control over the propagation path we only make use of a very small fraction of the transmitted signal being used by the receiver. The economy for reception of such a small proportion of signal energy is regarded significant in the industry of mobile telephony, mainly due to the fact that the mobile industries dominate the market and under extremely low cost mass production they are able to claim the required productivity. However, the case is very different for WSNs where in the long run any waste of energy contributes to the deployment paradigm and from an operational point of view the nodes should manage their own energy requirement with most of energy being used. So through wireless communication the smart sensors encounter serious energy crisis and could die out in the middle of an operation. The case becomes more critical in a shared environment where too many interfering signals a further drop of efficiency by increasing the transmission power.

The certainty factor stems from the chance of a user or node being preset within a signal saturated area, where in a classic wireless communication environment the coverage supports the natural random mobility as a superiority feature for the cellular users for accessing the service. However, in the case of WSN nodes the sensors are already located in their required points; all they need is to find a way to transfer the collected information to the nearest database or control center without being delayed or lost due to a lack of proper connectivity or insufficient local storage. In most WSN design cases instead of using the wasteful classic system of chance discovery and searching for a good reception area/direction we use an ad hoc style relay connectivity, which should enable a process in which a node could be enabled with a small amount of intelligence dynamism to maintain connectivity upon some pre-established direct links.

The degraded trust factor is a natural wireless side effect product not noticed by many but it can easily give be vulnerable to intruders and jeopardize user's confidence. In general due to the possibility of losing data or an unauthorized manipulation of sensitive information' trust can be compromised. Although all wireless systems inherit such a natural side effect as a worrying element, in common cellular applications/systems its impact could be significantly reduced by employing/using some costly embedded security measures within the user device.

The above factors can easily accumulate and grow into much higher deficiencies for some applications in extreme environments. For example, due to the scale of distance and low-density usage, in outer space, classic transmission techniques may struggle to perform without the proper use of beamforming. In such cases the disappearance of signals can be due to the received signal's drop of energy or insufficient condensing before transmitting the signal.

For monitoring, manipulation, and control of underwater vehicles remotely, the use of mixing technologies for example has been very popular. This solution, however, can become very complex, slow, and inefficient in many ways, while it is subject to deterioration of wireless signals and natural interruptions. The UWSN systems were expected to resolve many of these problems, which they may do but only if all the links in the network are properly connected. We know for sure that such ideal connectivity is not always possible, thus hindering cost effective practical solutions for a long distance underwater environment.

Other harsh and extreme environments such as those underground have their own unique situations so that practical solutions could be only feasible by conquering most of issues related to the three main deficiency factors mentioned earlier.

The rest of this chapter is organized as follows. Section 6.2 provides the basic beamforming modeling for regular arrays of antenna elements. Section 6.3 discusses the importance of directivity and the benefits of being able to control the beams. Section 6.4 discusses development of beamforming in selective areas while examining the state of the art and potential for future uses of beamforming in design, development, and deployment of the WSN in the extreme environments.

6.2 ANTENNA ARRAYS

Ideally the best directivity is the functional feature of conductors and cables. One way to make use of wireless directivity is to use some of the existing antennas, arranged in the form of an array so that their phased, manipulated combined beams result in a sum in a certain unified direction. That is, a phased array of some isotropic antennas, each propagating waves in a media (uniform 3D pattern) are combined to deliver a much stronger beam (ideally a linear sum of all elements) in the expected direction only. Figure 6.1 shows an ideal *directed beam lobe*. Each lobe's precision of directivity is characterized by its beamwidth, which is measured as the angular distance between two points on the lobe measured at 3 dB in power below the peak. The narrower the beamwidth better the beamforming performance of the array. For example a half-wavelength power beamwidth for a single dipole with 78° (degrees) can be reduced to 32°, 15°, and 7°, using 2, 4, and 8, cascaded arrays, respectively. Using reflectors is very useful for directivity because it doubles the power in the desired direction without doubling the number in the array. The directed beam pattern shown in Figure 6.1 is an ideal view as it can only be provided through being specifically designed for a media upon the characteristics of the elements of the array being used. Also, due to its reciprocal property the directional efficiency of antennas applies to receiving signals exactly in the same way they do as they were transmitting them.

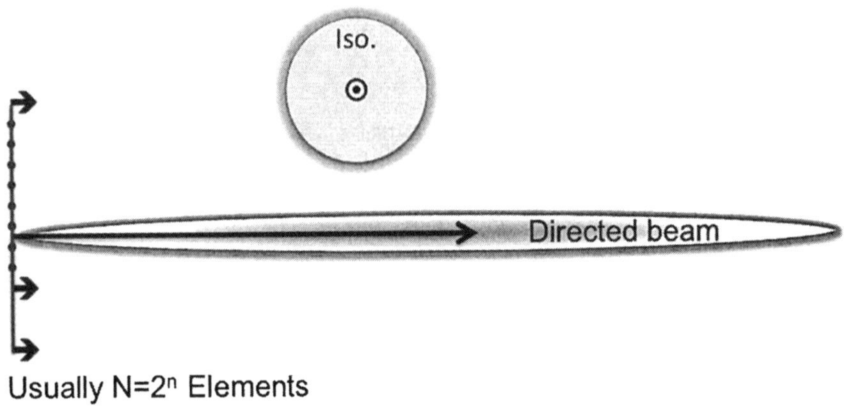

Figure 6.1 Two-dimensional (2D) image comparing two 3D beam patterns: in the upper, the sphere pattern of an ideal isotropic antenna; and in the lower the 2D image an ideal conic beam pattern that could be generated from a simple array of sensors.

Most efforts of the last few decades, however, have been devoted to cellular telephonic applications, where spread of the propagation energy from some cooperative base stations are directed at their prime objective of coverage. That is, although a true beamforming-enabled WSN design for extreme environments could benefit from using existing commercial off-the-shelf (COTS); common practice today for a rapid provision of low cost components, this could however discourage expert designers and innovative industries from creating new designs.

Modeling Array Antennas

Basic array beamforming has a much longer history than most recent research publications indicate. The term "arrays" here not only indicates the mathematical ability of manipulations to explore new ideas through modeling but also guides a visionary penetration of scientific investigation. It can actually guide engineers to solve their technical problems systematically, reliably, and step by step. The lossless signal appearing at a distance r from an isotropic source, s as shown in Figure 6.2, is given by,

$$s(t,r) = s_0 \cdot e^{j\Psi(t,r)} \tag{6.1}$$

Where, $\Psi(t,r) = \omega t - k \cdot r$, $k = 2\pi/\lambda$ with λ as the wavelength is the directional angular phase change of the wave over one wavelength of travel. This means that variable Ψ dictates the energy pattern of the propagation (e.g., 1 for the isotropic antenna in all directions that means zero beamforming).

Figure 6.3 shows one- and two-dimensional array elements uniformly distributed at a space distance a from the neighboring antenna elements.

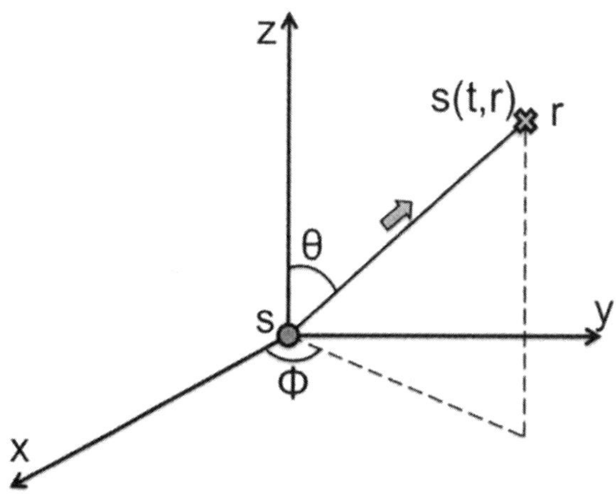

Figure 6.2 Signal appearing at a distance r from an isotropic source, no beamforming.

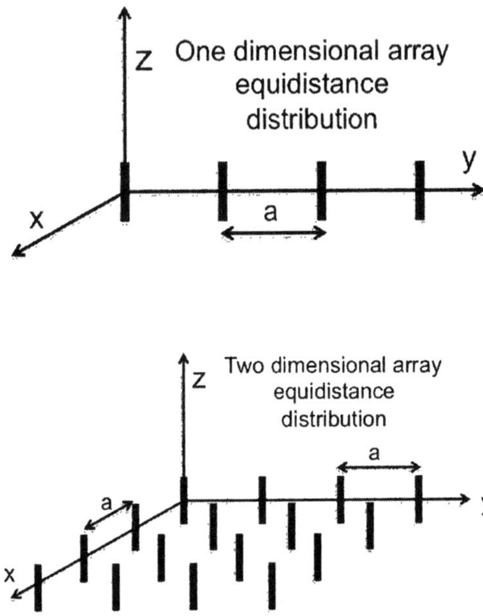

Figure 6.3 Typical antenna spacing distributions of 1D and 2D uniform arrays.

Then, upon (6.1), we can establish an estimation of the *normalized signal distribution* for any set of N, $n = 0 \cdot\cdot (N-1)$ distributed sources of Ψ_n for a relatively far away distance r (r is assumed the same for all elements) as,

$$s(\varphi) = \frac{1}{N}\sum_{n=0}^{N-1} e^{j\Psi_n} \tag{6.2}$$

where, $\Psi_n = k \cdot na \cdot \cos(\varphi) - \zeta_n$ with φ and ζ_n indicating the angle from the reference axis and element's feeding phase, respectively. The process of controlling the phase and adjusting its contribution from element n, Ψ_n and its associated properties create the fundamental beamforming functions, which can be also called BF rules.

The parameter $k = {}^{2\pi}/_{\lambda}$ brings the influence of frequency where, a designer can adopt one or a combination of the four following methods: (1) varying N; (2) varying the length of array $L = a \cdot (N-1)$; (3) varying the frequency; and (4) varying the amount of delay to each signal being fed into the elements. The last method is new and normally requires precision signal processing capabilities, which many devices may not posses.

Therefore, the simplest mathematical model for an 1D array the *total relative energy gain* of an isotropic set of N equally a-spaced array elements with the same signal strength s at a relatively far away distance r becomes,[1]

$$s(\varphi) = \frac{1}{N}\sum_{n=0}^{N-1} e^{jkna\cos(\varphi)} \tag{6.3}$$

For the energy distribution of (6.3) considers the real parts only resulting in a sinc-function gain formula to become,

$$|G(\varphi)|^2 = \left|\frac{\sin(N\cdot ka/2)\cos\varphi}{N\cdot\sin(ka/2)\cos\varphi}\right|^2 \tag{6.4}$$

Figure 6.4 shows two symmetric distributions of energy and corresponding radiation patterns of a single-array for quarter-wavelength and half-wavelength element spacing.

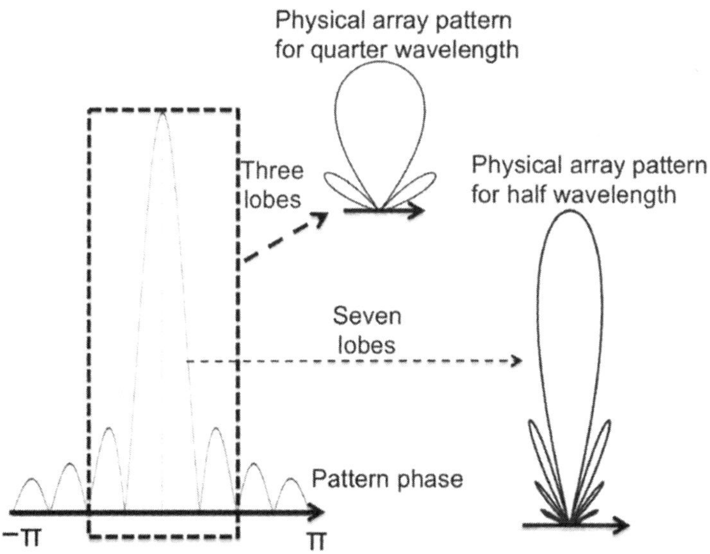

Figure 6.4 Distribution of energy and corresponding 2D radiation patterns of 1D arrays for two cases of quarter-wavelength and half-wavelength.

[1] For far away distances (large r) the effect of differential r becomes insignificant.

Nowadays with the increasing capabilities of devices for using software-enabled digital signal processing, the integration of steering functions to the antenna array is feasible. Therefore, antenna arrays can adopt static and dynamic steering and navigation functions, electronically. This most welcomed function is making new devices superior to traditional mechanical systems. Figure 6.5 shows 60° deflection for the main beam.

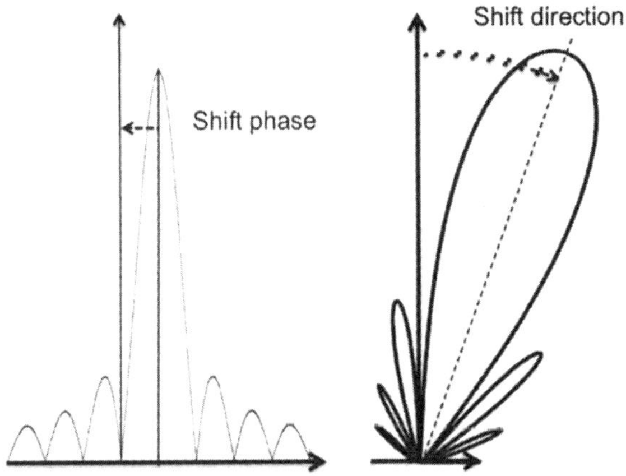

Figure 6.5 Electronically rotated beam produced by an array.

We can extend the configuration of the array antenna (Figure 6.3) into a third dimension (3D) for some applications; however, due to its required extra complexity and specific incremental benefits it has rarely been used so far. However, with ongoing developments in integration technologies could become a common practice in few decades.

Also, under common equal-distance distribution of the transceiver element in the array on either one or two Cartesian dimensions, we encounter serious design problems, especially for lower frequency arrays. The good news is that due to the nature of the beamforming process most of the design problems can be moved from hardware into software harnessing suitable, agile, and cost effective algorithms. This stems from the fact that most common beamforming applications can perform under almost any distribution of elements as long as the correct phase changes are applied to the elements to comply with the correct result. One typical example is use of circular (2D) or sphere (3D) arrays. Figure 6.6 shows an equally spaced 2D circular array.

In general any wirelessly networked system can create beamforming as long as the basic rule (phase adjustment) can be applied to all contributing elements of the antenna. The antenna in such cases does not need to be actually part of a device (physical, physically fixed, or connected) to cooperate. The beamforming antenna then becomes a virtual process, built upon cooperative propagation of the signals produced at a correct set of phases maintaining the BF rule of compensating the differential paths. However, to implement the beamforming idea in practice we should include two complementary functions to the operation: (a) a centralized cooperation with a powerful computing capability to organize the whole process and (b) all individual contributing smart sensors possessing sufficient computing capability to maintain their contribution for the required length of communication.

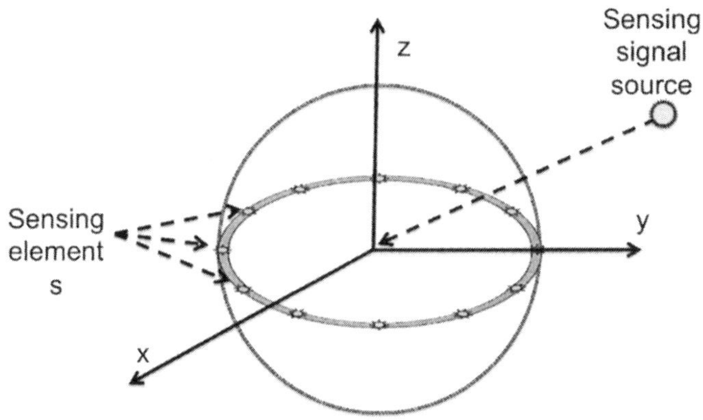

Figure 6.6 An equally spaced 2D circular array beamformer.

6.3 WHY DIRECTIVITY?

Due to the nature of sensing applications we have realized that when using the WSN concept in most extreme environments we often need unstructured and dynamic ad hoc networking. Along with it we need the best way to design a network making use of direct connections between nearby nodes, cluster heads, and data stations (sinks) of the network.

Before answering the question of beam directivity (in WSN), let's explain the difference between near-field and far-field signals. Recalling the assumption we made for long distance from the source estimation (6.3) of signal power, the near field effects (NFE) signal estimation of the array requires detailed physical aspects of the array. From beamforming design considerations such details cannot help with the far-field effects (FFE) while injecting unnecessary complexities. That is, near-field signals vary tremendously from one point to the next, which makes the task of modeling both complicated and unreliable. For these reasons we cannot rely on any design process for NFE regions. The best practical recommendation is avoid near-field issues or take your pick.[2]

The demarcation between NFE and FFE regions is usually blurred; therefore for every design we consider d_{min} as our measure for avoiding the near-field issues. This minimum distance normally depends on the structure of the array, its materials, the size of the elements, their spacing, and the surrounding environment. We generally use some estimation based on practical evaluation, which can be simplified to an empiric formula $d_{min} = 2L^2/\lambda$, with L the largest dimension of the array antenna. We, therefore, should limit our use of WSN cooperative beamforming to applying the following considerations: (a) in order to avoid NFE it is best that we not use cooperative beamforming between subgroups of the nodes (often many to one) unless the d_{min} restriction does not apply to most; (b) we should use cooperative beamforming for a cluster to the cluster head unless the average distance falls to around or above d_{min}; (c) we should use beamforming for almost all long distances within WSN; and (d) we need to understand that direct cooperative communication between WSN and very far objects such as satellites and HAPs may not be practical as their energy and resources are not designed for long distance communications.

Directivity represents an important but somehow overlooked feature for implementing the WSN. This feature could become significantly more effective when it comes to extreme environment applications. From a performance point of view we consider: (a) the point of interest sensing instead of coverage of an area (i.e., apart from special multi-sensor sampling cases the sensors used in most WSN applications collect data from a chosen sampling point, where one point of the media is as good as its surrounding area); (b) spreading a signal's energy means weakening a good signal; (c) all unused signals missing the intended receiving antennas are a waste of useful energy and often interfere with other nodes'

---

[2] The pick in here means choosing between 'rule-of-thumb' guesswork and a systematic approach minimizing the risk. In practice, systematic approach is recommended. Here one can test performance of a near-field beamforming design using a case study avoiding its possible pitfalls and limitations, where choices are: (a) use a suitable computer simulation with a few variables to cover all possible working conditions and/or (b) setup an experimental measurement in a laboratory or field representing the case under the specific conditions.

communication; (d) desired communication paths in the extreme environments encounter heavy losses (e.g., due to the scale of distances in space and underwater and underground any propagation rays with potential bandwidths are subject to heavy losses); and (e) unplanned multipath rays for communication in extreme environments are not as useful as terrestrials due to the fact that reflections aren't usually as many or of the required quality; therefore they normally require significantly more costly signal processing, which may also add extra internal shot noise, weakening the receiver in making use of such indirect signals.

For most practical WSN applications desirable directivity solutions can be met with using *dynamic beamforming*. In this case the common feature of directivity falls in the accuracy of the beamforming and the selectivity of the target receiver in the midst of neighboring transceivers. Figure 6.7 shows an example of such a case when a WSN node is trying to communicate with a further away node at a known direction while avoiding three surrounding nodes.

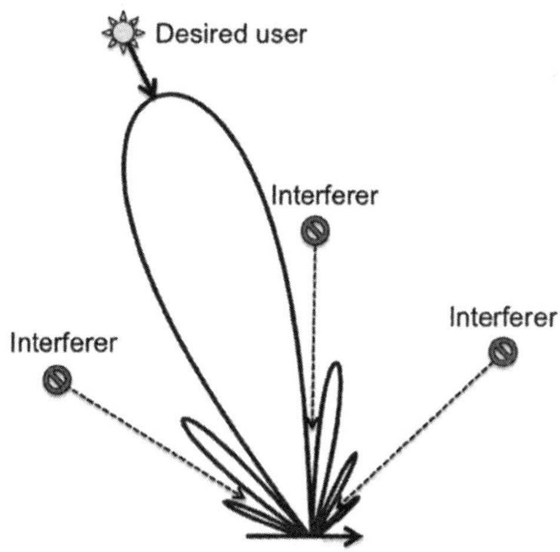

Figure 6.7 Directivity use for communicating with a desired node amongst many undesired transceivers.

In general three common cases of using beamforming in WSN are: (a) cluster-to-cluster transmission, where the cluster heads use their own antenna arrays to communicate with others; (b) when whole cluster of sensors are organized to generate an effective cooperative beam for directly communicating wirelessly with the base stations upon each sensor following its own BF rules as a contributing element of the array; and (c) making use of extra information such as baseband or channel conditions for generating a multibeam propagation pattern to communicate with more than one node. Figure 6.8 shows a third case example, where a WSN uses a complex (e.g., industrial site) environment to make use of baseband information such as the direction of arrival (DOA) for an available channel to make a connection.

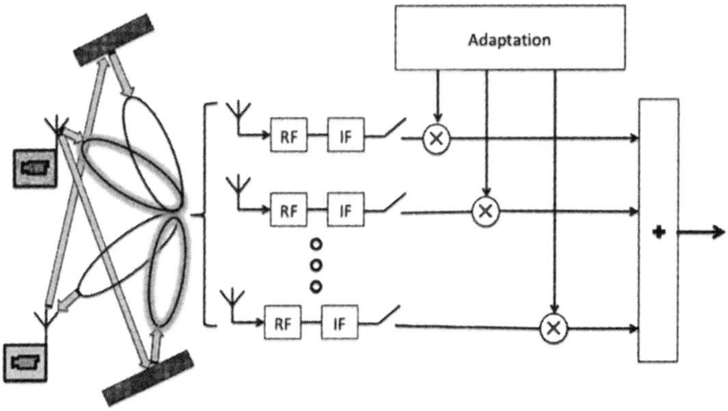

Figure 6.8 Multibeam propagation case of multibeam environment.

6.4 WSN BEAMFORMING

Before we look into various possible WSN beamforming design ideas we need to analyze the ways various beamforming approaches made their way into using this feature in different industries. We have four different lines of development, each with its own specific industrial requirements. That is, although all should follow the common design base they become different as they follow their own specific development routines for solutions to match their specific deployment in the field. For example, due to commonality in the early historical developments of WSN with mobile industries many beamforming techniques have been used in both systems, which in some cases can help with the WSN solutions but in many cases could be harmful. Another example is MIMO technology, in which, due to the domination of generic wireless technology, the WSN application scenarios have been overlooked. In effect the integration of beamforming with classic MIMO technology should make great impacts in design and implementation of beamforming enabled large WSN systems.

6.5 MOBILE BEAMFORMING SOLUTIONS

WSN has been expanding by using the potentials of beamforming for various aspects of energy and performance; however, being relatively young and operating at lower scales for many industries its independent world market developments have not yet been proven and therefore their specific design ideas have not been integrated into other technologies such as mobiles and IoT.

Mobile telephony as the main force behind the wireless industries have shown their own way of promoting beamforming. That is, all beamforming ideas coming in for improving mobile and generic wireless technology can be shared with WSN and some can be retailored and used for WSN deployment. There are many such cases in the wireless propagation, networking, energy conservation, beam management, and unique infrastructure of mobile technology and of course in the development of standards. Now let's look at one familiar case of using beamforming in contemporary WiFi standards.

In the process of developing mobile technology we have seen a great deal of technical discussions of the design of smart antennas. A smart antenna is a generic name that has been used for a mobile antenna with some integral adaption capability to help a device alter its configuration relative to the changing propagation media. The changing media could be part of the service or integral nature of the application. A single element smart antenna can be also called smart if it can automatically reconfigure its signal parameters by adjusting to the field [e.g., a mobile telephone moving from one standard media to another mainly for being able to recognize the receiving signals (modulation system, frequency, etc.)]. The term smart antenna has been also applied to many other mostly wireless systems including remote scanners "object trackers" and radio telescopes.

The title of "smart", however, is more commonly used for arrayed antennas that are able to adjust their reception and/or transmission patterns automatically. When designed to change the direction of their main beam, in order to pick up the maximum energy signal in a field (via altering their differential phase applied to the elements in the array), then, for example, the smart signal processing algorithm behind the array makes them algorithmically use the DOA of the field (generally called the spatial signal signature) to adjust their own elements' internal differential phase to move gradually towards it and capture the main lobe of the transmitting antenna.

Another interesting beamforming related area of mobile technology is based upon its earlier integration with WLAN, also called WiFi. Using an ever-increasing WLAN and associated WiFi, function of sharing the spectrum at the access point, boosting the 3G/4G technological development. However, due to the limited power of such a shared service they had to make use of a new multichannel BF-MIMO, which enabled longer access using reciprocal array antennas. It may also be worth mentioning that in principle MIMO technology has not yet been designed for using beamforming as well as sharing multichannel applications (e.g., the IEEE 802.11n standard) [3]. Therefore, in order to enhance WLAN with WiFi capabilities the operators had to make use of antenna arrays for two cumulative enhancements: (1) at access points to increase the power adaptively and (2) to enable basic beamforming for directivity and reaching some intended nearby receivers while remaining transparent to other receivers.

One of techniques commonly used in wireless systems for improving performance and extending the range is called *complementary beamforming*. The IEEE 802.11 WLAN systems have served today's wireless communications and internet access extensively for decades. They, however, come with some weaknesses including power deficiency and limited access range mainly due to their use of classic antennas. Using conventional power beamformers can increase the transmitted power in specific directions while reducing the radiated power in other directions. This, however, can cause some other users in the area to experience lower *signal to noise ratios* (SNRs). Another issue that had to be resolved at the earlier stages was related to original basic beamforming propagation. That is, when a channel is busy in a different direction it may also to be assumed idle by some who wrongly start transmitting their packets at a loss, causing unnecessary extra back-offs, and excessive interference. This is commonly called a hidden beam problem and it can become more serious for heavily loaded wireless networks using hot spots to increase a system's range. Using complementary beamforming on the other hand enables some control mechanism to maintain the benefits of basic beamforming features for the beamforming users while watching other users (called silent users) in the network [4].

Another developing beamforming technology coming from the mobile arena is the advent of automatic reconfiguration of an array antenna, which is more of art and design than engineering. Let's look at some interesting pieces of work on an electronically adjustable and steering antenna.

The first one called the "real-time reconfigurable pixel antenna" is claimed as superior to the classic reconfigurable antennas due to (a) its flexible modular style matrix of switches, which makes them cost effective and suitable for the global market, (any 2D matrix size); (b) their monolithic switch integration features, now under rapid production due to availability of the infrastructure technology; and (c) their addressable dynamic reconfiguration, which enables us to overcome many programmable design issues for their use.

Due to dynamic reconfiguration features the reconfigurable pixel antenna can be adopted to a wide range of applications including radar and *ground penetrating radar* (GPR). Although the complexity of such antennas sounds high but at the production stage specific designs could minimize the complexity leading to a low cost product [5].

Another more interesting work on reconfigurable matrix structured array antenna design, *reconfigurable aperture*, comes as a multifunctional programmable supporting switchable MEMS for deploying many antenna based applications including functional beamforming, pattern synthesis, dynamic spectrum sharing, MIMO communications, and physical layer wireless security. The concept is to be able to build wide range applications for any propagation environment using a rather simple hardware supported by a manageable software complexity. Figure 6.9 shows typical samples of switchable settings for a 9x9 matrix antenna [6].

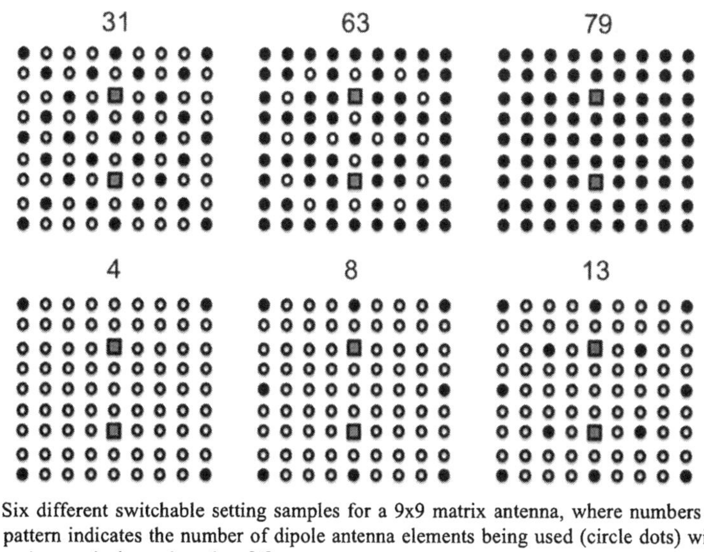

Figure 6.9 Six different switchable setting samples for a 9x9 matrix antenna, where numbers on the top of each pattern indicates the number of dipole antenna elements being used (circle dots) with two feeding points (squares), drawn based on [6].

The sample techniques we have mentioned above are regarded as the *generic mobile and wireless beamforming* mainly because their design is oriented towards high usage statistical performance instead of being suitable for WSN's specific common directional application scenarios. They are also mostly designed and demonstrated for the terrestrial wireless applications that stem from their convenience of access rather than specific device functions. We therefore envisage that manipulating the radiation patterns in the propagation of wave energy in such working media will be far easier than in the cases of extreme environments, where limited choice of propagation make the case far more challenging.

6.5.1 Generic WSN Beamforming Solutions

Following our discussions on directivity in Section 6.3 and the early part of this section the potential differences between a WSN with a mobile system is twofold: (1) wireless nodes can behave like an antenna array element and (2) the main stream of data within the WSN normally flows one way. Therefore, if the propagation energy managed to take advantage of these features then WSN would provide a superior performance. That is, the beamforming-lead WSN process carries virtually untapped potential to be utilized. Examples provided in Sections 6.4.2–6.4.4 show some successful sensor-based beamforming works.

For generating a beamforming function our elements need to be located at a specific place or set in a regular physical array (i.e., any systematic or randomly distributed set of WSN nodes can be employed for enabling a beamforming process and producing a desirable beam pattern to suit the required circumstances). But due to the importance of a relative location for accurate estimation of the differential phase a regulated placement helps with the complexity of calculation. Such a purely random distributed beamforming technique features freedom from obligations over choosing the location for active elements. This over simplified distribution, however, would require a very complicated cooperative process from the cooperating elements or appear with unwanted annoying side lobes. That is, it would necessitate synchronizing the potential nodes (elements) for generating a correct form of contributing signal. Often side lobes of a powerful lobe interfere with neighboring nodes where removal of (nulling) an interfering sidelobe requires integration of additional information such as the *channel state information* (CSI) [7].

Similarly, we use so-called collaborative null-steering beamformers designed to improve the overall network performance by incorporating a specific task of removing a particular sidelobe. This simple process helps in practice due to two interrelated facts: (1) naturally the main lobe is improved as it makes use of saved side-lobe energy and (2) the sidelobes are dynamically reduced for all other receiving stations. In this case the overall collaborative gain improved in proportion to the number of collaborating nodes results in 6.6 dB improvements over conventional collaborative beamforming [8]. Use of WSN for medical purposes is a common practice, where due to the noninvasive requirement of nonsurgical treatments the body can be counted as an extreme environment. Tomography of the brain and endoscopy of the digestion system are two interesting cases that use array antenna style processing. Here, the movements of the wireless endoscopy unit (capsule) that carries our investigation sensors ought to be identified and analyzed along with the collected information. The 3D localization, tracking, and movement control of the capsule is achieved by an array of wireless sensors located on a cylindrical unit enabling the beamforming, signal processing, and other data management [9]. Figure 6.10 shows three different but complementary placement configurations of 16 selected wireless sensor arrays for locating the digestive path using a wireless capsule endoscopy (WCE) enabled.

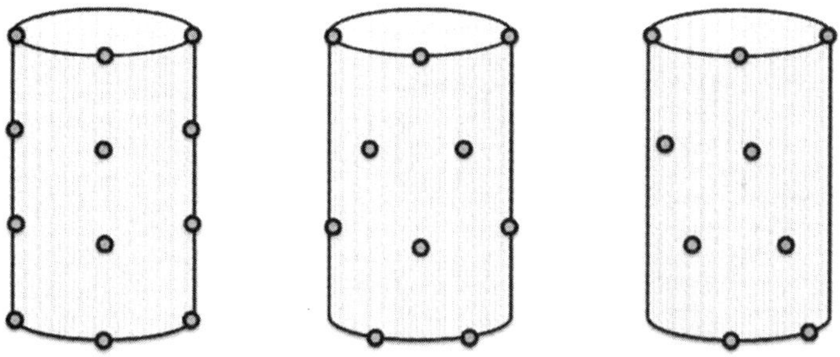

Figure 6.10 Three different location configurations of active array elements on the cylinder.

6.5.2 Low Frequency Beamforming Solutions

As mentioned in Chapters 9 and 10 unfortunately underwater and underground propagation environments do not allow the proper use of radio frequency communications and limit our bandwidth capabilities for exploring such harsh and inhabitable media. Sounds and micromechanical vibration signals, however, are able to propagate but at the cost of considerable losses and certain limitations such as bandwidth and obvious inconvenience of access. Despite all these difficulties we have been making considerable progress in developing some practical low frequency solutions for using our smart sensors to manage our urgent tasks and limited explorations.

When it comes to the question of progress in using smart sensing we realize that our smart system capabilities provided by low frequency signals for such extreme environments are neither exciting nor sufficient; why?

There are two main obstacles hindering progress: (1) cost per bandwidth meter for UG and cost per bandwidth per kilometer for UW are too high to cross any meaningful threshold for low frequency transmission of the signal and (2) under the existing solutions the propagation waste of transmitted signals is incredibly high and rapidly growing with the distance.

Before we try to find alternative and more viable technologies we look further to see if there are any missing solutions upon the existing technologies. We then see that what possess more than enough signal and information processing capabilities to call upon beamforming. That is, it is easy to see how beamforming can help with both parts of hindering obstacles, yet making better use of directivity requirements/features of WSN and beamforming combined could help us with new design solutions. We therefore examine weather any recent innovative ideas can be used to solve this problem. Specifically we examine samples from three main areas of technology, namely acoustic, seismic, and sound.

6.5.2.1 Acoustic

An active extension to sonar is called *directional frequency analysis and recording* (DIFAR) normally used in hydrophones. It stands for a small array of three acoustic sensors to help with measuring properties of incoming acoustic signals. One of them is called the *omni bender element*. This is located on the bottom of the hydrophone measuring the acoustic waves' amplitude of the pressure. Then two remaining sensors cooperate onto a wobbler resting on four ceramic discs to measure the direction of the signal. With acoustic pressure waves detected by the hydrophone the pressure moves over onto the ceramic discs. With the piezoelectric element embedded in the discs the vibration energy is converted into a voltage signal. Under the orthogonal geometry arrangement this results in two sine and the cosine lobe patterns. These three then together provide the directional information as well as the power spectral density of the incoming acoustic waves [10].

That is, after having this directional device for well over two decades we see very little use for this method to be upgraded for our "beamforming-WSN" adoption. One useful work is saving interference between humpback whales and the offshore operations. The coastal shelf north of Western Australia is known as the winter resting grounds for the Western Australian population of humpback whales, normally populated from early June to late October. Unfortunately, this part of the continent can be affected by intense offshore operations for oil and gas explorations. Therefore some urgent measures are required to prevent physiological and behavioral impacts on the marine environment as well as the possibility of collisions with vessels and other side effects of man-made underwater noise. In order to assess the possibilities and find a good solution for this particular problem we need to monitor, detect, and localize the whales. Therefore the operation sites, ports, and vessel corridors have to be sensed for warnings. At this stage a sparse set of DIFAR sensors arrays are deployed. For this experiment some arrays with three to six drifting buoys are located at distances, between 1 and 15 km apart, so far results are promising with some new types of sound recorded from the whales [11].

Another idea that could be used for beamforming-WSN development is related to unmanned autonomous vehicles (UAVs) is using antenna based on retrodirective principles, which uses an antenna that transmits the signal back in the same direction it came from. This is based on the principle of self-phase conjugation of the input signal. Directivity depends on relative phases of signal radiated from different elements. The array antenna consists of some multiple identical radiating elements [12].

Microphone-Array

For developing any beamforming-WSN solution for sound systems, one may directly look into arrays of the source elements. Here we examine how beamforming for such arrays can be established. The combined sound generated from the array provides expected but very interesting features for applications of WSN for surveillance, audio/video conferencing, and tracking objects that may lead to new trends in the design of sonar systems.

Let's begin with the very basic form of a signal waveform generated from an ideal single source located at the origin of our coordinates, as shown in Figure 6.2, and propagated away in the media at the speed of v. The expression we normally use is for the decaying signal power through the time t while crossing the microphone at a distance r from the source the following wave equation applies,

$$\nabla^2 s(t,r) = \mathrm{v}^{-2} \frac{\partial^2}{\partial t^2} s(t,r), \tag{6.5}$$

where, for a wave-front plane the signal energy can be simplified further to a frequency dependent complex lossless[3] formula,

$$s(f,r) = s(f)e^{-jkr} \tag{6.6}$$

Where, f is the frequency of the wave, $s(f)$ is the signal power, and k is the same being defined in Section 6.2,

$$s(f,r) = k = 2\pi f \cdot v^{-1}. \tag{6.7}$$

The receiving of the signal, however, depends on the way this signal is detected. The technical term used for the process is called aperture and its frequency dependent impact on the received energy at point r can be denoted by a weighting factor $w(f,r)$ as,

$$d(f,r) = s(f) \cdot \mathrm{w}(f,r) \cdot \mathrm{e}^{-\mathrm{j}kr}, \tag{6.8}$$

and when integrated over the whole valid small area (θ, ϕ) of the aperture at the receiving point, distance r from the reference point, it normalizes it to $s(f)$, we then have,

$$D(f) = \int_{\theta,,\phi}^{A} w(f,r) \cdot \mathrm{e}^{-\mathrm{j}kr} dr. \tag{6.9}$$

Figure 6.11 shows two sets of typical example patterns for using two popular beamforming techniques known as *delay-sum* and *superdirective*, respectively [11]. These methods used in DIFAR in the UAVs [12].

[3] Here due to isotropic source.

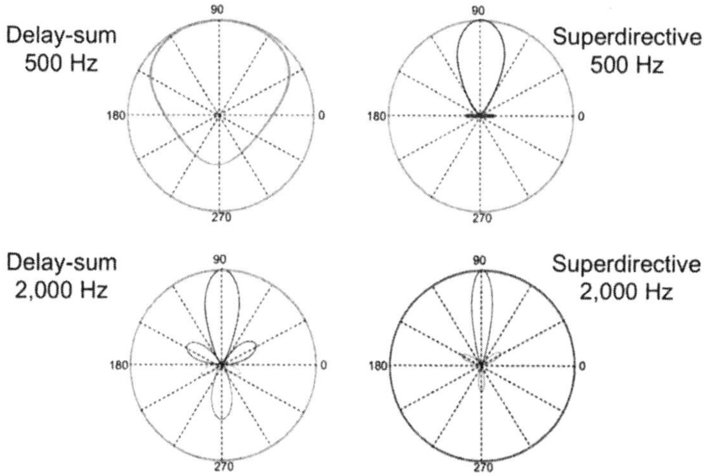

Figure 6.11 Typical patterns sketched for using two beamforming methods: (a) delay-sum beamforming and (b) superdirective beamforming. The antenna consists eight elements distributed over a 20 cm diameter circular array.

Sonar

Another area of development in acoustic domain is for shrinking sized sonars. As the usage increases the higher volume production helps with a natural exponential surge of technology as long as the two main components of the market and developments continue to respond. New trends and technologies for developing light, smart, and portable sonar devices are promising. Beamforming is now, with additional new capabilities, particular useful for precision recognition and extended range using new higher frequencies.[4]

For example, a recent portable sonar design uses an arc-shaped scanning unit consist of 10 uniformly distributed elements. Its receiving hydrophone uses an 80-element array. The two dimensional beamwidths for this design are 45° horizontally and 15° vertically. It works in a frequency range of 550 kHz to 600 kHz with a primary power level of 190 dB providing hydrophonic beamforming sensitivity of −195 dB. The signal conditioning system it uses the main amplifier shown in Figure 6.12 [13].

[4] Exceeding the traditional sound range of 10 kHz.

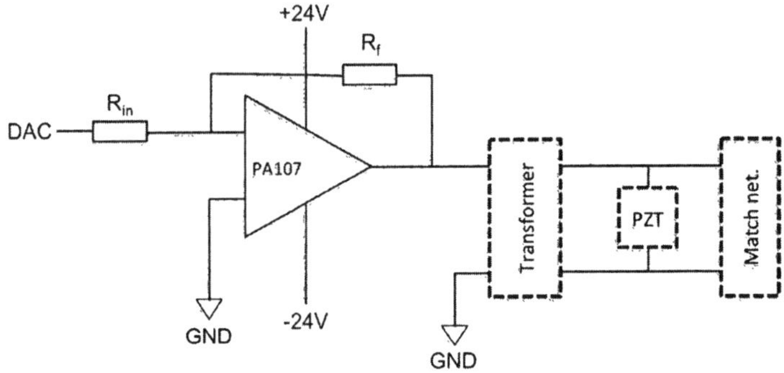

Figure 6.12 Conditioning circuit designed for the portable sonar, drawn based on [13].

6.5.2.2 Seismic

Natural seismic waves and their properties have been introduced in chapter 10 for some of the global scale UWSN applications. There is also growing interest in seismic waves generated from devices using technologies such as PZE.

Most applications of seismic waves require intensive signal processing; therefore deployment of WSN using seismic waves for many applications is neither easy nor cost effective. This is particularly true for any using the natural seismic waves for investigating natural disasters such as earthquakes. Such deployments normally require gathering sufficient sample signals from cooperative sensors to make a decision upon the total energy above a certain threshold. The gathering of the signals from cooperative sensors is regarded as a beamforming process (reverse style). This process involves estimations for the source (location and intensity), path (distance and propagation sensitivity of the material), and sensitivity of the receiver.

However, when using new devices and equipment for common geographical and industrial applications such as exploration, localization of objects, and investigation of the materials we use artificial seismic waves with known properties, and therefore the signal processing associated with the beamforming process can be simplified extensively and reduced to the path and its conducting properties. Figure 6.13 shows a typical example of localizing an object under three layers underground.[5]

[5] Using common devices such as GPR have little chance to operate effectively under such underground conditions.

Although the beamforming function of the source beam provides higher efficiency for the system and naturally much fewer possible interfering signals, the beamforming process of the received signals through smoothing, filtering, and associated signal processing functions counts towards the final task of localization.

The localization consists of the preprocessing of any relative time differences in the arrivals of recorded signals happening at the seismic arrays, correcting any noncoincidence arrivals of the seismic waves from different seismic sensors, and estimating the energy of overall seismic signals. For doing this normally the location of the source and identification of its peak output power is computed using the elastic wave equations where the model parameters can be analyzed and applied to subsurface geology and the locations of seismic sensors and sources. In this experiment the beamforming function is applied at the receiver sensors that use traditional geophones and then geophones only measure the vertical component of the seismic waves through the ground vibrations [14].

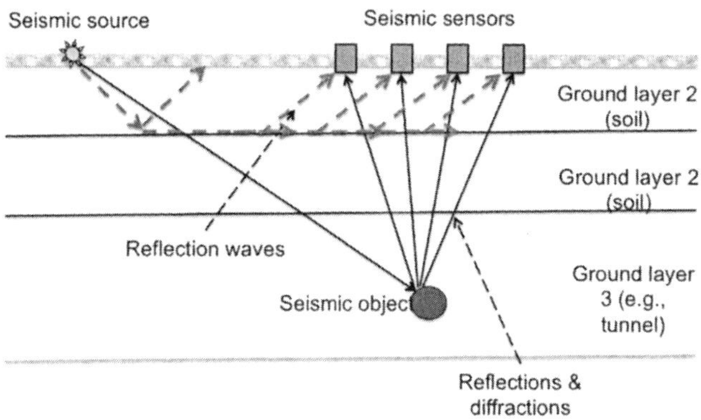

Figure 6.13 Typical localizing scenario for detecting an object located three layers underground, drawn based on [14].

6.5.3 MIMO Beamforming Solutions

One of prime wireless communication technologies that can be integrated with beamforming is the multiple-input-multiple-output (MIMO) technology. Due to their multichannel features and structural similarities BF and MIMO share some common configuration properties suitable for a new advantageous design.

MIMO is a generic multichannel multiplexing system featuring a wireless link to model, use, and enhance its properties. It can be easily configured and used for sharing a common high-speed channel between some users to increase the power efficiency or to widen its bandwidth of link using multiple antennas in multipath propagation. The beamforming function, however, works in energy propagation to handle and manipulate the propagation patterns of a controllable multi element antenna.

As mentioned earlier recent developments are showing more prospects for combining beamforming with MIMO. This development may not look as serious for overpopulated wireless applications (such as UMTS) but for WSNs, especially those to be used in extreme environments, where energy is scarce and propagation of signals face very heavy losses, it can be regarded as a critical factor. Some reconfigurable array antennas, for example, can be used for integrating MIMO functions with beamforming for the WSN [6].

Also, under the advent of data tolerant networks it is clear that on the MIMO style multiplexing side of WSN applications that we have significant untapped flexibility features to enhance the systems' performance in both sides of the MIMIO-BF technology. MIMO side sensors with higher data priority or speed could be assigned to more array elements. In order to make our discussion clearer we denote the number of users by m and the number of array elements in the array by n. Obviously under normal practice, we have $n \geq m$.

Then a generic configuration for the combined MIMO-BF would be as given in Figure 6.14. In this m users are making use of n elements of the antenna. Functional blocks in Figure 6.14 are: (a) "Mux," a special multiplexor for distributing elements to the user channels whereas in normal practice $n \geq m$ is strongly valid and it is possible to have some elements idle, so some users may have their own arrays (i.e., more than one); (b) "AAC," the *antenna array controller*, which passes on the right signals with the right level and phase; and (c) "BFM," the *beamformer manager*, brings MIMO and beamforming functions together for an overall smooth and optimized configuration.

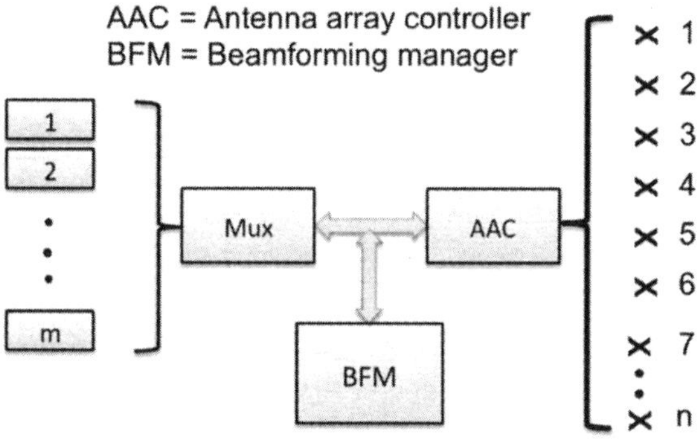

Figure 6.14 Generic configuration of combined MIMO-BF.

In order to understand the system, we divide the MIMO and beamforming functions of a complex system into two separate parts. We have two separate parts that are normally overlapped for an integrated design: (1) multiplexing the common high speed channel for the users to have some share of elements of the antenna in the way required by each user, which represents the basic MIMO function of the process; and (2) manipulation of the transmitted rays propagated from the array elements in a form suitable for calculated directivity as the BF factor to make best use of the beams in the media.

When these two functions are handled separately and adequately, the MIMO controls the beams for the required multiuser operation, and then the beamforming capability of the system is compromised. The same can be applied to the case when we have a single user (no MIMO); then beamforming can be easily managed for its directivity and avoiding disturbing nearby transceivers. For example using DOA in the MIMO-BF system can enable the main lobe to be directed with maximum beam power towards the intended receiver while a null of the beam pattern is directed at the interfering transmitter. Similar cases apply for a MIMO using channel state information (CSI) to enhance the WSN lifetime further to automatically adjust the transmitter power (CSIT) or the receiver (CSIR).

As various parts of MIMO-BF system can be designed following modular rules, the extension of this system can make use of COF modules, and building any size system with any capability is expected to be easy and cost effective. To that effect, design of massive MIMO products, though still complex, can become easier then a design form of MIMO-BF for its futuristic applications in WSN solutions [15].

6.6 CONCLUDING REMARKS

As originally intended this chapter takes the viewer through different aspects of design. The long journey of looking for the missing technological potentials of beamforming for the WSNs working in extreme environments requires an intensive examination of various areas of the technology, including basic arrays of antennas; the importance of directivity; and seeking new solutions through the various technological domains of mobile, TWSN and low frequency applications, leading us to the great untapped potentials of beamforming, unique to wireless sensor networks. This guides us to finding both centralized and distributed beamforming for various energy efficient solutions.

In doing this we have come across many fundamental design ideas for the core process of generating desired beamforming through calculating and applying the BF rules to control both the magnitude and angle of the lobes, as well as nulls of the beam patterns. This design approach should guide engineers to configure, design, and deploy any required patterns, static or dynamic, at the cost of accurate values of the phase being provided to each contributing element. Applying these along with sophisticated control of the MIMO technology is expected to provide a new platform for promoting large-scale WSN paradigms through properly designed beamforming-enabled networking.

References

[1] http://www.antiquewireless.org/uploads/1/6/1/2/16129770/45-did_marconi_receive_transatlantic_radio_signals_in_1901.pdf (10 Jan 2018).

[2] https://en.wikipedia.org/wiki/Guglielmo_Marconi.

[3] https://en.wikipedia.org/wiki/IEEE_802.11n-2009.

[4] Tarokh, V., Y.-S. Choi, and S. Alamouti, "Complementary Beamforming," *IEEE VTC 2003 78*, 3136–3140.

[5] Lopez, D. R. "Real-time reconfigurable pixelled antennas," Polytechnic University of Catalonia, July 2010.

[6] Mehmood, R., "A Study of Reconfigurable Antennas as a Solution for Efficiency, Robustness, and Security of Wireless Systems," PhD thesis, BYU 2015, https://scholarsarchive.byu.edu.

[7] Berbakov, L., Anton-Haro, C., and J. Matamoros, J., "Distributed beamforming with sidelobe control using one bit of feedback," in *IEEE 73rd Vehicular Technology Conference* (VTC Spring 2011), May 2011, pp. 1–5.

[8] Zarifi, K., Affes, S., and Ghrayeb, A., "Collaborative Null-Steering Beamforming for Uniformly Distributed Wireless Sensor Networks," *IEEE Transactions on Signal Processing*, 58(3): 1889–1903, 2010.

[9] Nafchi, A. R., S. T. Goh, and S. A. R. Zekavat, "Circular Arrays and Inertial Measurement Unit for DOA/TOA/TDOA-Based Endoscopy Capsule Localization: Performance and Complexity Investigation," in *IEEE Sensors Journal*, Vol. 14, No.11, pp. 3791–3799, November 2014.

[10] Desrochers, D., "High resolution Beamforming Techniques Applied to a DIFAR Sonobuoy," in Physics Department. Kingston: Royal Military College of Canada, 1999.

[11] Gavrilov, A., et al., "Passive acoustic monitoring of humpback whales in Exmouth Gulf using a sparse array of DIFAR sensors", UA2014 - *2nd International Conference and Exhibition on Underwater Acoustics.*

[12] Sun, J. S., D. S. Goshi, and T. Itoh, "A sparse conformal retrodirective array for UAV application," 2008 *IEEE MTT-S International Microwave Symposium Digest*, Atlanta, GA, 2008, pp. 795–798.

[13] Li, Y., "A design philosophy of portable, high-frequency image sonar system", UA2014 - *2nd International Conference and Exhibition on Underwater Acoustics.*

[14] Oh, G. L., "Model Based Beamforming and Bayesian Inversion Signal Processing Methods for Seismic Localization of Underground Source," PhD thesis, Technical University of Denmark, DTU Electrical Engineering, 2014.

[15] Larsson, E. G., et al., "Massive MIMO for next generation wireless systems," in *IEEE Communications Magazine*, Vol. 52, No. 2, pp. 186–195, February 2014.

Chapter 7

Role of the Overlay Technique for Deploying Wireless Sensing

Water, water everywhere, not any drop to drink
from "The Rime of the Ancient Marina" by Samuel Taylor Coleridge

7.1 INTRODUCTION

In this chapter, we discuss a new way to *manage overlay networking* to help to promote developing new WSN application paradigms through *extreme environments* (EEs). Networking as a concept has a long and rich history of providing services by connecting users of the service. The path goes through various stages including electromechanical, electronic, and computer-controlled switching providing communication channels for connectivity and sharing common resources. In the early stages of the telecommunication explosion it provided complicated services through highly congested systems, which were using the very poor quality channels of the time (radio, wires, or optical); this substantiated the need for a sophisticated way of controlling the network and managing the services in order to meet ever-growing user demands for quality and speed in a very competitive market.

However, since the second half of the 20th century the exponential changes in computing and communication technologies pressurized operators to respond to the ever-growing demands for quality and grade of service, where the need for better controlled services supplemented with some supervisory systems created an opportunity for a new culture of experts and design professionals to maintain full control of the networking and associated management aspects of the service.

These continuous efforts led the experts and visionaries then to improve their network infrastructure to cope with the overwhelming demands for providing the qualifying services using noisy channels and fault prone equipment, devices, systems, cables, and unstable radio channels, which then triggered the new age of *network management* (netman for short), a *critical* part of the service infrastructure that then one could not afford to be overlooked.

With the arrival of digital communications followed by ISDN, a convergence of physical media, and then the internet we see some 30 years of technological fervor with diverse expansions in many new fields such as wireless connectivity and smart devices. Subsequently; a thick cloud of confusion emerged between the three main bodies of industry, policy makers, and advanced academics thereby weakening the process.

In the ICT arena network management as a viable technology was almost lost for few effective reasons: (1) due to their wasteful complexity the telecom standards could not support any new superior services and *cross-layer technology* had to be called in for saving some [1]; (2) lightweight services grew rapidly spreading all over the internet and private networks ignoring the network management; (3) an extended battleground between new software contenders (high-level, middleware, and web programming) and hardware (programmable devices, MEMS, and nanotechnologies) made businesses shallow and often improper; and (4) booming new ideas emerged with little to show for the growing global markets.

Therefore, the field of network management was overlooked by a series of chaotic industrial mismanaged results caused by the random nature of stray decisions arising from new ideas such as ad hoc and unstructured networks, hindering proper solutions to the persistent problems of insecurity, lack of trust, and poor sustainability. The result has been sparkles of too many new ideas with too little to show. Then, arrival of many sporadic poor and rushed applications crippling new markets for young technologies (such as WSN) and creating a daunting future for all new technological paradigms.

Now with the hope of promoting new WSN application paradigms we are on the verge of developing new services for the deployment of WSN in extreme environments. In order to generate such applications a designer should make use of any feasible overlays at every stage of development from planning, analysis (lightweight simulation), deployment (trials and pilots), and control supervisory (lightweight monitoring) of the service.

Many near future applications such as early warning systems, disaster prediction, effective surveillance, pervasive monitoring, and controlling unpredictable events through a better use of intelligent and smart devices would significantly benefit from such overlays.

The rest of this chapter is structured as follows. Section 7.2 describes key aspects and recent developments of overlaying technologies by dividing them into five viable groups. Section 7.3 provides a brief on P2P technology for the IP services and their "overlaying" requirements. Section 7.4 discusses traditional network management systems and analyzes their limitations to cope with new services using 5W1H. Section 7.5 introduces the key aspects of advanced network management as a way to handle managing upcoming services. Here we offer our new network management overlay system under the dynamic design flagship, *dynamic network management* (DNM), for developing the network controlling process. Finally, in Section 7.6 we discuss the extreme environments as disaster-prone working media, where we identify an urgent need for the implementation of dynamic early warning systems using DNM overlay solutions.

7.2 OVERLAY MANAGEMENT

The word "overlay" indicates the provision of a specific form of service supervisory or management mechanism over a field of activities or networked resources enabling operators to provide a clear set of specialized management functions with the required degrees of trust, reliability, and quality of service. In scanning the literature for new ideas and information on the products, we had a rather hard time identifying a proper use of overlaying techniques. Here are some examples.

The traditional SCADA (supervisory control and data acquisition) system as the wired ancestor of WSN, which uses a master terminal unit (MTU) to control some remote terminal units (RTUs), is a well-established technology so its use can save a great deal of capital resources, but if it is not upgraded with new technologies to interwork properly we could face some serious problems including: (1) a lack of compatibility that makes systems slow and in some cases jeopardizes the stability of interworking the system; and (2) SCADA is capable of alarming security concerns. In case of the traditional power grid, the system is normally separated into two different subsystems of power generation and long distance transmission before being distributed amongst the end users. An RTU simply measures power and voltage to issue commands to intelligent control electronic devices to send appropriate signals to particular breakers and generators over long distances through the Internet. The modern distributed SCADA systems in practice require continuous adjustments and cannot stop any attackers, outside or inside, compromising the security of remote substations and therefore they weaken the system seriously [2–5].

It may be worth stressing that due to the importance of security new standards are integrating some protection. For example we use the IEEE 802.15.4 MAC protocol in the beacon-enabled mode for low rate wireless personal area network (LR-WPAN) in secure mode (IPsec). However, for short distance high rate applications such as wireless body area networks (WBAN) we can make use of the protocol in non-beacon mode to enjoy the security provided by the new 5G. That is, if SCADA is upgraded using new technologies such as wireless, and with remote platforms over the internet, we can enhance it further with an overlay system categorized as basic *supervisory overlay monitoring* (SOM).

Due to the environmental issues and energy requirements, nodes of a WSN working in extreme environments frequently fail. This common problem could become a bottleneck for the provision of smooth service. Finding the faulty node is quite critical as the cause could easily get accelerated beyond control. Being categorized as SOM there are three basic methods for detecting failed nodes: (1) using the *heartbeating technique*, where nodes systematically check on each other; (2) using structured pinging messages and waiting for a response; and (3) adopting a gossiping procedure. For the first method each node produces a three-columned table of entry with (a) the address of a randomly selected node; (b) the heartbeat count; and (c) the last time of receiving the heartbeat.

As we will see in Section 7.3, by using the Internet Protocol layer within the traditional OSI standards for provision of internet services we have moved beyond classic telecom systems. For providing such networking services we had to develop what is commonly known as a "service overlay" for two main services, which are associated with two distinct networking protocols, TCP and UDP, called the *service overlay network* (SON) [6].

The original internet services were simple and straightforward; therefore they did not need to be managed. However, as they expanded rapidly into networkable sizes and numbers then they also needed lightweight service management. For improving IP services this supervisory function needed to grow into a service management layer to become another overlay on its own. We therefore categorize this management as the *service network overlay* (SNO).

We also have been using the "overlay" technique in complex computer system designs where the software can help with their natural growing process. The use of overlaying in here is mainly deploying the *distributed memory management* (DMM) framework and sharing cache memories. That is, when the option is enabled a processor makes use of other processors' memory space by extending the direct and indirect addressing methods to enable more complex virtual addressing domains. This complex memory sharing can become further simplified by adopting a dynamic look-up table (LUT) for borrowing and releasing the required memory.

We also have some other overlaying techniques particularly useful to extreme environment applications that can play significant roles in improving the overall performance of the system. Having a predefined objective as a goal or serving mission, it can be managing the process and watching its progress closely towards some achievable points of the operational goal. The goal of the mission then can be easily translated into some operational routines and managed using a specially designed overlay algorithm. These overlaying mechanisms are called *mission overlay management* (MOM), which in practice, means using a flexible structure that can be adjusted from one mission to the next without significant changes in the operation and resource requirements.

Due to their nature most networked applications of sensors are subject to faults, errors, and erroneous behaviors. These malfunctional behaviors are expected to happen far more in the extreme environments than in conventional environments, where requirements for dynamic functional alternation mechanisms become essential. Under the control of a powerful overlay then we should be able to allow them to reconfigure themselves to cope with such changing conditions. This overlay technique usually falls in our MOM category but as we will see in Section 7.5 such functions could be also amalgamated with other overlaying functions and designed under a generic category of our *network management overlay* (NMO) [7].

Some large-scale overlays can be designed to become a common platform. Some can be configured for monitoring the nature of the targets, their available characters, and their variable behaviors. For example we can construct an overlay to match the characteristic requirements of an application network through dynamic scaling of the relevant nodes. That is, upon the size and complexity of the network management we can design some optimum overlaying mechanisms for each of the five categories, DMM, SOM, MOM, SNO, and NMO. Figure 7.1 shows the five members of our overlay management systems. Their complexity is expected to increase from left to right.

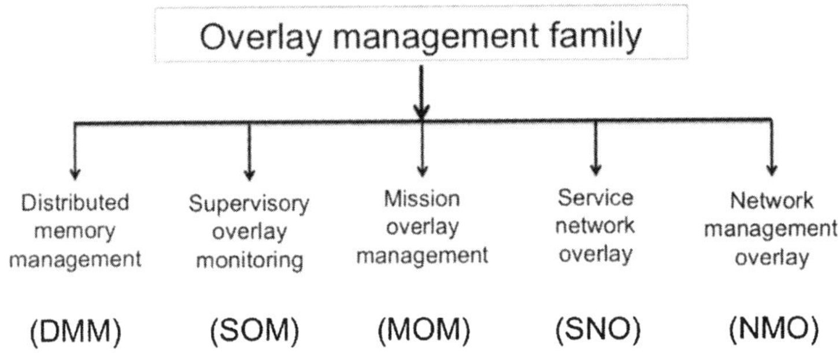

Figure 7.1 Five functional overlaying categories.

Here is a recap of our five functional overlaying management categories of Figure 7.1: (1) DMM, a lightweight and wideband overlaying controller mechanism to manage multiprocessing operations for handling very low complexity structures; (2) SOM, a replacement for traditional SCADA systems suitable for lightweight management of supervisory and monitoring applications of WSN; (3) MOM, for managing one out of a selective set of management operations. It can be designed for a specialized mission or be configurable to change the mission upon a command or changing environment; (4) SNO, a specialized overlaying service management such as the internet, where P2P services are managed through a layer on top of SON; and (5) NMO, our new generic network management overlay system to implement the advanced network management, DNM, which is flexible for managing wide range of applications such as WSN, IoT, and upcoming 5G services. See Section 7.5.

7.3 PEER-TO-PEER NETWORKING

The internet as a networkable service should remain as an unstructured network. The original idea of using it as part of the 1970s telecom industry began with introduction of the *Internet Protocol* (IP) in 1974 as a low cost alternative to traditional packet switching technology (ITU X25). It provides two networked services at OSI's network layer using complementary protocols: (1) TCP for a fully structured connectivity and (2) UDP for a semi or totally unstructured connectivity services. Figure 7.2 shows the IP stack used for typical structured services. In the early days this complete structure was considered feasible for slow internet services but these days its use is limited to connecting the legacy equipment [8].

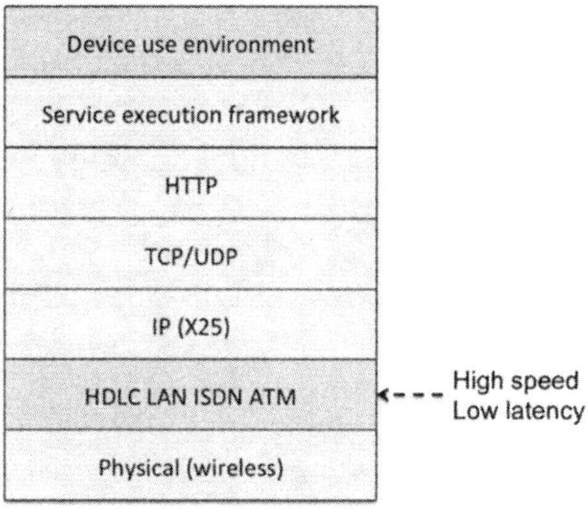

Figure 7.2 Full-layer stack for integrating legacy devices and gateways.

Out of two basic extreme networking topologies, star (totally centralized) and mesh (uniformly spread) most practical internet services build their applications closer to the first for two reasons of unlimited number of users and minimum networking complexities. It is therefore called P2P (peer-to-peer protocol) network. With this our communication technology is evolving one more step beyond digitization revolution. Further expected benefits achieved by reducing the cost, reducing the complexity, and improving the maintenance for smoother operation, where the new trend of spreading the networking complexity of the service onto user domains with new and powerful devices would help. Therefore, any further move towards uniform unstructured network applications is now achievable using the UDP technologies, which are far lighter and much simpler than the TCP.

Due to the above mentioned advantages arising from internet's unconditionally unstructured connectivity, the sudden growth of demand under the superiority of low cost internet broadband access encouraged many new P2P narrowband and broadband applications to emerge enabling service providers (ISPs) to maintain very competitive rates and to let many P2P network providers to offer more secure and more reliable services.

As we look at some P2P examples below we see that practical P2P companies had to rediscover their need for some dedicated overlay technologies to control and supervise their own services, and they then had to design it all from scratch. This requirement has been critical for maintaining their success, as many functions such as network security and trust could not be overlooked. We will discuss the details in Section 7.5.

One of success stories for providing internet services over structured networks (TCP to start) happened around the turn of the millennium. The NTT initiated its innovative idea of integrating internet technology into mobile telephony in 1991 i-mode (mobile internet-via-cell-phone). These mobile phones were regarded as the first step towards using internet services within the telecom industry. The technology was introduced in Japan in 1999 and within two years it grew into an over 10-million-user network. This success in effect enforced significant changes to mobile technology and triggered the move towards 3G. It is not surprising to see many innovative technologies for a wide range of new services, changing demographics, and seeking the demand. Being praised by the world media and industry analysts alike NTT made the top of the list market capitalization in Japan and second in the world of telecommunication businesses. By 2008, there were more than 50 million telecommunications companies [9].

In order to create viable businesses on the data business side of providing internet services, however, the companies faced many challenging problems. Building new services over the available bandwidths at a very competitive cost to the users of unstructured networks was more than a natural challenge. Many ideas emerged to solve the problem of providing emerging QoS-sensitive services under the pressure of costly bandwidths. The challenges included decoupling current internet services, provisioning a supervisory function-enhanced end-to-end connectivity, and designing new required value-added internet services. Quality services such as quality VoIP and high quality video-on-demand were characterized as bandwidth-hungry services and required tight management. That is, the integrator received bandwidth from various network domain providers according to service level agreements to build logical end-to-end service delivery infrastructures on top of existing data transport networks. The overlaying service then needed to do some serious formulation of dynamic bandwidth provisioning for the usage to manage various traffic demand distributions. Figure 7.3 shows how the SNO nodes control gateways of the complex *heterogeneous supplying networks* (SOM) [10].

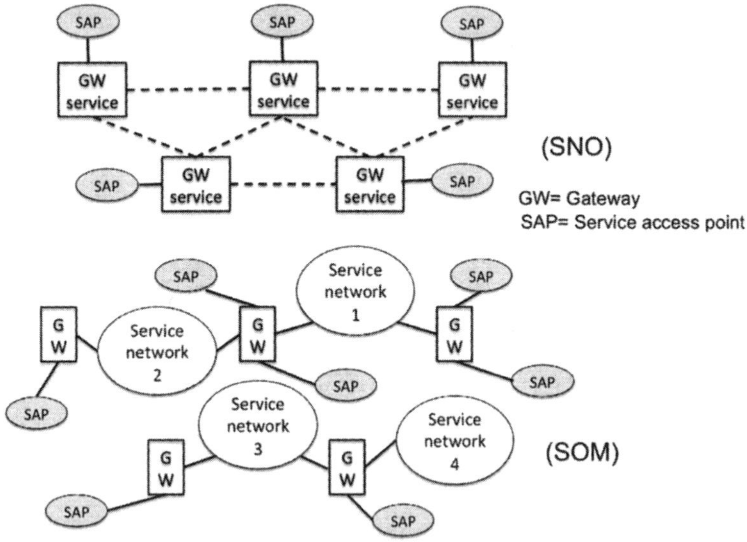

Figure 7.3 Demonstrating the requirement of an (super) overlay network to control heterogeneous resources.

One way to overcome the problem of market competitiveness in heterogonous P2P service networks (e.g., maintaining interworking with traditional standard devices) is to remove the interconnection rigidity of the service through designing a specialized overlay. In such cases one may bring a marketing service intelligence factor into the overlay management for optimizing autonomic service implementation. This function would help overall P2P system improvement in two beneficial parameters: (1) service providers' best use of resources due to adopting a self-configuration function (adaptive) and (2) enforcing a graded charging for the users, where regular and heavy users are differentiated from occasional users to pay less for their low grade required access. Provision of a successful unstructured internet services (web, data) over the internet has been resolved. Provision of P2P service for individual users is simple and easy especially if the end users trust one another (as in private networks). However, this is not a common case for businesses [11, 12].

The earlier services of SON divided into two groups: (1) structured ones such as Chord, CAN, Tapestry, Pastry, Viceroy, and Kademlia, which due to their lack of proper network management overlay virtually all suffer from insecurity problems including man-in-middle and Trojan attacks; and (2) unstructured ones such as Gnutella, BitTorrent, and Freenet, security measures have made them more popular [13, 14].

The tendency towards using online virtual activities such as design, development, training, gaming, and shopping with various groups of users around the world is increasing everyday. Users of these applications need to interact with each other and collaborate upon some shared virtually enabled environments. Also, such applications need to use some distributed processes, 3D graphic representations called *collaborative virtual environments*, to enhance their services [15]. With IP services promising best-effort end-to-end guaranteed QoS using low-cost Internet resources, we envisage the provision of an SNO overlay that could enable such service with no need of support from the underlying network resources [16].

7.4 NETWORK MANAGEMENT

Traditionally, telecom services were controlled using in-service supervisory functions. The concept of network management started with the control of signaling systems in electronic exchanges, before becoming the controller of *supervisory services* for various reasons including reduced complexity, supporting heterogeneous functions, reduced cost, and managing the ever-increasing volume of traffic. Examples of the controller functions include basic device maintenance, regular service signaling, monitoring, maintenance, fault recovery management, user location identification, tariff management, traffic control, and maintaining the quality of service.

As the computing capability of the main controllers and stations gathered higher capabilities in various dimensions of speed, memory, and intelligence along with network components that were becoming more reliable, more resilient, and less error-prone, then the network management roles became simpler. The older and more primitive functions were replaced by more sophisticated functions such as location identification and mobility. Then, with the internet and high speed internet many new service support capabilities, under the flagship of LTE & NGN, demanded much better use of the new computer functions to satisfy new user appetites. Network-managed location identification (NMLI), for example is a basic service supported by WiMAX (IEEE 802.16) supporting 3G. This network management function supports many control and supervisory services including tariff, traffic, and mobility parameters of the services supported by WiMAX. The measurement of the location requires significant calculation of various signal attenuation and distorting errors due to various propagation disparities such as multipath, fading, and non-sight of light. The main components for such measurements were the time of arrival (TOA), the time difference of arrival (TDOA), and angle of arrival (AOA) [17, 18].

Managing a network, a service, or an application scenario requires making many business decisions based on facts, figures, and reliable predictions. A traditional method, evolved in management boardrooms, is trying to find out the details and shaping a solution by questioning all possible effective and critical aspects of the process under investigation. The 5W1H is the most popular model covering location (where), time (when), people/users/clients (who), the process/product/service/network (what), and the main objective (why) for five dimensions of "W" followed by one "H" for the design and plan of action (how) to deploy the *analyzed process*.

Although it sounds complex and primitive to some, professionals involved in strategic decision-making appreciate this method as they can easily tailor it and slim it down to an optimum, agile, and practical deployment of the process. Like all classic management methods, which are usually built, upon practice and experience, 5W1H despite frequent criticisms is shown to be applicable to many management cases. Its practical effectiveness becomes clearer when a complex case (e.g., integration of many unrelated fields) gets involved where "nothing looks obvious." Due to the continuously evolving nature of dominating technologies and prime demands, "the strategic business designers" won't know where to start, where systematically designed network management is the one that has been a *key success factor* in fast moving telecom industries. The sequence for the queries normally start with "Ws" to answer followed by "how" at the end. To save time and improve the effectiveness of the meeting the sequence could be decided prior to a discussion session. Do not forget that this sequence could be adjusted to suit any particular requirement. In order to visualize the process objectivity a typical sequence is shown in Figure 7.4 as a stack of functional layers. Figure 7.4 shows how these six layers can be used for fault recovery [19].

In some cases when using the 5W1H management tool some of the components could be omitted when exercising the tool. Special care should be taken to avoid missing any components critical to the process. Normally "why" should not be omitted. Why normally represents the main objectivity of the process; even if it is obvious or well known to all parties involved, it should be included on the list. As a key component it requires a proper definition and often goes through a number of refinements throughout the meetings. Missing the "why" could result in missing the aimed objective instead of achieving it.

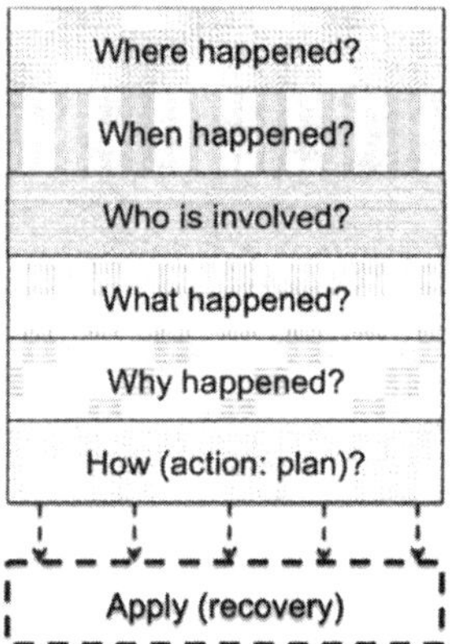

Figure 7.4 Simplified example demonstrating the 5W1H top-down management queries for a fault recovery process.

For example, a project using 4W1H for designing some robotic system features with no human functions involved can miss the middle W (objective, who) but a similar project from the same team of researchers uses 5W1H due human to functions (interactive design sensor clusters and self-organizing maps) [20, 21].

Another use of 5W1H is for predicting faults. This is far superior to classic design methods, where a piece of software can set some statistical rules for learning about sources of the faults and their processing mechanism. Using the 5W1H contexts the designer produces some fault diagnosis rules to predict a particular fault pattern, which can be used to generate a warning signal [22].

Similarly the 5W1H method can be adopted onto a client's usability analysis process leading to some interactive design procedures and generating some classes of user groups. Usually "why" is associated with the main factor of the goal and sometimes normalized to a cost factor in the process, while other factors of the set are closely associated to real system parameters [23].

Despite various work and solutions to improve traditional network management to make it suitable for new services, there are still too many unanswered questions and practical issues remaining to substantiate our claim that the traditional network management system is too rigid for substantial changes to be adopted today as a viable solution. Chapter 9 of Sohrabi et al. is a good source for analyzing this case [24].

7.5 ADVANCED NETWORK MANAGEMENT

With the impact of the internet and mobile technologies following significant technological improvements in their connecting links, in terms of error rates and reliability, traditional network management started to look like a vintage idea and started to fade away. This radical change of perception, however, could not be justified in practice, as all networks, structured, ad hoc, or unstructured need to deal with supervision of the services [25]. Then IP (P2P) services could not progress without a proper centralized layer of control. Service providers are now marketing this as the real comfort factor behind the market boom. Now with the introduction of 5G, we are integrating internet services under the IoT and of course the WSN with the ever-evolving telecommunications. In conjunction with an increase of insecurity and distrust (inherent to internet services) we see a clear path for the return of traditional network management but advancing into more proactive system architecture; we therefore call it *dynamic network management* (DNM).

The tendency to advance the network management in a unified format has been around for decades, helping with the better integration of internet services with 5G and its earlier generations. Most of the work has been towards promoting new IP networking to make it easier for it to be supported by 5G standards. Due to their independent roots, managing the internet networks come with many limitations and challenging issues facing 5G.[6] However, managing complex networks can be helped with adoption of a lightweight overlay using the following ideas: (1) extending the emerging service overlay to cover critically sensitive areas of the network; (2) making use of gateways, and other interface devices to gain compatibility; (3) adopting basic packet switching for virtual connectivity; and (4) adopting the convergent network architectures leading to the *delay tolerant networks* (DTN) [26–28].

---

[6] Network management aspects of integrating the internet and 5G during the transient period is beyond the scope of this book.

7.5.1 5G Initiatives

Due to the rapid growth of P2P and the dawn of extended internet services such as WSN and IoT, along with the smartphones of NGN, the fifth generation of mobile technology is once again challenging the traditional layer structure. As shown in Figure 7.5 the conflicting views of telecom operators dispute accepting the internet as part of the standards that have been resolved by new IEEE initiatives, enabling 5G agreements to support and incorporate at the physical and application platform levels. The four main service layers in the stack, the 5G version of NGN, are then required to be redefined to be able to support the three ever-evolving technologies of wireless technology, internet, and mobility. Then, incorporating some functions of network protocols and applications, enables the "application platform" to perform mainstream Internet services under Web5G, which includes traditional internet services, P2P, and other web applications. Due to the fact these are mostly designed to be independent from the carrier infrastructure, these applications do not require further network management services. To this effect we see the Web5G project being set to enable some return from the heavily invested infrastructures [29].

The new standards, therefore, should accommodate *five new service capabilities*: (1) new applications with higher bandwidths (5G) and very low latencies; (2) further integration and higher interconnections between the application network and service scenarios; (3) greater controlling functions at the network layer for supporting web applications; (4) provision of new capabilities to support future services for further impacts; and (5) provision of enhanced control mechanism capabilities network management attached to the control plane.

Three main enhancement components for supporting the 5G new functions in Figure 7.5 are: (a) Web5G, (b) a new layer of advanced computing functions over the control column to support control and deployment of DNM, and (c) the "5G and low power wide area networking (LPWAN)" group of physical connectivity.

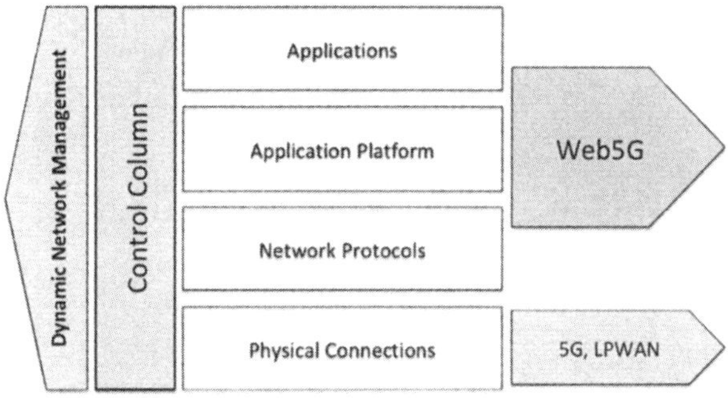

Figure 7.5 New model of layering structure to integrate dynamic network management (DNM) into internet-enabled 5G standards.

The functional group (a), Web5G, explained earlier, is for 1, 2, and 3 capabilities. The functional group (b) is a new layer of control and supervisory to enable better use of advanced computing and networking [e.g., discovery, recovery, security, augmented reality (AR), cloud computing, enhanced virtual private network (VPN) services, and narrow band (NB-IoT)]. The functional group (c) was expected to enable new physical layer support for the IoT and smart device applications. It was mainly expected to provide extended and long-distance capabilities (e.g., remote sensing, global, localization, and target identification).

Figure 7.6 shows the state of technology range versus distance for 5G low power wide area networking and associated applications. Three main activities in this area are: (1) high data rate long-range services using IEEE 802.11ah WiFi style short distance wireless connections supported by P2P-style internet low cost and long distance services. These services are expected to have a large number of applications in extreme environments as well as classic applications such as remote maintenance and surveillance; (2) support of the long term evolution (LTE) process towards services on the move including robotics and other unmanned vehicles for moderate data rates and moderate distances under machine-to-machine (M2M) projects for smart transport applications (see Chapter 12); and (3) new low power long distance products to support IoT-style applications (LPWAN, NB-IoT).

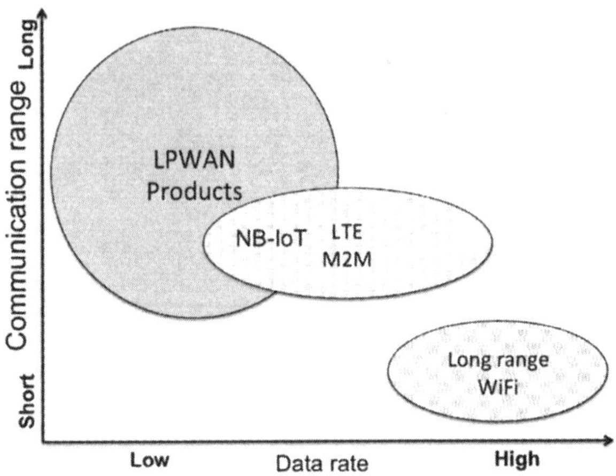

Figure 7.6 The spread of products for developing new products supporting the NGN. The network range versus data rates for low-power, short-distance, and wide-area networking.

7.5.2 Mobile Edge Computing

One of the new core control functions of 5G (see Figure 7.5) is *mobile edge computing* (MEC), which is included in the new design to make better use of *cloud computing* in the provision of new services and to give mobile devices such as smartphones a better use of open standards and application programming interfaces (APIs). New standards developing by the European Telecommunications Standards Institute (ETSI) are expected to create a new industrial paradigm. Supporting the virtual appliance applications as a new operating system and virtual machine platform, means it can provide any middleware applications and DNM services.

In conjunction with other functions of the DNM (SDN and NFV), by creating multicloud application delivery (MCAD) platforms and efficiently amalgamating resources, MEC is expected to help in critical parts of managing (trillions of) devices and smartphones. These changes would be achieved through better management of, the so called dumb pipe radio access network (RAN), converted to "intelligent" pipes by overlaying distributed edge cloud computing onto the RAN. With virtualization at the edge of the RAN, mobile network operators can allow multiple third party tenants at the base station. For example intensive computing processes such as AR and image processing of applications sensitive to latency or energy can be hosted at the edge of the network, usually at the base stations and net data can be delivered to the user devices [30].

7.5.3 SDN, NFV, AR

Traditionally, network devices contain both control and forwarding functions that are configured per device, creating problems of limited intelligence, excessive complexity, and tedious device management. However, if the network is controlled by a central software management platform and it is kept separate from the traffic forwarding hardware then we have a *software-defined networking* (SDN) solution [31].

In practice, building and operating a network is both time consuming and expensive plus it requires massive amounts of equipment and specialized expertise to (1) design the infrastructure; (2) set up and run that equipment; and (3) to get new ideas and services to market. For today's world of internet connectivity, service providers are always looking for cost-effective ways to meet the growing demand for more bandwidths, which pushes business development and is very tense and tedious. These pressures in conjunction with progressive world of computing, encourages the use of *network virtualization* (NV) and *augmented reality* (AR). Their potential is in dramatically lowering the costs by using network infrastructure from the network control and application layers in the form of abstracts for enabling centralized auto-provisioning of network resources, through a virtual network model. Such an abstract process would allow a service provider to expand the capacity and the access range, along with flexibility across a multi-network solution for a choice of devices from a centralized management center. Users also will benefit by gaining a better experience as a result of rapid network adaptation and the introduction of new services that previously could not have been provided. *Network function-virtualization* (NFV) therefore can be regarded as a new tool to enable the better use of enhanced flexibility in designing network applications. This is regarded as one of few key complementary computing technologies, enabling programmability of control and network functions for the eventual migration of key constituents of the network to the cloud.

AR is a view of a physical or real-world object or environment where some of the objects are augmented through computer-generated sensing data such as image, sound, video, graphics, and GPS data. It is in contrast to *virtual reality* (VR) where computers simulate the real world. AR is defined as a technology that combines real and virtual objects to provide simultaneous interaction between them. The additional sense of reality between virtual and real worlds is created through the manipulation of overall perception. The effective illusive content in AR created by injecting additional objects (sounds, images, and videos) originally sourced from other physical environments. This leads us into another definition of AR: "*a real-world, real-time interactive system combining various real and virtual objects in a three-dimensional environment.*"

A simple example is inserting the scores on a TV screen during a football match. Unfortunately, this prime technology has been diluted due to excessive use of VR objects. DNM has the potential to make effective use of AR for enhancing the usability of its tools and algorithms. AR as a service can also be used in an application for excursions or sightseeing where accurate data is required. The net data of the analyzed data deducted from the device, such as an image from a camera, a precise location, and the user's position, means that after analyzing the information, the application is able to provide additional information in real time [32].

In order to clarify AR from VR we look at an analogy of our observation considering two extremes of real and virtual shown in Figure 7.7. There is another complementary variant in the process of mixing (mixed reality) called *augmented virtuality* (AV).

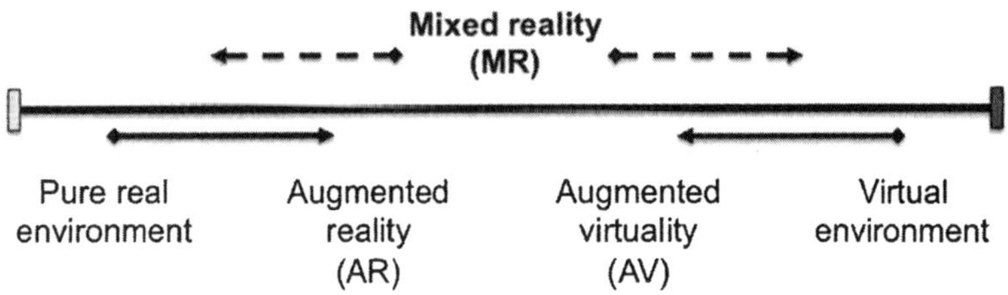

Figure 7.7 Four distinct status of the (Milgram) mixed reality [35].

AR, for example, can become an important feature of innovative designs implementing a method traditionally called *trialability* (as in product exhibition centers). In this stage of marketing innovative products the customers try a new design so that they get the proper feel by using it before they decide. Creating such a stage of demonstration for large machineries, such as *locomotives* or complex *software* if feasible, requires costly samples and preparations make fair negotiation almost impossible. However, AR can make the task much easier if customers (or company representatives) were able to try and see for themselves the outcome of combining various available parts and systems, in virtual products, where AR would enable users to generate realistic combined images [33].

7.5.4 DNM Four Basic Factors

For the smooth deployment of a sustainable and reliable DNM we need to consider some orthogonal stability factors to help our developers throughout the lengthy rocky path. Naturally a minimum of *four basic factors* (as four foundation columns of traditional buildings) are most relevant to the overlaying process, namely *trust, objectivity, security, and stability*.

Trust from a measurement point of view is a subjective element, or nonmeasureable factor that can change its impact upon people's state of mind and the surrounding environment. This management factor is something to watch out for in untrustworthy events. Similar management components have many tools to measure some aspects of trust, but no decisions should be taken about them without a fresh open-minded review of possible outcomes.

Before introducing our new simple model we look at few aspects of trust: (1) Trust is a complex human perception or feeling. (2) It therefore comes in a two-dimensional subjective perception, factual trust and virtual trust. These two components are expected to be orthogonal (i.e., independent in all aspects influencing our evaluation). (3) A factual component of the trust is a logical evaluation or perception of an expert from the trustee. (4) The virtual component is a *common perceptive trust value* of the trustee and therefore builds the visibility and quality of the presentation. (5) Also, as in here we are involved with our systems in use; we therefore need to extend our definitions to the impact of systems on our trust. The analysis in this area should carry very little human feeling, and therefore the outcome of trusting a user in here is limited to the system part of the trust, in which we should include the content trust (received information by the trusting user from the trustee user) at a later stage, if required. (6) For the system level evaluation the system receives two components from the trustee namely, "system usable factual feature" and "system usable virtual feature." Having these two components received by the media, if features such as reliability and security then influence it, it could enhance the system before being passed over to the trusting user. Intelligent systems might be able to enhance the factual component but basically are expected to enhance the virtual component more. In cases involved with the extreme environment the limited access capability of the other users could increase the trust.

Now, we can set our model using Figure 7.8's trust variables of the trustee ($\propto_f$, $\propto_v$), for the originator of information as viewed by the system

(γ_f, γ_v) being converted to a complex trust vector of information with a trust vector of (β_f, β_v) for the trusting user, as below.

$$\begin{cases} \beta_f = \alpha_f + \gamma_f \\ \beta_v = \alpha_v + \gamma_v \end{cases}$$

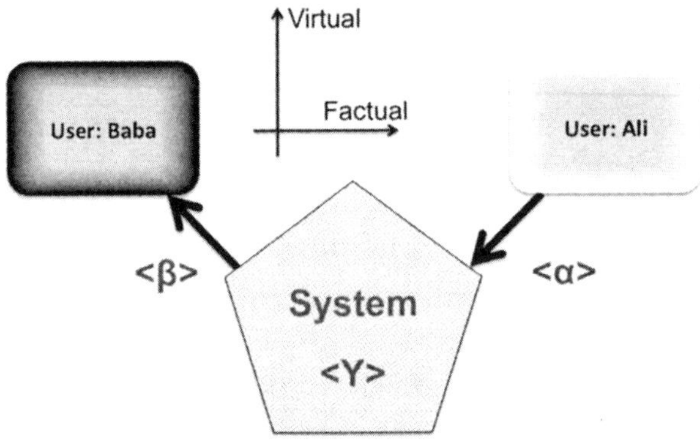

Figure 7.8 The trust factor, its two-dimensional perceptive components, and its manipulation through the system.

Objectivity, as a prime factor, means a decisive plan for achieving a manageable set of precisely defined goals. Stemming from our lengthy management experience, in conjunction with a vast number of management tools, the generation of a dynamic plan would enable designers to achieve their complicated goals in minimum time. Being open to cooperation from other teams would help (1) to visualize the outcome far ahead of time and (2) to minimize overheads and wastes. We will use this element in our early warning system cases in Section 7.6.

Security[7] is a prime factor for maintaining any meaningful operation and therefore should be included in the overlaying process of DNM. Therefore system resilience against the *insecurity threats*, like trust, can be modeled in a two-dimensional format, for measureable factual components in physical and upper system levels of DNM, where some figures indicating the probability of risks should be evaluated for the degree of insecurity. Due to strong dependency of resiliency on the design and control processes, the insecurity measurements should be integrated in the DNM process.

[7] We do not have a measure for security but there is a possibility of measuring the security risks for a system. Details are out scope of this book.

It is interesting to note that due to the special case of "security" there is the possibility of systematic and well-calculated security attacks on or via the service overlay networks using some weakly designed operators. Examples include jamming transmissions, cutting wires, distributed denial of service (DDoS) attacks, and objectively eavesdropping. Although some users can protect themselves by applying cryptographic techniques against monitoring and eavesdropping, or devising their own prevention once under attack, for many that is too late. However, there are more complex cases being developed by disturbing attackers everyday and any weakly designed overlay could expand towards vulnerable users [34].

Despite the general perception of higher security in extreme environments, we are convinced, due to the natural changing conditions expected to dominate in such environments, that the degree of insecurity could easily rise in many susceptible areas making our WSN systems and associated components vulnerable and weaker against systematic attacks.

Finally, our last prime element of DNM comes under "stability" of the service, which is expected to cover the overall dependability, reliability, and sustainability or an amalgamation of these terminologies, with some service and environmental variables such as continuity, availability, and predictability to clear one's mind of any possible unpredictable events. Therefore, under different circumstances our basic term of stability could be simplified to one or two of these terms.

As systems sustainability is hard to measure, for this case we make use of its measurable feature of system reliability and its related feature, the recovery factor (QoR). As both humans (users, experts, and managers) and systems (devices, hardware, and software/middleware) are involved in the measure of reliability it should be expressed in a two-dimensional format of *objective* for robots and *subjective* for humans. The system reliability is often called the system's resilience against design issues such as faults, operational misbehavior, and abnormality (such situations happen more often in extreme environments) as well as human errors of those involved. This two-dimensional reliability could be simplified to system reliability if weaknesses of the network against the second dimension (human errors) could be converted into faults and errors at the vulnerable points of system. Despite the possibility of designing devices with much higher accuracy and precision, the DNM reliability could be easily compromised working under extreme environment conditions, a quality assessment for system reliability is also called *quality of recovery* (QoR). The system quality enables operators to build a relevancy measure of the recovery work for customers' satisfaction upon their expectation from the service.

Hence, we measure the value using the classic exponential definition of reliability. R_u, as a function of system's un-interrupted service period, u, $R_u = e^{-u^{-1}}$ [35].

7.5.5 Core Overlaying Processor

Due to the importance of long-term *global market viability* success in marketability of the DNM, the "design trust" factor plays a critical role. Trusting a design heavily depends on its "working life without modification". This factor then dictates specific attention to the design process as a prime requirement. Further into the design requirements the details should be addressed per each application scenario. Although the overall bandwidth for processing the DNM is quite low, whatever the overlay, it is required to manage a wide range of activities each with different speeds of response and reliabilities. The sensitivity to the energy-versus-performance factor can be addressed through some flexibility at the core via programmable embedded self-configuration.

The design procedure for a core processor to manage the highly critical process of DNM could at thc samc time integrate a heterogeneous network which may impose: (a) extra burden on the gateways and node processors; (b) an specialized, cost barriers due to high-speed, and expensive service resources; (c) significant extra complexity to networking; and (d) serious performance and quality risks that could endanger the whole process if the service network grows fast beyond its practical limitations. Although small and fixed WSS and WSN applications would survive any such complexity and its extra overheads, any large and evolving networks such as global seismic networks, integrated climate monitoring networks, large WSN, future P2P services, and the upcoming IoT services would not be able to deal with this burden of wasted resources. We recommend that since today core process of DNM is designed to deal with a set of specialized processing functions, in the form of software, then all that remains is to build a set of modular, lightweight dedicated processing functions to be handled by a core processor as the *core overlay processor* (COP). The following paragraph may shed some light on our design process.

The nature of an overlaying process dictates preparedness to handle one or two out of a vast number of independent and accountable small selectable processes at any one moment. We, therefore, need to find a practical way of amalgamating all these functions including the discovery of a new event, activating reconfiguration tools, refining service adjustment knobs, and supervising monitoring processes by converting them into a ready-to-use sack of operating subprograms (e.g., a Pandora's box). The box can contain all sorts of functions varying in their nature of process and demand, but they all need to share the one common feature of very low bandwidth usage. That is, though there might be tens of functions waiting to be acted upon, only a few urgent functions at the top of the queue and only one highly urgent task need to be processed immediately. Therefore through the overlay, each function does not need any significant amount of bandwidth.

This process of bringing in and integrating all the small, but extremely important, network management elements together under one operational platform requires the introduction of two new subprocesses of prioritization to be tagged at entry. We therefore use a flowchart to visualize our DNM algorithm called *convergence divergence ubiquitous-access algorithm* (CDUA) [36,37].

For using CDUA we generate a long list of tasks associated with all input functions to be checked and used for making decisions. They are processed in order of priority based on their degree of urgency in managing the network. To do this we first pile up all DNM objectives under a long-objective-list management (LOL) as shown in Figure 7.9. LOL can be produced at the early stages of the deployment phase and updated as required. In the case of a major change in the WSN operation, a total upgrade is required where a new set of objectives is piled up to accommodate the changes, especially if most of the significant changes are in the target WSN. The service requirements in the application scenario can be implemented gradually under refinements of DNM system, or during the recommendable annual refreshment as in normal progressive WSN technologies.

In CDUA, tasks are units of DNM operation usable by the processor. They are different than the objectives, which are for planners and maintenance engineers to use for implementing their requirements. Then, another DNM operation is objectives-to-task conversion (OTC) shown as "LOL to LOT", which generates a set of tasks each carrying its own priority class. For example, for many NMO applications, having three classes could be sufficient. Classifying them as 1, 2, and 3, then "class 1" tasks could be representing regular checking, maintenance, and basic tasks. If "class 3" is reserved for special and critical tasks we will be left with "class 2" for important tasks, which could entail regular but not critical extreme environment-specific tasks. The LOT table plays as important role in the DNM process.

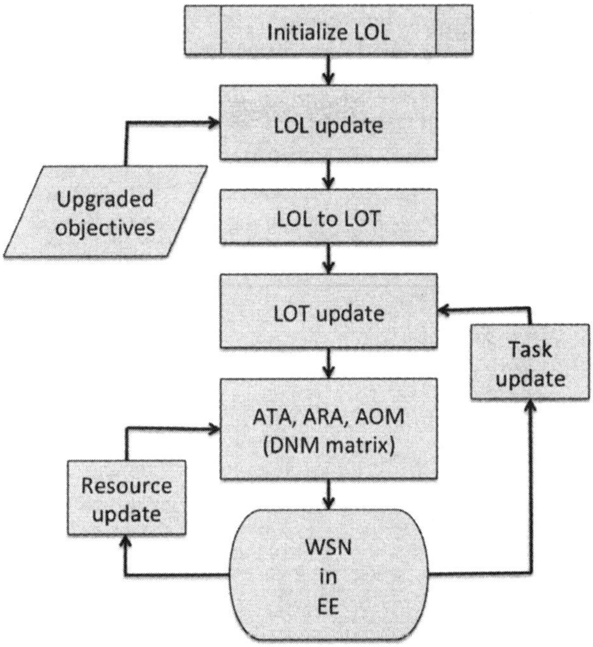

Figure 7.9 Typical CDUA flowchart for implementing DNM algorithm (COP).

At the heart of the DNM process lays a controller and a task allocation matrix with some DNM agents to take control of a task unit from the table and manage its operation with a report on its completion; see Figure 7.10. Inside the matrix we use three main operating agents: *agent task allocation* (ATA), *agent resource allocation* (ARA), and *agent operational management* (AOM). The DNM resources are WSN and environmental agents.

Our network resources are of two kinds, internal and external. The internal resources are those located within the WSN, and external resources are located outside the WSN, either within the surrounding extreme environment or other environments, usually near by.

Two types of agents are directly involved in DNM operation. The master agents (MAs) for initiating and monitoring task implementation and environment agents (EAs) for executing the tasks. The MAs are located inside the matrix (e.g., embedded or in a form of middleware) controlling the EAs. Then EAs are scattered around in the physical monitoring area to enable the required resources, located in sensitive and strategic places for controlling the key resources.

Under normal uses, internal resources (within the WSN) controlled by internal agents are operational to monitor and maintain the network's dynamics. Use of external resources can be made available to use through other environment agents. For further information on DNM see Section 7.6.

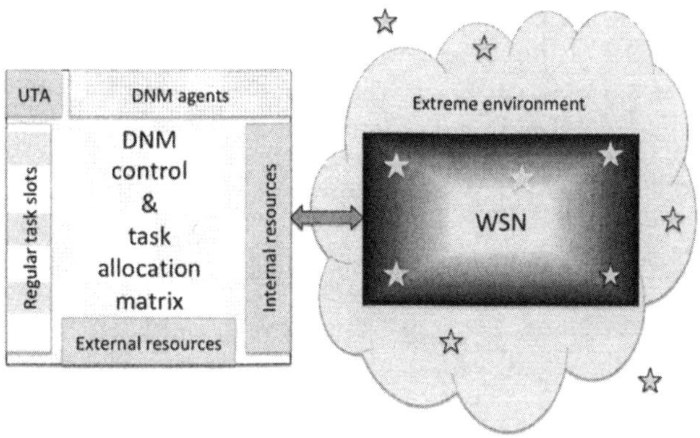

Figure 7.10 DNM architecture, the matrix structure, and its internal and external components.

7.6 DYNAMIC EARLY WARNING SYSTEM

To look for the meaning of "disasters," we find some vague and confusing definitions appearing in most dictionaries, including the two categories of "tragedies" and "failures." Tragedies or catastrophes are viewed subjectively a by human mind as a misfortune. Failure, however, sounds more logical and proactive, which leads us to our definition of disaster as *massive losses due to unexpected and unpredictable events.* Being man-made or a result of human error, including bad designs, unavoidable high-risk operations, and carelessness, we all share the resulting *scars and bad memories*. We, therefore, should do our best to reduce such possibilities and to save lives before saving our goods and other dispensable belongings. Our main reaction to disasters therefore remains to save lives, especially where we are dealing with extreme environments where we have far less control, access, and capabilities.

Now, by using our smart sensors either as a small scale or a large scale sensor network we have far more confidence in being able to save lives and valuable resources from disasters and other unpredictable events through providing reliable information from the disastrous environments and beyond. It is true that our technologies are still too naïve to be able to control natural disasters such as earthquakes and storms, but we do have reliable information systems and sensor technologies that save lives. On the sensors side of technology we are able to collect extensive readings from the disaster's source and we can filter out any useful data to use for timely actions thus minimizing the casualties. These days we have a very large variety of low cost sensors measuring earthquakes (seismic), imaging (satellites), processing (computing), and intelligence (agents) as a complementary devices using micro and nanotechnology to build an extensive number of warning and predicting systems to save people from disasters. Unfortunately, in comparison with the number of cars on the road, we see very little of such technologies being made available in vulnerable places on earth.

We really need to urge our industries to seek life saving systems. To develop meaningful solutions to help with disasters and emergency events we need to analyze individual cases by using reliable fundamental data (e.g., recorded facts and figures for a specific disaster). Unfortunately, most of the old and ancient figures are more of a guesstimate rather than true measurements of the relevant features. Sometimes we may need to rely on the words and claims of those directly, or indirectly, involved. There is also an eye-catching factor with high volume figures tending to be more accurate for large-scale events. In the case of small scale and seminatural disasters such as the collapse of old mines and bridges, which are sporadic, local, and subjective, the overall figures are often inaccurate and hard to be traced.

Design information, also called data, is required for the development process. This is due to two aspects of viability (market: for mass applications and global use) and availability (product: for maturity of the required design technologies to deliver proper and lengthy solutions). These two aspects could share some data but those used for the products are expected to be more accurate than those used for the marketing.

The design must be objective and we ought to decide precisely how to make use of WSN technologies in form of better early warning systems and professional rescue management to save life caused by natural and man-made disasters. Saving lives requires a reliable source of data including the history of events, clear design objectives, and resources to enable timely deployment. For statistical data we can use Rose et al.'s recent data [38]. Their classification of absolute annual deaths due to natural disasters includes nine categories: (1) wildfire, (2) volcanic activity, (3) storm, (4) landslide, (5) impact, (6) flood, (7) extreme temperature, (8) earthquake, and (9) drought. Their 1900-2015 figures, in 12 columns, indicate huge changes from one decade to the next, which makes us to believe that, due to lack of a proper reporting system and upheavals caused by the century's world scale wars, figures for first five decades may not carry any significant weight. We, therefore, for our analysis look at the last seven columns from 1950. This is shown in Figure 7.11.

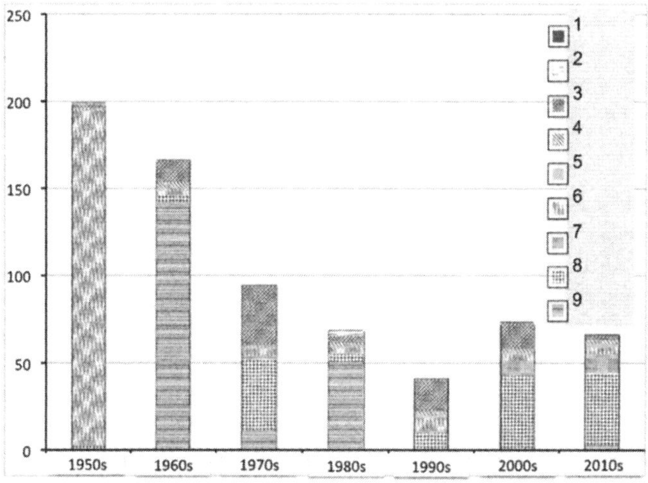

Figure 7.11 Absolute annual deaths due to natural disasters for five and half decades configured in nine categories. Redrawn from [38].

As shown in Figure 7.11 two historical alternating disasters with major human casualties are drought and flood. The patterns in recent decades are very different from earlier ones. For example, drought has not returned as before in the last decades, and, global scale flooding has been significantly reduced. The reduction in drought is due to the better use of technology in farming, agriculture, and transportation. The flooding reduction is from better housing, style of living, and telecommunications. That is, though disastrous events are happening and are totally out of our control, our loss of lives due to them have been reduced significantly. However, the loss of properties, goods, and belongings (not shown in Figure 7.11) remains persistent. That is, in the case of flooding, events have been happening much the same as before but better general management and telecommunications combined with public awareness enabled by the internet have been effective contributors. For human flood safety see Section 9.2.

Our improved life saving in the last five decades indicates that though we have progressed towards saving lives through better dwelling and transport technologies, not as much visible professional and direct progress has been made as expected from smart sensors and the internet.

Earthquake, as a natural disaster, has always been our worse enemy. In Chapter 10 we look at seismic waves and new predictive methods including a suggestion for building a GSTN, which has been collecting information about the earth's plates and their movements in new databases. Sadly these have rarely been used to save lives.

Having the unknown effects of global warming on the plates, and now knowing that their movements are the source of earthquakes it would be wise to prepare a better prediction system and advance the design of smart early warning systems under a more practical title of *dynamic early warning systems* (DEWS) as an important application of DNM. That is, using DNM it should be possible to harness 5G and internet technologies to achieve *two* main objectives: (1) better prediction of forthcoming earthquakes through further advancements of the GSTN idea using recent developments of big data for a more precise understanding of the earth's plates forces and the main causes of their movements and crashes; (2) enhancing the GSTN with objective use of better designed DEWS placed in all sensitive, strategic, and earthquake-prone locations on earth. As DEWS carries a natural flexibility it can be used in any advanced 5G compatible WSN applications of any size (small or massive) and for any environment.

Now, in order to design the DEWS as part of the DNM to manage disasters and gain a better understanding of the sources, we adopt our earlier design of DNM in Section 7.5. One of the functions not properly discussed in Section 7.5 is UTA. UTA is an active part of the control matrix reserved for urgent operations. Here urgent is used for any extraordinary situation within the environment where WSN is triggered externally through the DNM operational system. When activated, UTA takes the highest priority, and the agents take control of the operation. It first overrides all routine DNM functions, and then reactivates a new set of tasks and resources through a relevant subset of the agents. The new agents, upon the control of a new CDUA process, maintain the required sequence of urgent state function of control, specific monitoring, and dynamic expansion of the network.

Section 7.5 describes the DNM structure under its normal operational condition, where the DNM uses internal resources and internal agents. The use of external resources through external agents is possible, but not essential. In the case of receiving an urgent and unexpected disaster warning information activates the UTA controller to take on the role of managing all, internal and external agents assigned to the special emergency operations. This urgent "event" warning information usually received directly (via news and traditional sources) or detected through a combination of three possible DEWS routes: (1) internal agents; (2) external agents; and (3) some emergency news through public channels via the gateway securely connected to G5, the internet, etc.

As shown in Figure 7.12, any unexpected event under an *event* or *public emergency warning* can trigger an alert for a UTA to wake up for a watchdog routine analysis of the event to be classified as false, possible, and urgent. The possible case would lead the DNM into further investigation and to prepare some resources towards better and more reliable information, while urgent cases would convert virtually the whole of the DNM system into a ready to act status in case of a possible disaster.

A possible disaster case then initiates three special processes:

(a) enhanced processes to increase certainty of a disaster (e.g., an earthquake); (b) kind of disaster (e.g., earthquake's strength); and (c) identification of the disaster (e.g., location). Under the possible disaster condition the UTA activates the agents to expand the sensing capability by providing new resources and allocating much higher computing capabilities to the event identification. Also, under possible disaster and true disaster conditions DNM generates some early warning messages and alarms, previously agreed on by local authorities to prelisted disaster warning responsible bodies, to cascade and store data.

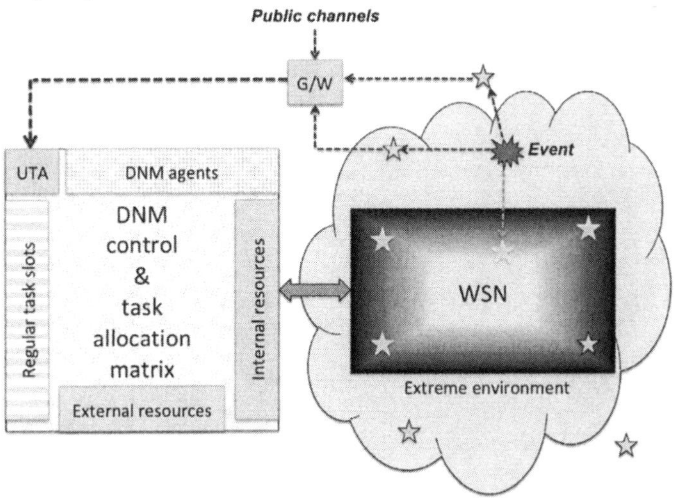

Figure 7.12 DNM core structure showing basic connections reacting to an urgent event–case activating appropriate routes of the dynamic early warning system.

7.7 CONCLUDING REMARKS

The excitement of visualizing the development of a new dynamic advanced network management by using up to date technologies once again raises the hope of an end to the last four decades of technological disparity and confusion. We head towards a new set of paradigms to harvest the extensive efforts of dedicated experts trapped in the midst of a circle of confusion. Using the overlaying concept for managing the wireless sensor networks and for designing suitable applications for the extreme environments encourages us to devise a new set of practical solutions, which lead us towards our missing paradigms.

Overlaying as a concept and its five identified categories have been analyzed, challenged, and configured to bring together many varieties into one dynamic format as a solution for all upcoming services. By using agile distributed intelligence in the form of interactive agents, along with the serving network to carry the load based upon the optimum use of valuable resources, to achieve an almost fully automated supervision of the service we cover the main set of objectives of this chapter. For analyzing these objectives this chapter examines a selective set of technologies to develop our flagship approach of NMO and its suboptimum overlaying solutions: SOM, MOM, and SNO.

Amongst the main topics discussed in this chapter are the limitations of traditional network management and the P2P service overlay networking for WSN applications, while the dynamic overlaying functions provided by ANM and its associated NMO can suitably support all sensor rich extreme environment applications, such as DEWS.

References

[1] Rashvand, H. F. (Ed), and Y. S. Kavian (Ed), *Using Cross-Layer Techniques for Communication Systems*, IGI Global Press, 2012, ISBN: 9781466609600.

[2] Internet of Things Architecture (IoT-A), WP3 – Protocol Suite, Seventh Framework Programme, http://www.meet-iot.eu/deliverables-IOTA/D3_1.pdf.

[3] Formby, D., S. S. Jung, S. Walters, and R. Beyah, "A Physical Overlay Framework for Insider Threat Mitigation of Power System Devices," *2014 IEEE International Conference on Smart Grid Communications (SmartGridComm)*, 3-6 November 2014 Venice, Italy.

[4] Babita, S. I., "Survey on Performance of IEEE 802.15.4 Low Rate- Wireless Personal Area Networks (LR-WPAN)," *International Journal of Advanced Research in Computer Engineering & Technology*, Vol. 3, Is. 4, April 2014.

[5] Surender, R., and S. Srinivasan, "Performance Analysis of Non-Beacon Enabled IEEE 802.15.4 based Secured Wireless SensorNetwork," *International Journal of Advanced Research in Computer and Communication Engineering*, Vol. 3, Is. 3, March 2014.

[6] Capone, A., J. Elias, and F. Martignon, "Optimal design of service overlay networks," *2008 4th International Telecommunication Networking Workshop on QoS in Multiservice IP Networks*, Venice, 2008, pp. 46–52.

[7] Miyazaki, T., et al, "Robust Wireless Sensor Network Featuring Automatic Function Alternation," *2011 Proceedings of 20th International Conference on Computer Communications and Networks (ICCCN)*, Maui, HI, 2011, pp. 1–6.

[8] Rashvand, H., F., "Wireless Communications II," MSc Lecture Notes, ES433 and ES9R8, School of Engineering, University of Warwick.

[9] Copéré, L., "i-mode: wireless service a la Japanese," 3rd June 2004, http://www.88pdf.com/search.php?req=i-mode%20-%20KTH.

[10] Duan, Z., Z-L. Zhang, and Y. T. Hou, "Service overlay networks: SLAs, QoS, and bandwidth provisioning," in *IEEE/ACM Transactions on Networking*, Vol. 11, No. 6, pp. 870–883, December 2003.

[11] Wang, W. and B. Li, "Market-based self-optimization for autonomic service overlay networks," in *IEEE Journal on Selected Areas in Communications*, Vol. 23, No. 12, pp. 2320–2332, December 2005.

[12] Annexstein, F. S., K. A. Berman, and M. A. Jovanović, "Broadcasting in unstructured peer-to-peer overlay networks," *Theoretical Computer Science* 355 (1), 25-36, 2006. https://doi.org/10.1016/j.tcs.2005.12.013.

[13] Lua, K., et al., "A survey and comparison of peer-to-peer overlay network schemes," in *IEEE Communications Surveys & Tutorials*, Vol. 7, No. 2, pp. 72–93, Second Quarter 2005.

[14] Chopra, D., H. Schulzrinne, E. Marocco, and E. Ivov, "Peer-to-peer overlays for real-time communication: security issues and solutions," in *IEEE Communications Surveys & Tutorials*, Vol. 11, No. 1, pp. 4–12, First Quarter 2009.

[15] Boukerche, A., A. Zarrad, and R. Araujo, "A Smart Gnutella Overlay Formation for Collaborative Virtual Environments over Mobile Ad-Hoc Networks," *2006 Tenth IEEE International Symposium on Distributed Simulation and Real-Time Applications*, Terremolinos, 2006, pp. 143–156.

[16] Galán-Jiménez, J., and A. Gazo-Cervero, "Overview and challenges of overlay networks: A survey," *Int J Comput Sci Eng Surv* (IJCSES) 2, 19–37: 2011.

[17] Rashvand, H. F., Keynote Speech, "WiMAX Cybercity & NGN," *5th International Conference on Mobile Technology, Applications and Systems 2008* (Mobility 08), Ilan, Taiwan, 10–12 September 2008.

[18] Chan Y., and S. Hsu, "A Network Managed Location Identification Scheme for IEEE 802,16m Networks," IET, WSN 2010.

[19] Yang, L., Z. Hu, J. Long, and T. Guo, "5W1H-based Conceptual Modeling Framework for Domain Ontology and Its Application on STPO," *2011 Seventh*

International Conference on Semantics, Knowledge and Grids, Beijing, 2011, pp.203–206.

[20] Palafox, L., L. A. Jeni, and H. Hashimoto, "Using conditional random fields to validate observations in a 4W1H paradigm," *2011 4th International Conference on Human System Interactions*, HSI 2011, Yokohama, 2011, pp. 80–84.

[21] Palafox, L. F., L. A. Jeni, H. Hashimoto, and B. H. Lee, "5W1H as a human activity recognition paradigm in the iSpace," *2011 8th Asian Control Conference (ASCC)*, Kaohsiung, 2011, pp. 712–718.

[22] Lee, C-E., and K–D. Moon, "Design of the Autonomous Fault Manager for learning and estimating home network faults," *2009 Digest of Technical Papers International Conference on Consumer Electronics*, Las Vegas, NV, 2009, pp. 1–2.

[23] Yu Y., and Yun Bi, "A study on "5W1H" user analysis on interaction design of interface," *2010 IEEE 11th International Conference on Computer-Aided Industrial Design & Conceptual Design 1*, Yiwu, 2010, pp. 329–332.

[24] Wireless Sensor Networks: Technology, Protocols, and Applications, Wiley Online Library, 2007, http://onlinelibrary.wiley.com/book/10.1002/047011276X.

[25] Rashvand, H. F. (Ed) and Chao, H.-C. (Ed), *Dynamic Ad Hoc Networks*, Institution of Engineering and Technology Press, 2013, ISBN: 978-1-84919- 647-5.

[26] Burleigh, S., "Interplanetary Overlay Network: An Implementation of the DTN Bundle Protocol," *2007 4th IEEE Consumer Communications and Networking Conference*, Las Vegas, NV, USA, 2007, pp. 222–226.

[27] Macedo, D. F., A. L. dos Santos, and G. Pujolle, "From TCP/IP to convergent networks: challenges and taxonomy," in *IEEE Communications Surveys & Tutorials*, Vol. 10, No. 4, pp. 40–55, Fourth Quarter 2008.

[28] Baumgart, I., B. Heep, C. Hübsch, and A. Brocco, "OverArch: A common architecture for structured and unstructured overlay networks," *2012 Proceedings IEEE INFOCOM Workshops*, Orlando, FL, 2012, pp. 19–24.

[29] IETF, "Mobile Throughput Guidance Inband Signaling Protocol draft-flinck-mobile-throughput-guidance-04.txt,"https://tools.ietf.org/html/draft-flinck-mobile-throughput -guidance-04.

[30] Gupta, L., and R. Jain, (2016). Mobile Edge Computing – an important ingredient of 5G Networks. IEEE Softwarization Newsletter.

[31] Dixon, J., "Software Defined Networking: What it is and what you should know!," https://www.infosecwriters.com/Papers/JDixon_SDN.pdf.

[32] ETSI GS MEC-IEG 004 V1.1.1 (2015-11), "Mobile-Edge Computing (MEC); Service Scenarios," http://www.etsi.org/standards-search.

[33] Milgram, P., and F. Kishino. "A Taxonomy of Mixed Reality Virtual Displays," *IEICE Transactions on Information and Systems* E77-D, 9 (September 1994), pp.1321–1329.

[34] Keromytis, A. D., V. Misra, and D. Rubenstein, "SOS: an architecture for mitigating DDoS attacks," in *IEEE Journal on Selected Areas in Communications*, Vol. 22, No.1, pp. 176–188, January, 2004.

[35] Cholda P., and A. Jajszczyk, "Recovery and Its Quality in Multilayer Networks," *Journal of Lightwave Technology*, Vol. 28, No. 4, pp. 372–389, February15, 2010.

[36] Rashvand, H. F., "Distributed Intelligence: Convergence, divergence and ubiquitous access for future technological developments," Keynote Speech, Conference: *Future Tech 2011*, 28–30 June, Crete, Greece.

[37] Rashvand, H. F., and J. M Alcaraz-Carelo, *Distributed Sensor Systems - Practice and Applications* by John Wiley & Sons, 2012, ISBN: 9780470661246.

[38] Roser, M., and H. Ritchie, "Our world in Data: Natural Catastrophes," https://ourworldindata.org/natural-catastrophes/ (Visited 09/07/2017).

Chapter 8

Information Flow in Hard to Reach Areas: Source Estimation Using Correlated Streams

In a typical WSN, a number of spatially distributed sensors measure a set of desired environmental and physical parameters across a field and pass the collected data wirelessly to a central station for further processing. Various aspects of WSN systems including efficient data acquisition, optimal sampling rate and distributed compression, error recovery, information fusion, localization and tracking, routing and networking have been studied over the past decade. These studies have led to the design of novel algorithms to solve different technical challenges and alleviate implementation barriers. This chapter considers a specific variant of WSNs, where data sources are located in hard to reach areas and therefore additional challenges and issues arise. An important scenario is for field monitoring under extreme environmental conditions. Some real-world examples include jet engine temperature monitoring [1, 2], structural health monitoring [3], underwater passive aquatic listeners [4], and space exploration using composite sensors [5]. In some other applications such as navigation systems using battery-free sensors, placing sensors at exact data source locations is not feasible or economically viable since the source locations are not stationary or not known in advance. Figure 8.1 provides an illustrative example, where a set of autonomous drones track an animal of interest, collect intermittent and partially accurate imagery data and send it to a ground station for further processing [6, 7].

Figure 8.1 Habitat monitoring: A set of autonomous drones monitor the object of interest in an inaccessible area and collectively report their observations to a central ground station for further processing.

In these scenarios, an important challenge is the low fidelity of collected information not only due to communication errors, but also due to inaccuracies in the measurements of remotely deployed sensors [8,9]. To address this issue, distributed coding emerged as an efficient technique to collect information more elegantly by individual sensors such that reliable information is extractable by fusing partially accurate observations [10]. This chapter is devoted to exploring the state of the art in distributed coding techniques. In particular, a comprehensive review of the fundamental theoretical aspects of this problem is provided with a summary of recent developments. In this context, a novel and implementation-friendly solution is proposed to address this problem and answer a set of more specific technical questions including: (1) how many sensors are optimal in any given conditions? (2) what is the fidelity of collected information? and (3) what is the highest achievable information rate? We finally provide insights about the integration of the proposed method with modern networking approaches in order to extend it to a large-scale multihop communication infrastructure.

8.1 INTRODUCTION

A contemporary WSN consists of a multiple sensors mainly deployed in autonomous mobile ground or aerial vehicles with wireless transmission capabilities in order to collectively monitor environmental and physical parameters and transmit the collected data to a central data fusion unit for further processing. Wireless sensing is an integral part of the emerging Internet of Things (IoT) platforms with a broad range of applications including smart homes, smart grids, transportation and traffic control, aviation, wildlife monitoring, fire control, robotics and industrial automation [12, 13]. Due to the Adhoc nature of WSNs with dynamic and vibrant typologies, utilizing a central controller is costly, unreliable and inefficient. Therefore, distributed algorithm design and implementation has gained a lot of attention in the research community to realize agile, flexible and scalable data acquisition, compression and transmission in WSNs [14].

The majority of distributed algorithms aim at realizing a smooth and reliable data-flow from sensors to a central processing station by accommodating optimized compression, networking and error recovery techniques based on the assumption that sensor measurements accurately represent the actual monitored parameters. However, there are scenarios where placing sensors in exact data locations is not technically feasible due to a number of reasons including extreme environmental conditions (e.g., in a jet engine), lack of prior knowledge about source locations (e.g., a fire control systems in forests), and maintenance difficulty (e.g., space exploration). In such scenarios, an effective solution is to deploy a cluster of sensors in proximity of a putative data source and process the collected data. In this regard, distributed source coding (DSC) schemes are designed to exploit the intrinsic spatial correlation among sensors' observations to compensate for sensors observation errors. This solution brings forward a new set of questions such as (1) how many sensors are sufficient to achieve a certain level of fidelity, (2) how to compensate for sensing inaccuracies, (3) how often to collect sensor readings and (4) how to exploit the spatial correlation among sensors.

A comprehensive study of ongoing research in designing DSC algorithms during the past years reveals that most recent studies have focused on developing theoretically near-optimal algorithms. However, the high complexity of these algorithms prevented their commercial implementations in tiny sensors with limited capabilities [11]. Furthermore, most algorithms are designed and optimized based on the two assumptions of data sources being stationary and observation models being fully known. Both assumptions are far from reality. Therefore, the majority of currently available system on chip (SOC) sensor platforms such as Telos B Motes, Stargate, Mica2, Tmote Sky and IBM cricket stick with the conventional point-to-point coding techniques in their physical layer implementations following the IEEE 802.15 standard series and do not utilize recently proposed DSC schemes despite their proven superior performance [12–15].

This chapter is devoted to the study and design of practically feasible and self-tuning distributed algorithms for robust and adaptive data compression, error recovery and networking for WSNs with partially accurate sensors under dynamic environmental conditions. The main emphasis will be placed on a special case of remote sensing of a common data source with correlated sensors, the so-called the chief execute officer (CEO) problem. A novel implementation of distributed coding is discussed in detail; it is based on parallel concatenation of convolution codes (PCCC) with a novel bimodal decoder at the receiver.

8.2 THE CEO PROBLEM

The mathematical model of the CEO problem (Figure 8.2) was first proposed by Toby Berger in his seminal paper to capture the spatial correlation among the readings of a cluster of sensors that collectively monitor a common data source [16]. This model is inspired by a similar problem in business studies, where a CEO of a company interviews his untrustworthy employees individually and processes the interrogation results to extract reliable information about his plant.

The CEO problem is a special case of a more general problem formulation, where L sensors collectively monitor K sources with correlated data, as depicted in Figure 8.3. In this figure, S_i is the ith data source and U_j, E_j, X_j, N_j, Y_j and R_j, respectively, denote the observation, observation error, coded sample, channel noise, received sample and transmission rate of the jth sensor. The main goal is to design L distributed encoders with minimal rates R_i and one joint decoder in order to estimate K sources with a predefined distortion limit $d(S_i, \hat{S}_i) < D$, where $d(.,.)$ is an arbitrary distance measure. Here, we assume the additive observation error model $U_i = \sum_{S_j \in \chi_i} S_j + E_i$, where χ_i represents a set of sources observed by sensor i. For the sake of simplicity, we disregard the interference effect and assume parallel channel model with additive noise $Y_i = X_i + W_i$, where X_i is the coded sample and W_i is the white zero-mean Gaussian noise with variance N_i.

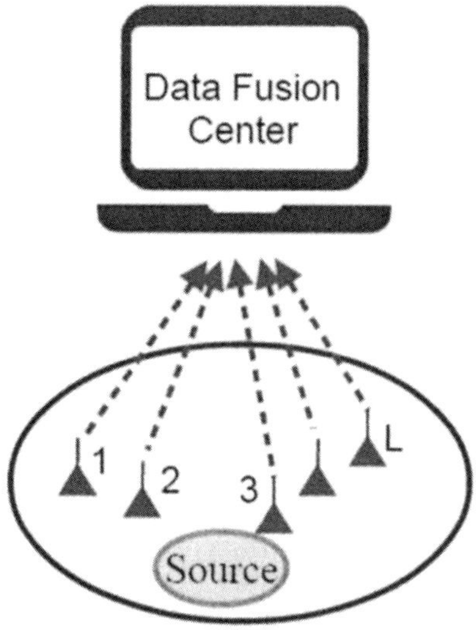

Figure 8.2 CEO problem: A cluster of L sensors collectively monitors a common source and sends their observations to a data fusion center.

This general model reduces for some important special cases if certain simplifying assumptions hold. For instance, we can set observation error to zero (i.e., $E_i = 0, \forall i = 1,2, \dots, L$), which converts the indirect observation scenario to a simpler case of direct observation [17]. Another special case arises if any of K utilized sensors monitors only one specific source (i.e., $\chi_i = \{S_i\}$), and the correlation among sensor readings is restricted to the intrinsic correlations among the sources [18, 11]. If full information about some of the sensors considered available to the receiver and the objective is to recover the other sensors' readings, the problem reduces to the coding with side information [19, 20]. Considering only one source $K = 1$, boils the problem down to the CEO problem, where the objective is to estimate the common source [16]. In all cases, the sensors are not allowed to convene and exchange information with one another; hence, distributed coding is utilized [21].

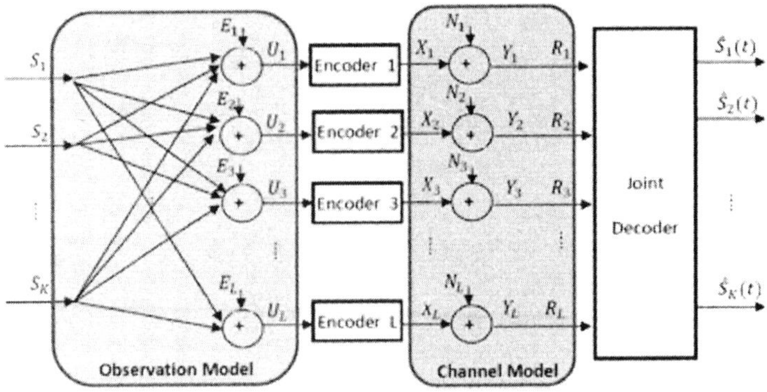

Figure 8.3 Multi-terminal source coding for sensors with direct and indirect observations.

8.2.1 Binary Correlation Model

The most commonly adopted model for correlation among sensor readings is joint Gaussian distribution, where the common source and the observations of sensors are jointly distributed according to a multivariate Gaussian distribution. This model is flawed in reality, since different data sources such light, temperature, humidity, stress, pressure do not follow normal distribution in general [22–27]. Furthermore, observation noise is not always normally distributed. The third drawback of this theoretical model is the ignorance of real-world sensor implementations. In practice, sensors read continuous-valued variables and convert them into binary streams after different stages of quantization (e.g., linear uniform PCM, Lloyd-Max quantizer or vector quantization [28, 29]), companding (e.g., A-Law and μ-Law for speech signals [30, 31]), and sample to bit mapping (e.g., Gray coding [32, 33]).

The recent developments of practical source codes are typically applied to binary bitstreams after a digitization step. To fill this gap between the theoretical framework and the practical implementation, we assume that the source generates binary symbols, which are observed by sensor i through a binary symmetric channel (BSC) with crossover probability β_i. Namely, we have $U_i = S + E_i, S \sim Bernoulli(0.5), E_i \sim Bernoulli(\beta_i)$, where $\beta_i = E[Pr(S \neq U_i)]$ is the expected bit flipping probability from the common source to sensor i for a given source type and sensor implementation. This model is very general and mimics the impact of noise and other nonlinearity sources at different stages of the path from source to the representative binary bitstream. Therefore, this model has been widely adopted in various studies [19, 34–38]. With this correlation model, the aforementioned CEO model converts to a special cased called binary CEO (bCEO).

8.2.2 Information-Theoretic Review

In the CEO problem, the goal is to find the rate region $\mathcal{R}_*(D) = \bigcup_{(R_1,R_2,...R_L)} R_D(R_1, R_2, \dots, R_L)$, which includes all combination of rate L-tuples. Here, R_i is the coding rate of sensor i and $R_D(R_1, R_2, \dots, R_L)$ represents all rate tuples for which there exist a joint-decoder capable of estimating the common source within a predefined distortion limit $D^{(n)} = \frac{1}{n} E[\, d(S^n, \hat{S}^n)] < D$, for a given distance measure $d(.,.)$, when the codeword block length n is chosen large enough. One may also be interested in solving a simpler problem of minimizing the sum rate, which is defined as $R = \sum_{i=1}^{L} R_i$, for a given target distortion threshold D.

This problem is intensively investigated from the information-theoretic perspective, yet the exact rate-distortion function for the CEO problem is not known in general [39]. The first information-theoretic result, reported by Toby Berger in [40], states that if the number of agents (sensors) approach infinity $(L \to \infty)$ and the agents are allowed to share their observations, then they are able to provide an accurate estimate of the common source symbol S with a limited distortion based on a distortion rate function $D(R)$. As a special case, if the sum-rate R exceeds the entropy of the common source $H(S)$, the decoder fully recovers the source symbol with an arbitrary low distortion (i.e., $R > H(S) \Rightarrow D(R) = 0$).

However, if the agents are not allowed to convene, which is the case in the CEO problem, there does not exist a finite value of the total rate R, for which even infinitely many agents can make the distortion D arbitrarily small. It was shown that for an infinitely large number of agents, the distortion decays exponentially as the sum-rate approaches infinity. This fact implies that in practical implementation of distributed coding with a limited number of sensors, we need to settle with approximate source estimation and not seek an error-free source recovery. This phenomenon is due to the inherent uncertainty of indirect observations. In contrast, for direct observation, correlated sources can be fully recovered within a rate region defined by the famous Slepian-Wolf theorem [18]. A concrete characterization of the rate-distortion region, also known as the Berger-Tung rate region proposed in [41] based on performing Gaussian quantization method on the observation bits and then applying the Slepian-Wolf coding to the quantized the bits.

An important variant of the CEO problem is called the quadratic Gaussian CEO (QGCEO) problem, where the source and observation errors are identically and independently distributed (i.i.d.) Gaussian variables. The problem of finding the distortion-rate for the QGCEO problem is studied from different perspectives including sum-rate distortion function derivation [42, 43], computationally tractable sum-rate calculation [44], sum-rate loss due to lack of inter-sensor communications [45], multiple source estimation [46–48] and robust coding to recover both the common source and the sensor readings with a certain fidelity [49]. Extensions of these results to the vector Gaussian CEO problem can be found in [50–53].

The CEO problem can be generalized in the sense that the CEO is interested in estimating both common source data and sensors' observations with arbitrary fidelity. If the maximum allowable distortion of observations' reconstruction approaches infinity meaning that the CEO does not care about the observation itself, this problem reduces to the classical CEO problem. On the other hand, if the maximum allowable distortion of the common source observations' reconstruction approaches infinity, the problem reduces to the two-terminal direct observation case. The achievable rate region for this general case is characterized in [49].

Another important variant with practical advantages is the binary CEO problem, as discussed earlier. Theoretical limits on the rate-distortion region for the bCEO problem are studied in [39, 54]. However, these results do not provide a closed-form expression for the rate distortion function and rather provide complex expressions for the upper and lower bounds on the rate distortion function. These expressions are complicated to quantify even for the special case of two sensors $L = 2$ and the equal observation error $\beta_1 = \beta_2 = \beta$, and consequently are not much useful in designing practical systems.

In this book chapter, we will show how our simplified calculations for the end-to-end coding rate proposed in [55] can be used to configure a cluster of sensors for real-world applications to achieve an acceptable level of accuracy. This method is based on finding a polymatroid region for the rates that fully recovers the correlated sensor readings instead of finding the rates required for recovering the common data source [56, 57].

8.3 PRACTICAL CODING DESIGN FOR THE BINARY CEO PROBLEM

8.3.1 Distributed Source Coding versus Joint Coding

In order to elucidate the concept of distributed source coding, here we present a simple example to show its contrast with joint coding. The main distinction between the two methods is the exchange of information among sensors. Consider two sensors with correlated binary inputs. Each sensor reading U_i is a sequence of three bits $U_i = [U_{i1}U_{i2}U_{i3}]$; therefore it takes $2^3 = 8$ options $U_i \in \mathcal{U} = \{S_0, \dots, S_7\}$ as depicted in Figure 8.4. Based on the observation model, U_1 and U_2 can differ at most in one bit (i.e., $d_{Hamming}(U_1, U_2) \leq 1$). The direct implication of this correlation model is that for each realization of U_1, there are only 4 possibilities for U_2, which include $U_2 = U_1 \oplus E$, where $E = \{[000], [001], [010], [001]\}$. Consequently, the entropy of U_1 is $H(U_1) = 3$ bits and the conditional entropy of U_2 given U_1 is $H(U_2|U_1) = H(U_1 \oplus E|U_1) = H(E) = 2$ bits. Then the practical implementation of joint coding becomes straightforward. Sensor 1 sends the plain reading U_1 with 3 bits to the destination. Sensor 2 receives U_1 and calculates $E = U_1 \oplus U_2$ and then maps it into two bits $[E_1E_2]$ and sends the result to the receiver. The receiver performs sequential decoding (also called onion pealing) by determining U_1 first and then calculating E based on received bits $[E_1E_2]$. Subsequently, it calculates $U_2 = U_1 \oplus E$ (Figure 8.4(a)).

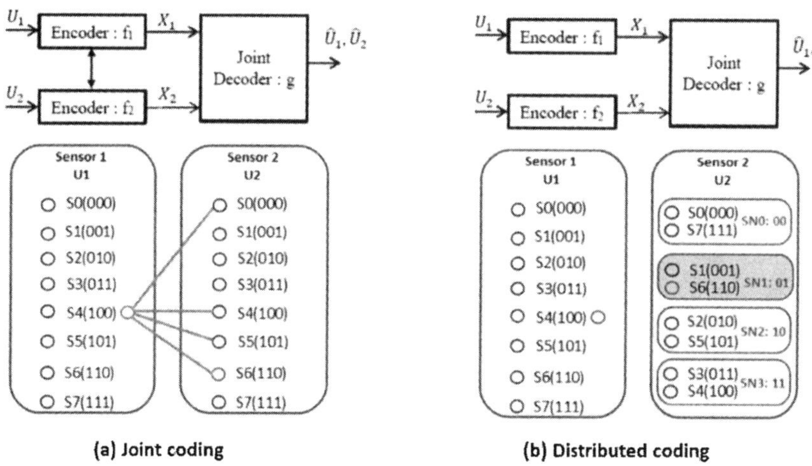

Figure 8.4 (a) Joint coding versus (b) distributed coding for correlated binary sources.

Lossless distributed coding relies on the famous Slepian-Wolf theorem, which suggests that we should be able to implement distributed coding of rates $(R_1, R_2) = \big(H(U_1), H(U_2|U_1)\big) = (3,2)$ without intersensor communication such that the receiver can perfectly recover U_1 and U_2. Consider the following implementation, where sensor 1 sends its plain reading U_1 using three bits as before. Sensor 2, however, divides the space of 8 possible outcomes into 4 disjoint partitions called cosets. Each coset SN_i encompasses two symbols with a Hamming distance of 3. Each coset is mapped to a 2-bit sequence $C = [C_1, C_2]$ called a syndrome as shown in Figure 8.4(b). The receiver first identifies $\boldsymbol{U}_1$ from the received bits $[U_{11}U_{12}U_{13}]$, and then identifies the coset SN_i based on the received syndrome bits $C = [C_1, C_2]$. Each coset includes two symbols and the one with a lower Hamming distance to $\boldsymbol{U}_1$ is mapped to $\boldsymbol{U}_2$. This simple technique is a variant of a more general technique called Syndrome-based distributed coding [58].

8.3.2 Review of Practical DSC Methods

Various practical implementations of DSC are proposed in the literature [11, 59, 60]. The first realization of DSC introduced by Ramchandran, et al. employing the concept of Syndrome-based coding [58]. Afterwards, different implementations of DSCs were developed based on the more powerful channel codes including low density parity check code (LDPC) [19], turbo code [35, 61−64], turbo code with pre-puncturing [65], turbo code with post puncturing [37], irregular repeat accumulate (IRA) code [66], and low density generator matrix code (LDGM) [67, 68].

Similar to point to point communications, combining the DSC stage with the subsequent channel coding stage to develop a single encoder called distributed joint source channel coding (D-JSCC) is more efficient in terms of complexity and implementation cost. The most reported realizations of D-JSCCs are based on LDPC and turbo like codes [69−72]. However, these methods have not yet been widely utilized in practice mainly due to implementation issues including the need for a prior knowledge of the observation model, the requirement of time-invariant observation noise, decoding complexity, sensitivity to a sensor failure, and complication of scalability to a large number of sensors [63, 71].

In this chapter, we propose a D-JSCC method, which provides a superior error recovery performance compared to the state of the art D-JSCC methods and enjoys a simple encoder and low-complexity decoder structure. The proposed scheme is robust to sensor failures and continues to operate if a sufficient number of sensors is still operational. Due to the linear complexity of the proposed decoding algorithm in the number of sensors, this method is easily scalable to a large number of sensors. The bimodal operation eliminates the unnecessary decoding iterations and reduces the complexity of the decoder by a factor of 10 to 20 in some cases. Finally, by employing a novel implementation of observation parameter extraction, no prior knowledge is required about the observation accuracy of sensors and the proposed system works even in time-varying observation models.

The following sections elaborate the details of the proposed method and offer solutions for finding the optimal number of sensors, analyzing the convergence of the decoding algorithm, and quantifying the performance of the system, when integrated with relaying techniques.

8.4 D-PCCC WITH BIMODAL DECODER

The proposed scheme is based on utilizing a distributed version of parallel concatenation of convolutional codes (D-PCCC) in the sensors and an adaptive bimodal iterative decoder in the receiver as follows.

8.4.1 Encoder Structure

Each sensor is equipped with a pseudo-random interleaver, a RSC encoder and a puncturing module, which collectively form a distributed version of a PCCC encoder as depicted in Figure 8.5. The rate of sensor i after puncturing the output codeword is R_i, where we set $R_i = R, \forall\, i \in \{1,2,..,L\}$ for homogeneous sensors unless otherwise specified. Here, the coding rate per sensor can be less than or greater than 1 in contrast to a typical individual encoder. An important distinction with the conventional PCCC scheme is that the input bits of each constituent encoder correspond to the noisy versions of common source bits, and hence may be slightly different.

We follow the BSC observation model, as described in Section 8.2.1. We note that the independence of observation errors for different sensors implies that the pairwise correlation between the observations of sensors i and j is modeled as a BSC channel with cross over probability $\beta_{ij} = 1 + 2\beta_i\beta_j - \beta_i - \beta_j$, which simplifies to $1 + 2\beta^2 - 2\beta$ for equal observation error $\beta_1 = \beta_2 = \cdots = \beta_L = \beta$. For small values of β, we can use the $\beta_{ij} \approx 1 - 2\beta$ approximation [73, 74].

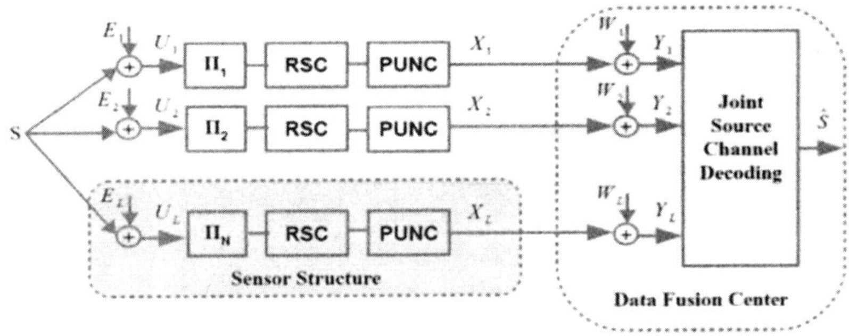

Figure 8.5 Coding scheme: Each sensor includes an interleaver, a RSC encoder and a puncturing module, which collectively form the D-PCCC scheme (Source: [83]).

8.4.2 Decoder Structure

Inspired by the encoder structure, we develop a multiple-branch turbo decoder (MTD) at the receiver with some modifications and additional features to accommodate the distributed nature of the encoder, as depicted in Figure 8.6.

The decoder includes L constituent soft-input soft-output (SISO) decoders operating in parallel. Each constituent decoder executes the BCJR Max-Log-MAP algorithm using the $log(e^x + e^y) \approx max(x, y)$ approximation for reduced complexity [75], which iteratively decodes the received bits in order to recover the relevant sensors observation bits [74].

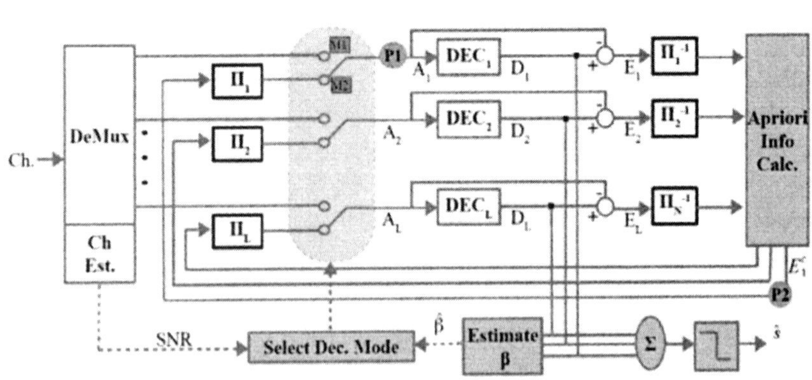

Figure 8.6 The proposed bimodal parallel-structure MTD decoder equipped with an observation parameter extraction module (Source: [83]).

Each constituent decoder receives a priori LLRs denoted by $A_i^{(t)}(k)$ for $k = 1,2,\ldots,N$ at iteration t and calculates the a posteriori LLRs denoted by $D_i^{(t)}(k)$. Here, N is the length of dataword (i.e., the number of information bits). The extrinsic LLRs are defined as $E_i^{(t)}(k) = D_i^{(t)}(k) - A_i^{(t)}(k)$. In contrast to the standard MTD, here the information bits of each constituent decoder represent the corresponding sensor's observation bits; therefore the extrinsic LLRs converge to the observation bits rather than the common source bits, which may cause convergence issues. The following is a list of modifications made to address this issue:

- *Decoder initialization:* In the first iteration, all SISO decoders are initialized with the LLRs that correspond to the observation bits of the respective sensor, which is in contrast with the classic MTD decoder, where only one RSC decoder is initialized with the LLRs of the common information bits.
- *LLR scaling:* In a classical MTD, the input LLRs for decoder i, $A_i^{(t)}(k)$ are calculated as the average of the extrinsic LLRs of all other RSC decoders (i.e., $A_i^{(t)}(k) = \sum_{j=1,j\neq i}^{N} E_j^{(t-1)}(k)$). In the proposed scenario, the extrinsic LLRs represent the corresponding sensor's observation bits. Therefore, we define a new set of extrinsic LLRs to represent the likelihood of the common source bits as follows:

$$E_{i,S}^{(t)}(n)\& = log_2\left(\frac{Pr(S(n)=1|\boldsymbol{U}_i)}{Pr(S(n)=0|\boldsymbol{U}_i)}\right),$$

$$E_{i,S}^{(t)}(n)\& = log_2\left(\frac{\hat{\beta} + (1-\hat{\beta})\, 2^{E_i^{(t)}(n)}}{(1-\hat{\beta})+\hat{\beta}2^{E_i^{(t)}(n)}}\right), \tag{8.1}$$

which is the nonlinearly scaled versions of the original extrinsic LLRs, $E_i^{(t)}(n)$. Here, $\boldsymbol{U}_i = [U_i(1),\ldots,U_i(N)]$ is the observation vector of sensor i and $\hat{\beta}$ is the estimate of β. The apriori LLRs of decoder i are then calculated as follows:

$$A_i^{(t)} = \sum_{\substack{j=1 \\ j \neq i}}^{N} E_{i,S}^{(t-1)}$$

$$A_i^{(t)} = \sum_{\substack{j=1 \\ j \neq i}}^{N} log_2 \left(\frac{\hat{\beta} + (1-\hat{\beta})\, 2^{E_i^{(t-1)}}}{(1-\hat{\beta}) + \hat{\beta} 2^{E_i^{(t-1)}}} \right)$$

$$A_i^{(t)} = log \left[\prod_{\substack{j=1 \\ j \neq i}}^{N} \frac{\hat{\beta} + (1-\hat{\beta})\, 2^{E_i^{(t-1)}}}{(1-\hat{\beta}) + \hat{\beta} 2^{E_i^{(t-1)}}} \right] \tag{8.2}$$

The details of the derivations are provided in [76].

- *Observation accuracy estimation:* In order to accommodate time-variant and unknown observation error parameter β, an observation error estimation module is developed to estimate β by the receiver. We first estimate the transmitted information symbol $\hat{x}_i(k)$ by hard-thresholding the output LLRs at the end of each iteration (i.e., $\hat{x}_i(k) = \mathrm{sign}[D_i(k)]$). Then, we use the following estimator:

$$\hat{\beta} \approx \frac{1}{2LN(N-1)} \sum_{i=1}^{N} \sum_{\substack{j=1 \\ j \neq i}}^{N-1} \sum_{k=1}^{L} \rho_{i,j}(k) \tag{8.3}$$

  where, $\rho_{i,j}(k) = \frac{|\hat{x}_i(k) - \hat{x}_j(k)|}{2}$, which is based on estimating the pairwise correlation followed by averaging over all pairs of sensors. It is easy to show that $\hat{\beta}$ is normally distributed with mean β and variance $\frac{2\beta(1-\beta)(1-2\beta+2\beta^2)}{LN(N-1)} \approx \frac{2\beta}{LN(N-1)}$. Hence, it approaches the observation error parameter β for large enough N and L values.

- *Decision phase:* In the last iteration, the hard decision is performed based on the average of a-posteriori LLRs, $D_{(av)} = \frac{1}{N}\sum_{i=1}^{N} D_i$ instead of the output LLRs of a particular SISO decoder D_i, for the similar reason of avoiding bias to a particular encoder/decoder pair.

- *Bimodal operation:* The last modification is to use a bimodal operation (i.e., by switching between the iterative and noniterative modes) as detailed in the following section.

8.4.3 Bimodal Decoding Operation

In a majority of the previously reported D-JSCC schemes for correlated sensors, iterative decoding is deployed in order to exploit the correlation among sensors' observations [11, 57, 63, 69, 71, 77–80].

In this section, we revisit the presumption of the superiority of iterative decoding for the proposed system by answering the following question, "Under what conditions does the iterative decoding improve the estimation accuracy?"

An intuitive conjecture is that iterative decoding is beneficial when the observation accuracy of sensors and consequently their pairwise correlations is high enough. In effect, the iterative decoding uselessness when the correlation among sensor observations is relatively low, has been noticed by former researchers in several studies including [37, 64, 65]. To the best of our knowledge, however, no study has been conducted to characterize this phenomenon. In this section, we propose a method to obtain the convergence region of the decoder, where iterative decoding is beneficial. Consider the block diagram of the decoder depicted in Figure 8.6. The idea is to verify whether the estimates of the common source bits provided by a constituent decoder improve during a full iteration (from point P_1 to P_2 in Figure 8.6) or not. The technique we use here is a modified version of extrinsic information transfer (EXIT) charts. The EXIT chart is a powerful technique to quantify the improvement of LLRs' relevance to the information bits during a decoding iteration by a constituent decoder [81]. A typical EXIT chart for a standard turbo decoder ($\beta = 0$) is depicted in Figure 8.7. However, its conventional form is not applicable to our system model noting the fact that the extrinsic LLRs of each decoder converge to the corresponding sensor's observation, and not to the common source bits.

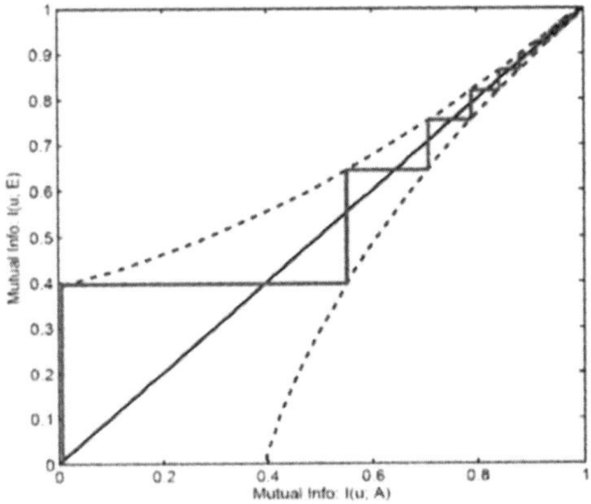

Figure 8.7 Conventional EXIT chart for the extreme case of complete observation accuracy ($\beta = 0,\ E_b/N_0 = 1$ dB).

Here, we provide the sketch of the modified EXIT chart technique that is used to find the convergence region. Interested users are referred to [82] and [83] for more details. To develop an EXIT chart, we note that the input LLRs received from the channel are *normally* distributed, hence we have:

$$A^{(1)} = log\frac{Pr(X = +1|Y)}{Pr(X = -1|Y)} = \mu_Y Y + n_Y,$$

$$n_Y \sim \mathcal{N}(0, \sigma_Y^2),\ \sigma_Y^2 = 2\mu_Y = 4/\sigma_N^2, \tag{8.4}$$

where X and Y denote the BPSK-modulated transmitted and received bits. It has been shown that if both channel observations and input LLRs follow Gaussian distribution, a MAP-family decoder with a fairly large frame length generates a posteriori LLRs, which tend to follow a Gaussian distribution [84]. The intuitive justification is based on applying the weak law of large numbers to the summands over the random-like decoder trellis structure. Moreover, extensive simulations confirm that equation (8.4) holds for the extrinsic LLRs as well [81]. Consequently, both apriori and extrinsic LLRs, A and E, can be written in the following format:

$$E = \mu_E Y + n_E, n_E \sim \mathcal{N}(0, \sigma_E^2),\ \mu_E = \frac{\sigma_E^2}{2}, \tag{8.5}$$

Now, if V denotes the BPSK-modulated version of the source bit, the input LLR (A) exhibits the following bimodal distribution:

$$P_A(\zeta|v) = \sum_{x=-1,+1} P_A(\zeta|x)P(x|v)$$

$$P_A(\zeta|v) = \left[\bar{\beta}e^{-\frac{(\zeta-\mu_A v)^2}{2\sigma_A^2}} + \beta e^{-\frac{(\zeta+\mu_A v)^2}{2\sigma_A^2}}\right], \tag{8.6}$$

where $\bar{\beta} = 1 - \beta$, which is called the m-order Binomial-Gaussian distribution with parameter set $(m = 2, \beta, \mu, \sigma^2)$ in this writing. This distribution for $\beta = 0$ boils down to a Gaussian distribution as expected. Consequently, the bitwise-mutual information (BMI), between the a priori LLRs and the source bits is calculated as:

$$I(A;\, V) =$$

$$1 - \frac{1}{\sqrt{2\pi}\sigma_A}\int_{-\infty}^{\infty}\left[\bar{\beta}e^{-\frac{(\zeta-\mu_A v)^2}{2\sigma_A^2}} + \beta e^{-\frac{(\zeta+\mu_A v)^2}{2\sigma_A^2}}\right] log\left(\frac{1+e^{-2\frac{\mu_A\zeta}{\sigma_A^2}}}{\bar{\beta}+\beta e^{-2\frac{\mu_A\zeta}{\sigma_A^2}}}\right) d\zeta. \tag{8.7}$$

For more than two constituent decoders ($m > 2$), an additional layer of averaging is required after calculating extrinsic LLRs. The resulting equations are more complex for $m > 2$ and are omitted here for the sake of brevity [83]. Note that $I(A;\, V)$ is a function of β and channel SNR. A typical $I(A;\, V)$ is depicted in Figure 8.8.

Similar equations hold for extrinsic LLRs. Note that in contrast to the conventional turbo decoder, here $I(A; V)$ and $I(E; V)$ do not approach 1, even for extremely large LLR values $\sigma_A \to \infty$; rather they approach $1 - H(\beta)$ [83].

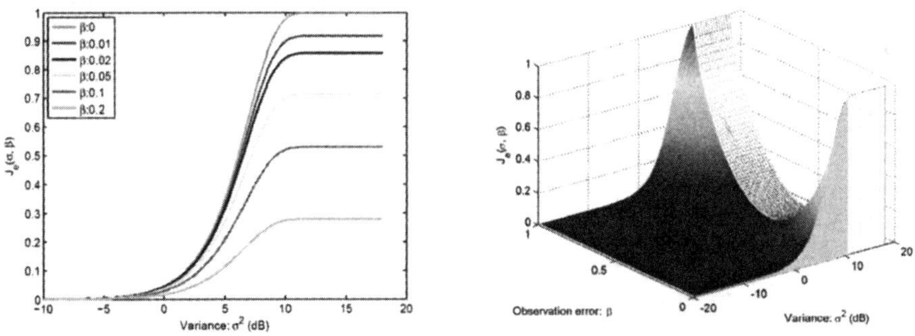

Figure 8.8 Bitwise mutual information between the a priori LLRs and the source data as a function of noise variance σ^2 and observation error β. The left hand side and right hand side figures present the 2D and 3D views (Source: [76]).

In order to derive modified EXIT chart curves, we generate random source bits, and then we pass the source bits through a BSC channel with crossover probability β to obtain the simulated observation bits. Subsequently, we generate a priori LLRs for the first iteration according to (8.4) for a given channel SNR. Then, we execute the decoding algorithm and obtain extrinsic LLRs (E) after proper scaling and averaging as mentioned in Section 8.4.2. The parameters of extrinsic LLRs (μ_E, σ_E^2) are estimated afterwards by fitting a Gaussian distribution. Finally, we calculate $I(A;V)$ and $I(E;V)$ using (7). Plotting $I(E;V)$ vs $I(A;V)$ as well as the reverse curves (i.e., $I(A;V)$ vs $I(E;V)$) presents the modified EXIT charts as depicted in Figure 8.9. If the direct curve falls above the unit-slope line and presents a positive initial slope, then iterative decoding is beneficial, since the BMI between the LLRs and the source bits exhibits improvement for a decoding cycle. The point at which the direct and reverse curves meet presents the final achievable error recovery performance and defines the number of useful iterations. The closer this point to the upper right corner $[I(A;V), I(E;V)] = [1,1]$, the lower the error floor. The general trend obtained from Figure 8.9 is that lower channel SNRs and higher observation accuracies are suggestive of more advantage for iterative decoding.

In order to find the convergence regions, we develop (β, SNR) grids and generate EXIT charts for all combinations of (β, SNR). Then, we identify the conditions that yield a positive initial slope of ϵ for the EXIT chart direct curve. A sample convergence region is shown in Figure 8.10.

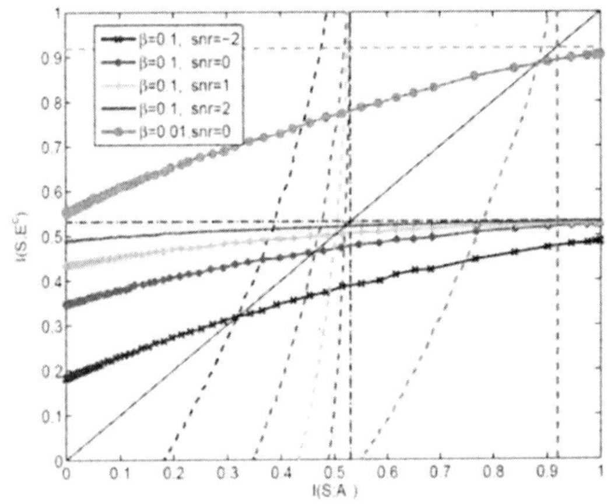

Figure 8.9 Modified EXIT charts for different observation accuracies. Number of sensors is L=2 (Source: [83]).

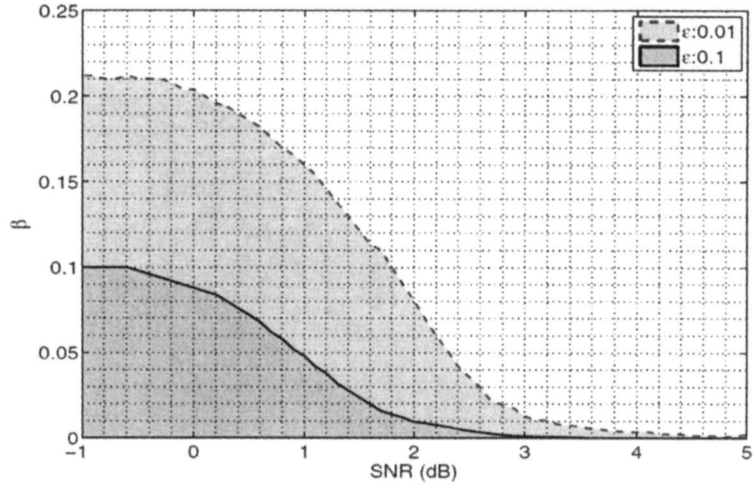

Figure 8.10 Convergence region of iterative decoding algorithm in terms of channel SNR and β for ε = 0.1 and ε = 0.01. Here ε is the level of minimum improvement due to iterative decoding. Outside the convergence region, the noniterative algorithm is selected (Source: [83]).

We note that developing the convergence region for the (β, SNR) grid, a time consuming task, but it is performed only once per system and is revealed to the decoder. Once the decoder receives a new multiframe from the sensors, it estimates the channel SNR and the observation accuracy parameter β. If the estimated (SNR, β) pair falls within the convergence region, the iterative mode is chosen; otherwise LLR exchange is bypassed to avoid unnecessary iterations.

8.5 PERFORMANCE ANALYSIS

To confirm the performance improvement obtained by the mode selection mechanism, we performed extensive Monte Carlo simulations for the proposed scheme (D-PCCC) with the following parameters ($M = 2000{\sim}bits, N = 8$ and $\beta = 0.1$). We compare the results against similar representative methods including distributed turbo coding in [71] and D-LDGM codes in [57] under equivalent conditions.

The error probability of AWGN channel $Pr(error) = Q(\sqrt{2SNR})$ is used to approximate the BSC channel used in [71] with an equivalent AWGN channel. The energy per source bit is calculated as $\frac{E_b}{N_0} = \frac{N}{R_i} E_s/N_0$.

The equivalent coding rate, $R_{eq} = \frac{N}{R_i}$ is used to make a fair comparison choices between different schemes, where N is the number of sensors.

The comparison results show that in the low observation accuracy scenario, $\beta = 0.1$, the proposed scheme performs close to the more complex LDGM codes, with performance degradation of less than 1 dB. The superiority of the proposed coding scheme over the classical DTC is shown as well (see Figure 8.11). This performance improvement is achieved with lower decoding complexity, when the decoder switches to the non-iterative mode. The order of decoding complexity reduction with respect to classical DTC can be as high as the typical number of decoding iterations, which is 10 to 20.

8.5.1 Optimal Number of Sensors

Apparently, using a higher number of sensors at each cluster reduces the estimation error, but on the other hand, it increases the bandwidth usage and the decoding complexity. A key question is "What is the minimum number of sensors that yields a desired performance?" In order to answer this question, we provide an approximate information-theoretic analysis. We note that the rate region of the bCEO problem even for the simplest case of two sensors is not yet characterized in terms of a closed-form expression [39, 54]. Therefore, the proposed simple method is valuable in practical system design.

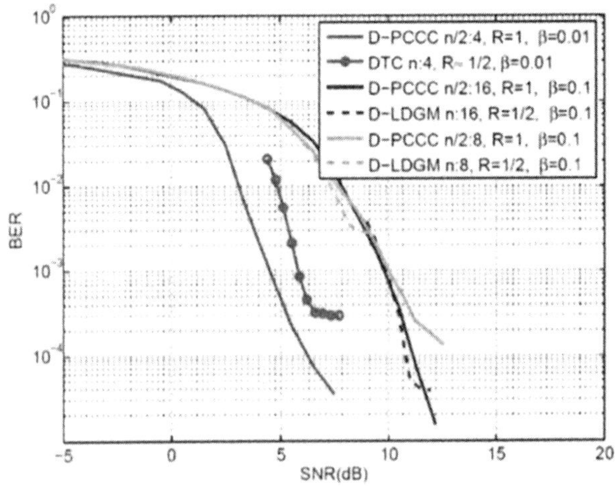

Figure 8.11 BER performance comparison of the proposed scheme with the state of the art D-JSCC methods.

In this regard, we first note that the virtual channel from the common source to the final destination is the cascade of a broadcast BSC channel and a set of parallel AWGN channels. It can be shown that for a cluster of N sensors with transmit power P, the observation error β and the noise variance σ_N^2, the capacity of this virtual channel is [55]:

$$C = \frac{1}{2(2\pi\sigma_N^2)^{\frac{N}{2}}} \int_{y^n} \left[(\gamma_0 + \gamma_1) \log\left(\frac{\gamma_0 + \gamma_1}{2}\right) - \gamma_0 \log \gamma_0 - \gamma_1 \log \gamma_1 \right] d_{y^N} \quad (8.8)$$

where, we have:

$$\gamma_\alpha = \sum_{k=0}^{N} \binom{N}{k} \beta^k (1 - \beta)^{N-k} \exp\left(- \frac{\sum_{i=0}^{k} \left(y_i + (2\alpha - 1)\sqrt{P}\right)^2 + \sum_{i=k+1}^{N} \left(y_i - (2\alpha - 1)\sqrt{P}\right)^2}{2\sigma_N^2} \right)$$

(8.9)

Therefore, the channel capacity is $C = f_c\left(N, \frac{P}{\sigma_N^2}, \beta\right)$, which is a closed-form function of the number of sensors (N), the channel SNR $\left(\frac{P}{\sigma_N^2}\right)$ and the observation error parameter (β). Now, note the fact that the BER floor is fully defined by the number of sensors in an error-free environment as follows:

$$P_{error}^{(min)} = \begin{cases} \sum_{k=\frac{N+1}{2}}^{N} \binom{N}{k} \beta^k + (1-\beta)^{N-k} & N \text{ odd} \\ \frac{1}{2}\binom{N}{\frac{N}{2}} \beta^{\frac{N}{2}} (1-\beta)^{\frac{N}{2}} + \sum_{k=\frac{N}{2}+1}^{N} \binom{N}{k} \beta^k + (1-\beta)^{N-k} & N \text{ even} \end{cases} \quad (8.10)$$

Based on the channel-coding theorem, there exists an encoder with rate $R \leq C$ that can be employed by the source for an error-free transmission of information bits to the common destination. Now, we conjecture that if we employ a proper coding technique of rate $R \leq C$ at the sensors, the final recovery error approaches the error floor, which is the best achievable error probability. Based on this argument, for a given set of system conditions $\beta, \frac{P}{\sigma_N^2}, R$, we can obtain the optimal number of sensors

N by solving $R = f_c(N, \frac{P}{\sigma_N^2}, \beta)$. This approach is demonstrated in Figure 8.12.

In Figure 8.12(a), the channel capacity is plotted for a fixed observation error parameter $\beta = 0.01$. The capacity is an increasing function of the channel SNR. If a certain value of capacity is desired, the minimum required SNR level is less for a higher number of utilized sensors. For instance, to achieve a capacity of 1/2 bits/transmission, the required SNR values for $N = 2,3,4,5$ sensors are −2.570 dB, −4.366 dB, −5.628 dB, and −6.621 dB, respectively. In another interpretation, for power-constrained sensors (limited SNRs), this graph can be used to determine the minimum number of sensors to achieve a certain level of capacity. For instance, at least $N = 4$ sensors are required to achieve a capacity of 1/2 bits/transmission at SNR=−5 dB. This is verified by extensive Monte Carlo simulations for ($\beta = 0.01, N = 256$ bits and $R_i = \frac{1}{2}$ for a different number of sensors as depicted in Figure 8.12(b), where the decoder reaches its error floor at SNR=−5 dB if at least 4 sensors are utilized in the system.

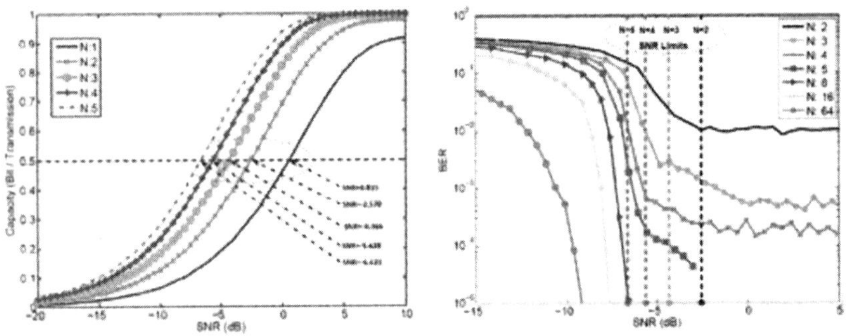

(a) Channel Capacity vs SNR, $\beta = 0.01$ (b) Decoder BER Performance

Figure 8.12 a) System information capacity versus observation accuracy (BSC crossover probability) and channel quality (SNR) for different number of sensors. b) The BER performance of the proposed coding scheme with a different number of sensors. In both figures, β=0.01 (Source: [55]).

8.6 EXTENSION TO MULTIHOP NETWORKS

So far, a collocated network scenario has been considered with all sensors directly communicating to a data fusion center. This scenario does not cover many applications, where the use of multihop communication is unavoidable due to a long distance between the source and data fusion center. The question here is how to extend the obtained results to such cases.

In order to address this requirement, we propose to use the idea of inner channels by replacing the multi-hop channel from each sensor to the fusion center as a BSC channel with an equivalent crossover probability. For instance, in [85], a two-tiered network, where the sensors at each cluster communicate to the base station through two intermediate relay nodes, is considered (Figure 8.13(a)). Utilizing two super nodes (also called cluster heads) as relay nodes has an obvious advantage of robustness to cluster head failure in practice. A demodulate and forward (DMF) relaying method assisted by a distributed version of space time block code (D-STBC) is utilized in order to improve the BER performance and overcome the potential fading effects.

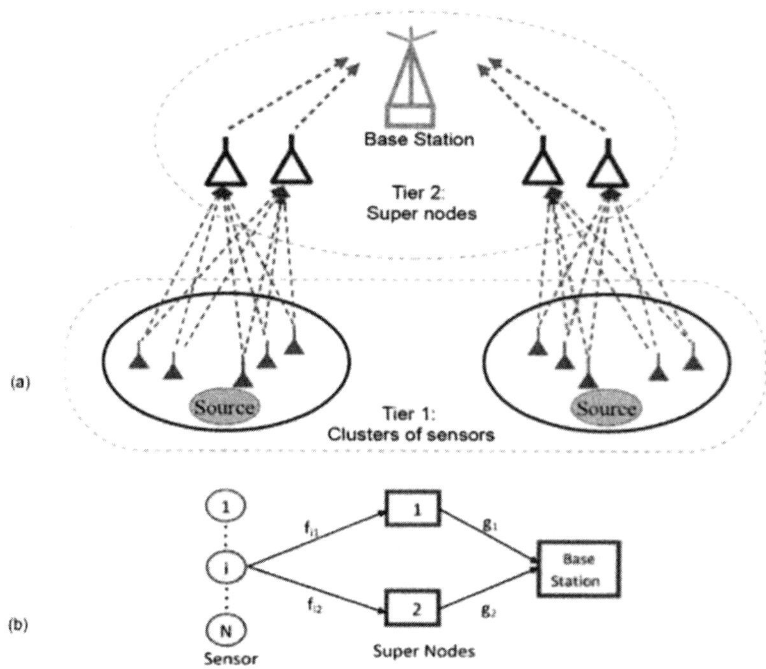

Figure 8.13 a) System model for two-tiered double-sink wireless sensor network, b) Channel coefficients for the communication links from sensors to the base station via two super nodes (Source: [85]).

Suppose that the total transmission power P is divided between the two tiers (sensors and super nodes) with a ratio of α. As such, each sensor consumes $\alpha \frac{P}{N}$ and each relay node consumes $(1-\alpha)\frac{P}{2}$. The inner channel error probability can be calculated as follows [85]:

$$p_e^{(in)} = \frac{1}{4(\alpha\,\gamma+1)} + \frac{1}{2(\eta\,\gamma+2)^2} - \frac{1}{4(\alpha\,\gamma+1)(\eta\,\gamma+2)^2}, \tag{8.11}$$

where we have:

$$\gamma = \frac{P}{N}, \; \gamma_1 = \alpha\,\gamma, \quad \gamma_2 = \eta\,\gamma, \quad \eta = \frac{3\alpha\bar{\alpha}}{2\,\alpha+1}. \tag{8.12}$$

Figure 8.14 presents the impact of power allocation on inner channel error probability. Despite STBC-assisted amplify and forward (AF) relaying, where the optimality achieved if the total power is equally divided between the two tiers (sensors and relays), the performance curve of DMF relaying mode is not symmetric. Dashed and solid lines in Figure 8.14 present analytical and simulation results for the resulting error probability as a function of the power allocation parameter α. The optimal power allocation scenario can be chosen based on the average SNR value of the system, $\frac{P}{N_1}$.

We complete this section by comparing the end-to-end error probability for the basis scenario with one and two relay nodes (cluster heads) with and without STBC coding in presence of the channel fading effect in Figure 8.15. It is shown that using two supernodes increases the BER performance by about 2~3 dB due to the space diversity gain. Also, about 1 dB additional gain is obtained using D-STBC due to the space-time diversity. All schemes ultimately reach the error floor for extremely large SNR values, which is corresponding to the error-free communication calculated using (8.10).

We finally note that the concept of the inner channel is general, and similar performance analysis and power allocation policies can be applied to other network configurations.

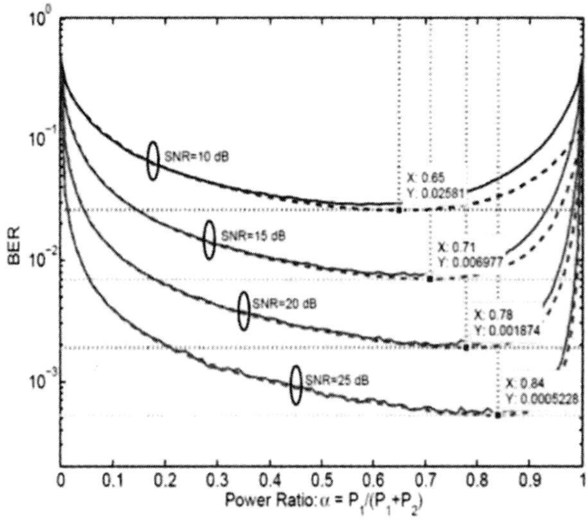

Figure 8.14 Inner-channel error probability versus power allocation parameter α (Source: [85]).

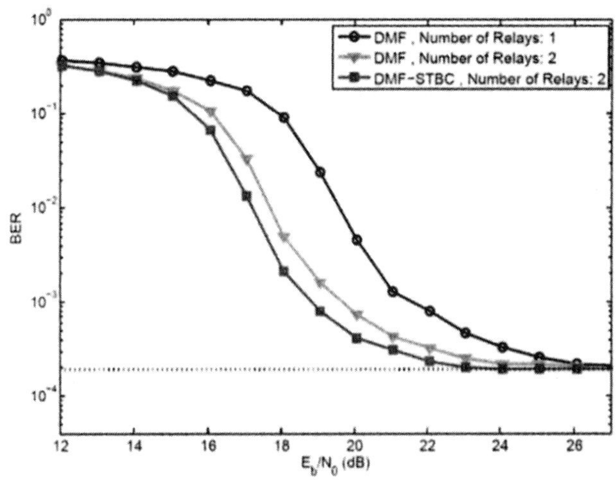

Figure 8.15 Comparison of the system performance for different numbers of supernodes, with and without the STBC coding at supernodes ($N = 4$) (Source: [85]).

8.7 CONCLUDING REMARKS

This chapter explores various algorithms designed by the coding research community to realize compression and error recovery in WSNs with partially accurate and remotely deployed sensors. Despite their promising theoretical performances, most of these algorithms are not well suited for practical applications due to a number of drawbacks. In this regard, a practical and implementation-friendly solution is reviewed for this problem. Previously reported algorithms, as discussed in Section 8.3, are not easily scalable to a large number of sensors. However, employing a large number of sensors at each cluster is unavoidable when the observation accuracy of sensors is low. Fusing multiple sensor readings compensates for the observation errors that occur prior to encoding. Therefore, there is an essential need for an easily scalable distributed joint source-channel coding algorithm. This solution should be flexible enough to accommodate dynamic clustering with a varying number of moving sensors at each cluster. Further, the developed algorithm should be capable of accommodating time-varying and unknown observation models and should be easily integrated with the current sensor structures and contemporary networking protocols. To address these requirements, in Section 8.4 an adaptive algorithm is presented for efficient data collection across a data field, where source locations are inaccessible due to either unknown source locations or harsh environmental conditions.

The commonly accepted presumption of the superiority of iterative decoding over noniterative decoding is revisited in Section 8.4.3. Indeed, the proposed convergence analysis, developed based on the EXIT chart technique, suggests that in some system conditions, the soft information exchange among constituent decoders at the destination does not improve the overall end-to-end error rate. Therefore, avoiding these unnecessary information exchange cycles reduces the average decoding complexity Further, in Section 8.5.1, an easy-to-evaluate information-theoretic function is derived to determine the minimum number of sensors to reach the highest achievable error recovery performance.

To generalize the proposed scheme to large-scale networks, a two-tiered clustered system model is considered in Section 8.6, where each cluster includes a single data source, which is remotely monitored by surrounding sensors. The sensors within a cluster collectively compress and transmit their observations to a data fusion center via two supernodes. This two-tiered network model is robust to supernode failure in contrast with the traditionally designed single supernode systems. The set of two-hop communication channels from the sensors to the data fusion center via two super-nodes is modeled as inner channels. This enables us to analyze the end-to-end performance of the whole system by evaluating the distributed coding operation over inner channels. The available total power is optimally split to the existing sensor and supernodes such that the overall data throughput is maximized. This system demonstrates the easy integration of our proposed coding scheme with desired relaying methods. This chapter suggests several directions for continuation of the current studies described as follows.

The first is designing and optimizing the coding parameters for the more general case of unequal observation parameters $\beta_i \neq \beta$ and unequal channel qualities $SNR_i \neq SNR$. The proposed scheme is robust to sensors' failure and stays operational as long as one or some of the sensors is active. However, each sensor failure degrades the overall end-to-end BER performance. This performance degradation compensated by changing the coding rate of the other active sensor. This method can be studied as another possible extension of this work, which involves quantification of the BER performance degradation due to a sensor failure and finding the new coding rate to compensate for this loss. Developing more elegant learning algorithms (such as reinforcement learning) in order to tune the required coding rates without the need for estimating the sensors' measurement accuracies and channel qualities can simplify the implementation of sensors. If all the passive sensors observe a common data source, similar techniques can be developed for passive sensors to harness the correlation among sensor observations to enhance the final estimation accuracy.

References

[1] Schmidt, R., J. R. Dickman, and D. R. Kiracofe, "Engine Health Monitoring Using Acoustic Sensors," U.S. Patent 15/063,582, September 14, 2017.

[2] Dudzik, E., A. Abedi, D. Hummels, and M. d. Cunha, "Orthogonal code design for passive wireless sensors," in *IEEE Biennial Symposium on Communications*, 2008.

[3] Zhang, J., et al., "A review of passive RFID tag antenna-based sensors and systems for structural health monitoring applications," *Sensors*, Vol. 17, No. 2, p. 265, 2017.

[4] Michael, Z., E. Anagnostou, J. Nystuen, and M. Anagnostou, "UPAL: Underwater passive aquatic listener," in *MTS/IEEE OCEANS'15*, Washington, 2015.

[5] Krishen, K., "Space applications for ionic polymer-metal composite sensors, actuators, and artificial muscles," *Acta Astronautica*, Vol. 64, No. 11, pp. 1160–1166, 2009.

[6] Berkcan, E., and Y. Lee, "Passive wireless sensors for turbomachines and method of

operating the same," U.S. Patent 9,909,443, March 6, 2018.

[7] Hodgson, J., et al., "Drones count wildlife more accurately and precisely than humans," *Methods in Ecology and Evolution*, 2017.

[8] Li, T., J. M. Corchado, S. Sun, and J. Bajo, "Clustering for filtering: Multi-object detection and estimation using multiple/massive sensors," *Information Sciences*, Vol.388, pp. 172–190, 2017.

[9] Gravina, R., P. Alinia, H. Ghasemzadeh, and G. Fortino, "Multi-sensor fusion in body sensor networks: State-of-the-art and research challenges," *Information Fusion*, Vol.35, pp. 68–80, 2017.

[10] Xiong, Z., A. D. Liveris, and S. Cheng, "Distributed source coding for sensor networks," *IEEE signal processing magazine*, Vol. 21, No. 5, pp. 80–94, 2004.

[11] Garcia-Frias, J., and Z. Xiong, "Distributed source and joint source-channel coding: from theory to practice," in *IEEE Acoustics, Speech, and Signal Processing Conference (ICASSP)*, 2005.

[12] Hill, J., M. Horton, R. Kling, and L. Krishnamurthy, "The platforms enabling wireless sensor networks," *ACM Communications*, Vol. 47, No. 6, pp. 41–46, June 2004.

[13] Cricket v2 User Manual, Cambridge, Lab, MIT Computer Science & Artificial Intelligence, 2005.

[14] Rowe, A., R. Mangharam, and R. Rajkumar, "FireFly: A Time Synchronized Real-Time Sensor Networking Platform," in *Wireless Ad Hoc Networking: Personal-Area, Local-Area, and the Sensory-Area Networks*, Auerbach Publications, Boca Raton, FL, 2006.

[15] Mangharam, R., A. Rowe, and R. Rajkumar, "FireFly: A Cross-Layer Platform for Wireless Sensor Networks," *Real Time Systems Journal, Special Issue on Real-Time Wireless Sensor Networks*, Vol. 37, No. 3, pp. 181–231, December 2007.

[16] Berger, T., and Z. Zhang, "On the CEO problem," in *IEEE International Symposium Information Theory (ISIT)*, 1994.

[17] Oohama, Y., "Indirect and Direct Gaussian Distributed Source Coding Problems," *IEEE Transactions on Information Theory*, Vol. 60, No. 12, pp. 7506–7539, December 2014.

[18] Slepian, D., and J. Wolf, "Noiseless coding of correlated information sources," *IEEE Transactions on Information Theory*, Vol. 19, No. 4, pp. 471–408, July 1973.

[19] Liveris, A., Z. Xiong, and C. Georghiades, "Compression of binary sources with side information at the decoder using LDPC codes," *IEEE Communication Letters*, Vol. 6, No. 10, pp. 440–442, October 2002.

[20] Wyner, A., and J. Ziv, "The rate-distortion function for source coding with side information at the decoder," *IEEE Transactions on Information Theory*, Vol. 22, No.1, pp. 1–10, January 1976.

[21] Li, Y., "Distributed coding for cooperative wireless networks: An overview and recent advances," *IEEE Communication Magazine*, Vol. 47, No. 8, pp. 71–77, August 2009.

[22] Breton, M., and B. Kovatchev, "Analysis, Modeling, and Simulation of the Accuracy of Continuous Glucose Sensors," *Journal of Diabetes Science Technology*, Vol. 2, No. 5, pp. 835–862, September 2008.

[23] Petrosyants, K., and N. I. Rjabov, "Temperature sensors modeling for smart power ICs," in *the 27th Annual IEEE Semiconductor Thermal Measurement and Management Symposium (SEMI-THERM)*, 2011.

[24] Robins, P., V. Rapley, and P. Thomas, "A review of wireless SAW sensors," *IEEE Transactions on Ultrasonics, Ferroelectrics, and Frequency Control*, Vol. 47, No. 2, pp. 317–332, 2000.

[25] Mendis, C., A. Skvortsov, A. Gunatilaka, and S. Karunasekera, "Performance of Wireless Chemical Sensor Network With Dynamic Collaboration," *IEEE Sensors Journal*, Vol. 12, No. 8, pp. 2630–2637, 2012.

[26] Tigli, O., and M. Zaghloul, "Surface acoustic wave SAW biosensors," in *the 53rd IEEE International Midwest Symposium on Circuits and Systems (MWSCAS)*, 2010.

[27] Hovakeemian, Y., K. Naik, and A. Nayak, "A survey on dependability in Body Area Networks," in *the 5th International Symposium on Medical Information Communication Technology (ISMICT)*, 2011.

[28] Bylanski, P., "PCM A-Law decoder using a circulating method," *Electronics Letters*, Vol. 12, No. 2, pp. 57–58, 1976.

[29] Gray, R., "Vector quantization," *IEEE ASSP Magazine*, Vol. 1, No. 2, pp. 4–29, 1984.

[30] Bylanski, P., "Signal-processing operations on A-Law companded speech," *Electronics Letters*, Vol. 15, No. 12, pp. 697–698, 1979.

[31] Dvorsky, P., J. Londak, O. Labaj ,and P. Podhradsky, "Comparison of codecs for videoconferencing service in NGN," in *ELMAR*, 2012.

[32] Linde, Y., A. Buzo, and R. Gray, "An Algorithm for Vector Quantizer Design," *IEEE Transactions on Communications*, Vol. 28, No. 1, pp. 84–95, 1980.

[33] Gray, R., and D. Neuhoff, "Quantization," *IEEE Transactions on Information Theory*, Vol. 44, No. 6, pp. 2325–2383, 1998.

[34] Krusevac, Z., P. Rapajic, and R. Kennedy, "Concept of time varying binary symmetric model-channel uncertainty modeling," in *International Conference on*

Communications Systems (ICCS), 2004.

[35] Aaron, A., and B. Girod, "Compression with side information using turbo codes," in *Data Compression Conference (DCC)*, 2002.

[36] Garcia-Frias, J., and Y. Zhao, "Compression of binary memoryless sources using punctured turbo codes," *IEEE Communication Letters*, Vol. 6, No. 9, pp. 394–396, September 2002.

[37] Liveris, A., Z. Xiong, and C. Georghiades, "Distributed compression of binary sources using conventional parallel and serial concatenated convolutional codes," in *Data Compression Conference (DCC)*, 2003.

[38] Tu, Z., J. Li, and R. Blum, "Compression of a binary source with side information using parallelly concatenated convolutional codes," in *IEEE Global Telecommunications Conference (GLOBECOM)*, 2004.

[39] He, X., X. Zhou, M. Juntti, and T. Matsumoto, "A Rate-Distortion Region Analysis for a Binary CEO Problem," in *IEEE 83rd Vehicular Technology Conference (VTC Spring)*, Nanjing, 2016.

[40] Berger, T., Z. Zhang, and H. Viswanathan, "The CEO problem [multiterminal source coding]," *IEEE Transactions on Information Theory*, Vol. 42, No. 3, pp. 887–902, May 1996.

[41] Tung, S. Y., "Multiterminal source coding," *Ph.D. dissertation*, School of Electrical Engineering, Cornell University, Ithaca, NY, May 1998.

[42] Viswanathan, H., and T. Berger, "The quadratic Gaussian CEO problem," *IEEE Transactions on Information Theory*, Vol. 43, No. 5, pp. 1549–1559, September 1997.

[43] Chen, J., X. Zhang, T. Berger, and S. Wicker, "An upper bound on the sum-rate distortion function and its corresponding rate allocation schemes for the CEO problem," *IEEE Journal Selected Areas Communication*, Vol. 22, No. 6, pp.977–987, August 2004.

[44] Courtade, T. A., "Outer Bounds for Multiterminal Source Coding based on Maximal Correlation," in *CoRR*, 2013.

[45] Yang, Y., Y. Zhang, and Z. Xiong, "On the Sum-Rate Loss of Quadratic Gaussian Multiterminal Source Coding," *IEEE Transactions on Information Theory*, Vol. 57, No. 9, pp. 5588–5614, 2011.

[46] Yang, Y., Y. Zhang, and Z. Xiong, "The generalized quadratic Gaussian CEO problem: New cases with tight rate region and applications," in *IEEE International Symposium on Information Theory Proceedings (ISIT)*, 2010.

[47] Yang, Y., and Z. Xiong, "The Sum-Rate Bound for a New Class of Quadratic Gaussian Multiterminal Source Coding Problems," *IEEE Transactions on*

Information Theory, Vol. 58, No. 2, pp. 693–707, 2012.

[48] Yang, Y., and Z. Xiong, "On the Generalized Gaussian CEO Problem," *IEEE Transactions on Information Theory*, Vol. 58, No. 6, pp. 3350–3372, 2012.

[49] Chen, J., and T. Berger, "Robust Distributed Source Coding," *IEEE Transactions on Information Theory*, Vol. 54, No. 8, pp. 3385–3398, 2008.

[50] Tavildar, S., and P. Viswanath, "On the Sum-rate of the Vector Gaussian CEO Problem," in *the thirty-ninth Asilomar Conference on Signals, Systems and Computers*, 2005.

[51] Zhang, G., and W. Kleijn, "Bounding the Rate Region of the Two-Terminal Vector Gaussian CEO Problem," in *Data Compression Conference (DCC)*, 2011.

[52] Lin, S.-C., and H.-J. Su, "Vector Wyner-Ziv coding for vector Gaussian CEO problem," in *the 41st Annual Conference on Information Sciences and Systems (CISS)*, 2007.

[53] Chen, J., and J. Wang, "On the vector Gaussian CEO problem," in *IEEE International Symposium Information Theory (ISIT)*, 2011.

[54] X. He, et al., "A Lower Bound Analysis of Hamming Distortion for a Binary CEO Problem With Joint Source-Channel Coding," *IEEE Transactions on Communications*, Vol. 64, No. 1, pp. 343–353, January 2016.

[55] Razi, A., K. Yasami, and A. Abedi, "On Minimum Number of Wireless Sensors Required for Reliable Binary Source Estimation," in *IEEE Wireless Communication and Networking Conference (WCNC)*, 2011.

[56] Del Ser, J., J. Garcia-Frias, and P. Crespo, "Iterative concatenated Zigzag decoding and blind data fusion of correlated sensors," in *International Conference on Ultra Modern Telecommunications Workshops (ICUMT)*, 2009.

[57] Zhong, W., and J. Garcia-Frias, "Combining data fusion with joint source-channel coding of correlated sensors," in *IEEE Information Theory Workshop (ITW)*, 2004.

[58] Pradhan, S., and K. Ramchandran, "Distributed source coding using syndromes (DISCUS): design and construction," in *Data Compression Conference (DCC)*, 1999.

[59] Qaisar, S., and H. Radha, "Multipath multi-stream distributed reliable video delivery in Wireless Sensor Networks," in *the 43th Annual Conference Information Sciences Systems (CISS)*, 2009.

[60] Esnaola, I., and J. Garcia-Frias, "Analysis and optimization of distributed linear coding of Gaussian sources," in *the 43th Annual Conference Information Sciences Systems (CISS)*, 2009.

[61] Bajcsy, J., and P. Mitran, "Coding for the Slepian-Wolf problem with turbo codes," in *IEEE Global Telecommunications Conference (GLOBECOM)*, 2001.

[62] Valenti, M., and B. Zhao, "Distributed turbo codes: towards the capacity of the relay channel," in *the 58th IEEE Vehicular Technology Conference (VTC)*, 2003.

[63] Daneshgaran, F., M. Laddomada, and M. Mondin, "Iterative joint channel decoding of correlated sources," *IEEE Transactions on Wireless Communication*, Vol. 5, No.10, pp. 2659–2663, October 2006.

[64] Garcia-Frias, J., Y. Zhao, and W. Zhong, "Turbó-Like Codes for Transmission of Correlated Sources over Noisy Channels," *IEEE Signal Processing Magazine*, Vol.24, No. 5, pp. 58–66, September 2007.

[65] Liveris, A., Z. Xiong, and C. Georghiades, "A distributed source coding technique for correlated images using turbo-codes," *IEEE Communications Letters*, Vol. 6, No. 9, pp. 379–381, September 2002.

[66] Stankovic, V., A. Liveris, Z. Xiong, and C. Georghiades, "On code design for the Slepian-Wolf problem and lossless multiterminal networks," *IEEE Transactions on Information Theory*, Vol. 52, No. 4, pp. 1495–1507, April 2006.

[67] Hernaez, M., P. Crespo, J. Del Ser, and J. Garcia-Frias, "Serially-concatenated LDGM codes for correlated sources over gaussian broadcast channels," *IEEE Communication Letters*, Vol. 13, No. 10, pp. 788–790, October 2009.

[68] Stefanovic, C., D. Vukobratovic, and V. Stankovic, "On distributed LDGM and LDPC code design for networked systems," in *IEEE Information Theory Workshop (ITW)*, 2009.

[69] Sartipi, M., and F. Fekri, "Source and channel coding in wireless sensor networks using LDPC codes," in *IEEE Conference Sensor and Ad Hoc Communication and Networks (SECON)*, 2004.

[70] Yedla, A., H. Pfister, and K. Narayanan, "Can iterative decoding for erasure correlated sources be universal?," in *47th Annual Allerton Conference Communication, Control, and Computing*, 2009.

[71] Haghighat, J., H. Behroozi, and D. Plant, "Joint decoding and data fusion in wireless sensor networks using turbo codes," in *the 19th IEEE International Symposium Personal, Indoor and Mobile Radio Communication (PIMRC)*, 2008.

[72] Garcia-Frias, J., "Joint source-channel decoding of correlated sources over noisy channels," in *Data Compression Conference (DCC)*, 2001.

[73] Razi, A., F. Afghah, and A. Abedi, "Binary source estimation using two-tiered sensor network," *IEEE Communication Letter*, Vol. 5, No. 4, pp. 449–451, April 2011.

[74] Razi, A., and A. Abedi, "Distributed Coding of Sources with Unknown Correlation Parameter," in *International Conference Wireless Networking (ICWN)*, Las Vegas, 2010.

[75] Ghaffar, R., and R. Knopp, "Analysis of Low Complexity Max Log MAP Detector

and MMSE Detector for Interference Suppression in Correlated Fading," in *IEEE Global Telecommunications Conference, (GLOBECOM)*, 2009.

[76] Razi, A., "Distributed Adaptive Algorithm Design For Joint Data Compression And Coding In Dynamic Wireless Sensor Networks," PhD Thesis, University of Maine, May 2013.

[77] Liveris, A., Z. Xiong, and C. Georghiades, "Joint source-channel coding of binary sources with side information at the decoder using IRA codes," in *IEEE Workshop Multimedia Signal Processing*, 2002.

[78] Haghighat, J., H. Behroozi, and D. Plant, "Iterative joint decoding for sensor networks with binary CEO model," in *the 9th IEEE Workshop Signal Processing Advances in Wireless Communication (SPAWC)*, 2008.

[79] Gunduz, D., E. Erkip, A. Goldsmith, and H. Poor, "Source and Channel Coding for Correlated Sources Over Multiuser Channels," *IEEE Transactions on Information Theory*, Vol. 55, No. 9, pp. 3924–3944, September 2009.

[80] Maunder, R., et al., "On the Performance and Complexity of Irregular Variable Length Codes for Near-Capacity Joint Source and Channel Coding," *IEEE Transactions on Wireless Communication*, Vol. 7, No. 4, pp.1338–1347, April 2008.

[81] Brink, S., "Convergence behavior of iteratively decoded parallel concatenated codes," *IEEE Transactions on Communication*, Vol. 49, No. 10, pp. 1727–1737, October 2001.

[82] Razi, A., and A. Abedi, "Adaptive Bi-modal Decoder for Binary Source Estimation with Two Observers," in *the 46th Annual Conference Information Science Systems (CISS)*, Princeton, 2012.

[83] Razi, A., and A. Abedi, "Convergence Analysis of Iterative Decoding for Binary CEO Problem," *IEEE Transactions on Wireless Communications*, Vol. 13, No. 5, pp.2944–2954, May 2014.

[84] Wiberg, N., H.-A. Loeliger, and R. Kotter, "Codes and iterative decoding on general graphs," in *IEEE International Symposium Information Theory (ISIT)*, 1995.

[85] Razi, A., F. Afghah, and A. Abedi, "Power Optimized DSTBC Assisted DMF Relaying in Wireless Sensor Networks with Redundant Super nodes," *IEEE transactions on Wireless Communications*, Vol. 12, No. 2, pp. 636–645, February 2013.

Chapter 9

Wireless Sensor Systems for Underwater Applications

Water covers over 71% of the Earth's surface and all the goodness below. Despite its life threatening storms and mighty waves, it is rich, with untouched resources ready to make new industrial paradigms enabling our smart sensor technologies. This chapter, therefore, presents a new analytical view of the potential opportunities, technological developments, and design aspects of *underwater wireless sensor networks.*

9.1 INTRODUCTION

An underwater environment signifies its own distinct evolutionary nature with specific properties and characteristics. In order to activate our new industrial momentum, therefore, we need to explore a few basic aspects of the challenge. One is to understand this media and its properties. Another is analyzing our ability against the stormy waters, and our technological capabilities for penetrating and controlling the media in making a better use of the vast and untouched underwater resources.

Traditionally we have used the seas as our shipping waterways, mostly for transportation and trading but also studying its nature for scientific and humanitarian purposes. However, expansion of new mining industries and new geographical excavations along with progressive underwater smart sensor applications are injecting new discoveries enabling new underwater wireless sensor network (UWSN) projects. One example is classic traditional acoustic technology, which has been dominating our underwater developments market for many decades in the past. The traditional sensors and associated equipment were bulky and very expensive with limited free market competition. More recent PZT and MEMS, however, are making UWSN new solutions both versatile and exciting. This chapter, therefore, carries design examples while discussing a progressive development platform of upcoming technologies with many typical and system level solutions.

We then address market opportunities and developing UWSN technologies along with building new possible industrial paradigms offering new trends to open our vision for more important future advanced industrial and humanitarian applications. In the case of reliably controlled humanity, preventive care aspects of disaster control UWSN can offer many new opportunities. These are not happening because such developments require the responsive support of new or restructured suitable organizations, nations, and human activists to organize, manage, and monitor. In the case of industrial opportunities, we need a new approach from management for full responsibility, where industries take responsibility for cleaning up the existing mess and abuse of the waters and demonstrate their own innovative capability through advanced products equipped with an adequate dis invention process so that all the loose cnds of their operations are controlled and closed up, preventing any future side effects from jeopardizing the quality of life on earth in the future.

This chapter is organized as follows. This Section follows the urgently needed aspects of UWSN including disaster technology management. Section 9.2 addresses the most challenging part of UWSN deployment, the technological developments of underwater propagation. Section 9.3 provides brief coverage of energy consumption. Section 9.4 provides an understanding of networking through remote sensing and remote control aspects of UWSN. Then Section 9.5 addresses the application areas where smart sensors and UWSN networking open up new opportunities for shallow and deep waters after which Section 9.6 discusses the challenging case of underwater oil and gas industry as a niche market for UWSN to thrive.

9.1.1 Technology Management

The best way to address "technology management" is to look at the way we perceive and apply technology to our environment. Due to their poor scientific understanding of nature, most ancient civilizations, desperate to feed their growing populations, after storms and floods rushed to pray as they believed in the role of invisible gods as a source of natural disasters.

For example, in the 5th century BC Chinese governing rulers demanded a yearly sacrifice of a *maiden virgin* to please the Yellow River gods to prevent flooding.

Flood has been at the top of the list of life-taking natural disasters in low ground coasts, and it is common in Europe. The lowlands of Holland have always been flood-prone disaster areas and are therefore very intense with disaster management. In some cases the occurrence of a disaster has triggered the improvement of a protective defense system. For example, the disaster protective systems in the Netherlands have been built mainly to control the floods. Floods of 1916 triggered construction of a dam and a flood disaster of 1953 (with over 1800 fatalities) initiated the delta works to protect the region [1].

These days smart sensors have shown their capability as the most effective core technology in many dimensions of disaster management, from prediction to prevention and operation. This becomes extremely important for saving lives. This life-saving management ensures that the technology is working at its best functional capability and is managed and coordinated with all involved parties, including private volunteers and the official members of the authorities.

Due to the possible loss of life major disaster management skills are becoming a number one priority. Building an appropriate technology infrastructure is a management task with a clear set of objectives. Without those objectives very little can be done to protect life, especially for the elderly or children who could be distressed and shocked by the event. On the management side of technology, it is vitally important to understand that ownership without possessing a sufficient grasp of its basic functions is useless and often antiproductive.

As they say, technology is a double-sided sword that can hurt the operation if it is unable to coordinate and manage the operation due to a lack of the required capabilities. That is why nations that are starved of new technologies should not be rush into adopting them without proper preparations and the required infrastructure for a full absorption of the operation.

In the case of a coastal disaster a UWSN installation requires an appropriate choice of technology, a suitable tactical management procedure, and clear knowledge of the environment. That is, a *smooth well-controlled operation* is far more effective in saving lives and economies than a last minute rush through a hectic, desperate evacuation, even with the best technology in place. We therefore advise that local responsible bodies take on the long-term preparation through a staged sequence of operations, such as the one shown in Figure 9.1.

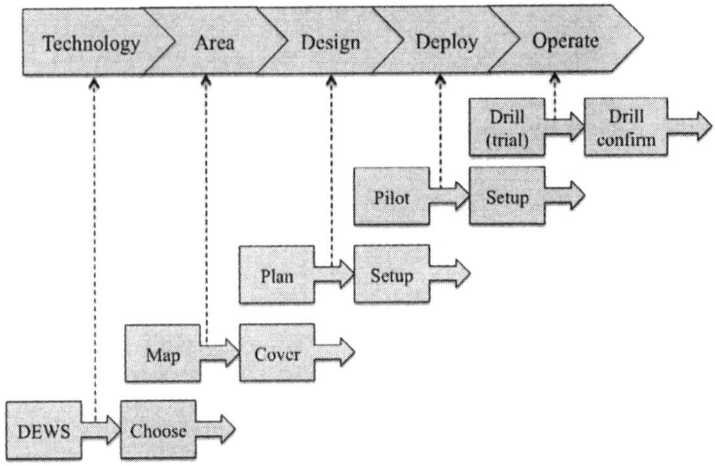

Figure 9.1 Sequence of selecting, preparing, and adopting a new technology in a disaster-prone area.

These days we have a wide range of sensors available for the right technological solution, once the critically important decision is made to build the most suitable integrated disaster warning and management system for the area. We propose a sensor-enabled dynamic early warning system (DEWS) in Chapter 7 as a tool to help with (a) living with peace of mind but being prepared during normal conditions and (b) saving lives and goods phasing through the disaster period when it happens. However, at the end of the day it is up to the management skills and effectiveness of the prepared and synchronized operation process to manage people whilE facing the impact.

Adopting a new technology to match a life-saving operation requires a smoothly controlled staged management. As indicated by Figure 9.1 we have five stages: (1) technology; (2) area; (3) design; (4) deployment; and (5) operate. Each stage comes with two complementary actions. In stage 1, DEWS represents choosing appropriate UWSN *technology* where, a basic understanding of the technology considers dependency, reliability, and ease of operation. One should appreciate that being new, or under development, may not necessarily mean the "best" for every scenario. The decision also requires a good knowledge of stage 2, *area*. For *area*, where a statistical density map of the effective area is essential for a professional *design*, which includes plans for the sequence of actions for the right configuration to be used in a follow-up stage of *deployment* for a pilot setup. Following the stage 4 setup, we need to take it through some drilling tests under the *operation stage* before being marking the system *ready to use*.

9.1.2 Human Instability in Floods

Due to flooding's potential for loss of life, for children in particular, early warning alarms are needed to prepare people in areas at higher risk to act and advise their children in advance. Such warnings should take into account the tendency of people to overestimate their strength against flooding mixed with storm, rain, and noise. Such dangerous overestimation is particularly dangerous for inexperienced people and frightened children, who forget the possibility of hidden objects and misjudge water level.

Out of many models in literature we look at one that measures individual's resistive strength, R, against the flood force, F. The flood force is a predictive statistical figure with a mean value of $F = h \cdot v$, where h is the estimated height of the flooding water and v is its running speed. The F is measured in m^2/s, which can be easily converted into Kg/s if a noneffective width of enforcing water is also considered in the estimation of R. The mass weight of trapped person m plays a main role in his safe passing. The height H also can be considered in building a model for R.

There are two models for R. One does not include the height as,

$$R_1 = 0.1\sqrt{m}, \tag{9.1}$$

but the other one includes the height, expressed as,

$$R_2 = 0.929(e^{0.001906 \cdot H \cdot m + 1.09})^2. \tag{9.2}$$

Figure 9.2 shows the changes of these two models against a person's weight, where for R_2 person's height H plays a significant role. As shown in Figure 9.2, the lowest curve, R_1 indicates a rather pessimistic view for the worse case scenario, warning that no one can be safe waddling through rough flooding waters but R_2 gives a more realistic view of those who might be able to run to safer spots and able to help others such as the elderly and small children. That is, under a sensible and guided rescuing condition, a flood strength of $F = 0.5$ m^2/s can be regarded as safe enough to take small distance hoping through the flood.

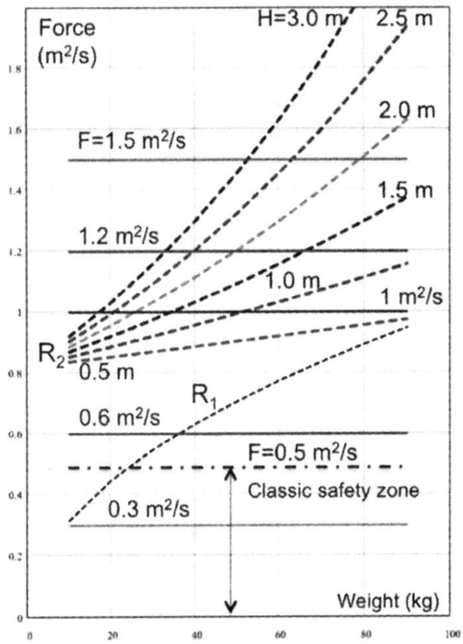

Figure 9.2 Force of flood versus weight and its flood safety margins.

It may also worth mentioning that the empirical model R_2 has been developed using some lengthy experiments by Abt et al. [2] and expected to be more effective for being used in the floods cause by coastal disasters. As shown in Figure 9.2 model, R_1 warns that anyone lighter than 25 kg is at risk for classic 0.5 m^2/s flood force. However, model R_2 provides higher risk margins such that everyone could be safe for floods up to 0.75 m^2/s floods but people (children) shorter than 0.5m are all at risk for 1 m^2/s floods. We also should not forget that coastal floods running through cities or industrial areas carry significant wild objects and debris with them and no one is 100% safe under such extreme conditions.

9.1.3 Shallow Water Earthquake

The worse case disaster scenario occurs when there is an earthquake along the coast not far from large coastal cities. Although every part of any natural disaster is sad, and we often feel helpless to control it, the most dangerous part is wild and strong rushing floodwaters running through habited areas. This imposes the most devastating impact on younger people; as mentioned in Jonkman's work [1] over 22% of the victims of flood disasters are children under 10, leading to over one third of the victims being under 20 years. This then leaves people with long-term regrets and sadness, causing further scars on society. For example, in the last seven years of the past millennium coastal floods were responsible for over half a million deaths with Bangladesh bearing nearly 90% of them [1]. Consider the following disaster parameters.

Hurricane: Hazard modeling has been developed estimate the strength of these storms' wind, surge, and rainfall flooding: (a) wind models estimate the storm wind field upon a parametric wind profile; (b) storm surge models for the storm surface wind and pressure are also sensitive to coastal bathymetry and topography; and (c) rainfall, due to large spatial and temporal variations is hard to model. Using local weather prediction models would simplify them [3].[8]

Tsunami: An impulsively generated sea wave as a result of large-scale seafloor displacement (earthquake). When running into the coast the tsunami wave-front, "bore," can pile up waters of 30m or higher sweeps up to several kilometers into nearby shallow lands. Another work is related to the tsunami caused by the Great East Japan Earthquake (GEJE). Japan, which is famous for its virtually everyday earthquakes and is a narrow island next to the greatest ocean of the moving crust's plates (see chapter 10), should have been always better prepared for such disasters. This tsunami was difficult to access for the following reasons:

(1) Having a source combination of two huge earthquakes made it very difficult to predict the magnitude, a moment magnitude of 9.0 Mw; therefore, it took everyone by a great surprise.
(2) Traveling along the ocean, it could not have been detected due to the complex topography of the ocean floor.
(3) The unique tectonic profile of the area (many small islands reflecting and distorting the seismic waves) added ambiguity to the detection.
(4) The traditional prevention design systems were set based on the tsunami style flooding height therefore most of the buildings were placed at a 4m height and nuclear reactors at 12m.

Later it was realized that the building designs should have been more dynamic and enhanced to deal with the following five parameters: (a) tsunami height, (b) tsunami speed, (c) inundation depth, (d) inundation

[8] Rainfall modeling is out of scope of this book.

height, and (e) run-up height, as shown in Figure 9.3. Figure 9.4 shows the affected areas by the tsunami [4].

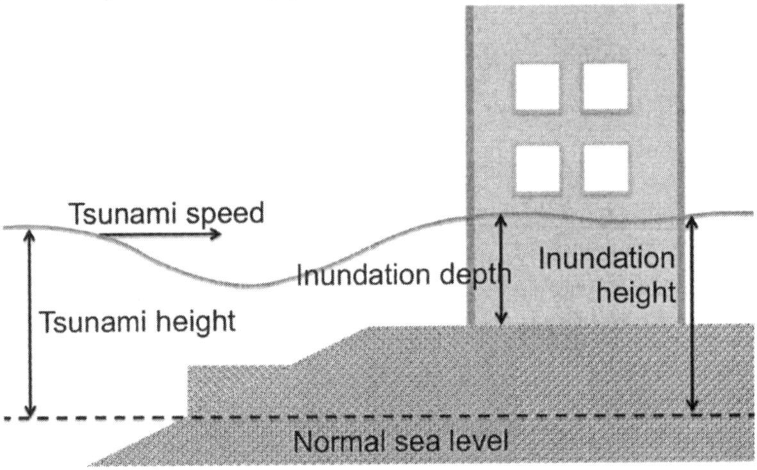

Figure 9.3 The run-up height model of the incident, drawn from [4].

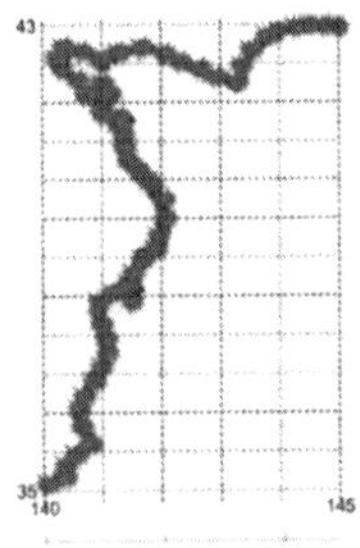

Figure 9.4 Results of the survey showing the affected shores, drawn from [4].

However, Japan has shown a unique capability to minimize the impact by building its disaster resilience. Building any disaster preparedness requires infrastructure development utilizing time, training, regulations, and the right technology. Japan's early warning system prevented a much larger death toll. Detecting seismic activities and broadcasting alerts on television, radio and mobile phone networks enabled people to minimize the impact [5].

The use of robotics has been a lifesaver in many cases. In fact Japanese advanced robotic technology was demonstrated in the special case of the Fukushima Daiichi nuclear power plant (NPP) where it handled radioactive devices and explored through the debris and half-collapsed areas [4].

9.2 UNDERWATER PROPAGATION TECHNOLOGY

This section briefly scans and analyzes the key UWSN technologies to identify those that require urgent development. This is essential, as we consider the world under water to be an extreme environment with huge potential opportunities for new applications, which can lead us to new discoveries for smart sensor and WSN technologies in the future.

By way of exploration areas we will emphasize our classic UWSN at large with an extensive coverage and those making use of already mature acoustic technology as *underwater acoustic networks* (UASNs). UASNs sometimes get extended beyond using sensors and actuators to include generic underwater communications for voice, and other telecommunication services. We call them *underwater acoustic networks* (UAN) to help with manned platforms and vehicles that have direct communications as well as connecting sensors and other devices through acoustic channels through the water.

Although underwater exploration opens up unique commercial and industrial opportunities for rapidly advancing sensor applications, we do not see much need for a swarm style networking of smart sensors. We do see much use for isolated sensors, clusters, bundles, platforms, and small-scale UWSN systems connecting scattered clusters equipped with agent style technologies, which we call connectivity for remote operations. In order to enhance the access to remote places most commercial and long-term projects make use of platforms, discussed in Section 9.4. Most autonomous underwater vehicles (AUVs)[9] accommodate a whole range of sensors, communications, and associated devices for special operations in the distant waters of the seas and oceans.

In general smart sensors require *three basic but key technological capabilities* to be accountable for building a UWSN: (a) communication, commonly wireless usually with the ability of an ad hoc style interconnect

[9] There are many abbreviations used in the robotic industry today, which could confusion in one's mind. In order to simplify our use of the terms we harmonize the book for two main abbreviations: (1) UAV for the unmanned aerial/air vehicle and (2) autonomous underwater vehicle (AUV).

for penetrating towards the point of interest; (b) computing as a lightweight intelligence as required for autonomy, house keeping, self-maintenance, task persuasion, quality control, and often with some kind of performance optimization; and (c) energy independence as an ability to support energy hungry communications and computing with a lengthy life span while functioning to their full capacity, without being dependent on short lived batteries or other power devices. Low energy smart sensors normally operate at their minimum energy consumption to help with the low cost mass market positioning, but this may not be feasible for underwater smart sensors.

For the communication aspect we first look at the commonly available acoustic propagation technology through the waters. Although acoustic is a good match for long distances of water, it suffers from one fundamental handicap in that low frequency carriers mean low bandwidth and associated long delays. We then look at other communication methods and examine their suitability to be used as an alternative technology for an underwater environment.

For the computing aspect we briefly look at the internal processing requirement of agent style with the limited autonomic capabilities of smart sensors for agile advanced UWSN applications. It is easy to see the vast amount of small but important processes at various levels of sensing, clustering, gateways, and networking but we need to add to them to make them to be efficient in designing viable new products. Building confidence with security and sustainability for a remotely operated vehicle such as AUV is essential and should not be compromised. Some of these capabilities have been generalized and already discussed in conjunction with overlying functions in Chapter 7. We discuss the energy aspects of smart sensors and their use in the UWSN in various parts of this chapter, but we address it in Chapter 11.

Now, in order to visualize these and other technical aspects of using smart sensors in UWSN we provide two sets of technical specifications for typical applications. Table 9.1 is for exploration and industrial applications, and Table 9.2 is for the habitat and the environment.

Table 9.1

Technical Specifications for using UWSN for Exploration and Industrial Applications

USAGE	ENV.	DEPTH	SENSING	COM.	NET.	SIZE
Industrial exploration (water)	River & Ocean	Meters to kilo meters	T, S, P, D	Acoustic	4D	Small
Industrial exploration (floor)	Ocean	1-10 kilo meters	Scan	Acoustic	3D	Small
Pipelines & Cables	Pool River Ocean	Meters to kilo meters	Vib Leak Localize	RF & Acoustic	2D 3D 4D	Small
Fish farms	Pond River Off-shore	Meters to kilo meters	T, pH, N, NH, Tur, O2, L, etc.	RF RF Acoustic	1D 2D 3D	Small & Large

Table 9.2

Technical Specifications for using UWSN for Habitat Applications

USAGE	ENV.	DEPTH	SENSING	COM.	NET.	SIZE
Water quality (man made)	Pools & reservoirs	Meters	T, S, P, D, L,O_2,N, EC, NO_3, pH, ..	RF Acoustic, RF	4D	Small
Water quality (natural)	Rivers, Lakes, Shores	Meters to kilo meters	Tur, pH, T, O_2, EC, etc.	RF & Acoustic	2D 3D 4D	Large
Marine life	Rivers, Lakes, Shores	Meters to kilo meters	T, P, S, V	Acoustic	3D & 4D	Small & Large
Marine life	Deep shallow waters	Meters to kilo meters	T, P, S, V, Tur, L, photo, etc.	RF & Acoustic	3D	Small & Large
Marine life	Ocean	kilo meters	T, P, S, V, Tur, L, photo, etc.	Acoustic	3D	Small & UAV

Abbreviations used in the sensing columns of Tables 9.1 and 9.2 are defined as follows: C= camera, Comp= magnetic compass, Cur= current, *D*= depth, EC= electrical conductivity, *H*= humidity, *L*= water level, NH_3-N= ammonium nitrogen, NO_3^- = nitrate, NH_4= ammonium, O_2= oxygen, Photo= light, *P*= pressure, pH= potential of hydrogen, *S*= salinity, *T*= temperature, Tur= turbidity, *V*= velocity, Vib= vibration.

9.2.1 Acoustic Propagation Technology

Today's traditional underwater systems make use of acoustic technology following da Vinci's notion of listening to the *echo of sound through water*.

As the water is still the same, the quality of connectivity should be mainly judged by the two main factors of *distance* and *unit distance loss of the signal*; this means that whatever technology we discover would have a huge potential for success due to the scale of applications. Now, we use acoustic technology as our *de facto wireless choice* for underwater communication as it is for the UWSN applications. Although in Section 9.2.2 we briefly discuss alternative propagation methods.

For a basic transmission of a signal from a transmitter '*A*' to a receiver '*B*' through the underwater environment a link between these two points is established. The link is normally complex consisting of a few significant components, commonly called multipath. The multipath could vary extensively from one location to another depending on the surroundings and the receiver's capability to receive those signals. If devices at *A* and *B* are installed and fixed and the surrounding environment stays steady, then multipath components remain constant, and it is easy to receive clear data. But any changes to these due to their movements or a rapidly changing environment then continuous variable phased components of the receiving signal have variable strengths, commonly called fading. For example, in Figure 9.5 we have a simplified multipath propagation model, where on the top of the direct path we have two additional sets of reflections, one from the surface of water and one from the seabed indicating a three-ray multipath reception. In many cases at the cost of dealing with the fading problems of multipath reception the stronger signal reception would help with communication capabilities (i.e., longer distances). But for deep waters and far away signals data reception fully relies on the direct path resulting in weaker data and therefore shorter distances.

For designing a UASN we have to consider four partly interrelated contributing parameters of: (a); speed (b) loss; (c) propagation environment; and (d) delay.

Due to the fact that acoustic waves travel through water based on the micromechanical waves principles, the energy transfer creates some molecular level vibrations resulting in two complementary effects. The first effect is that the temperature of the media governs the speed of propagation, and the second is that it consumes some of the acoustic traveling wave energy into so-called absorption loss.

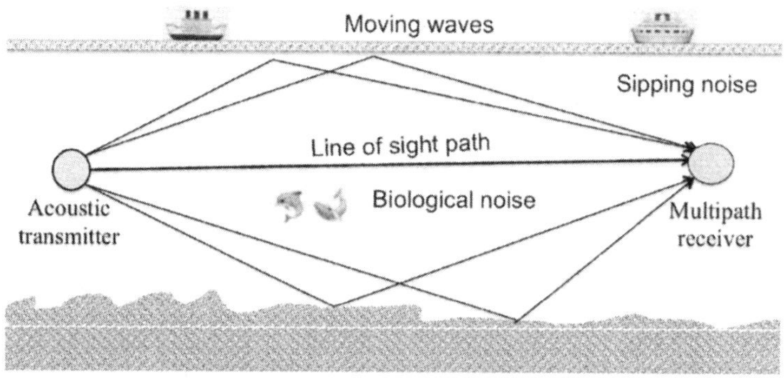

Figure 9.5 Three-ray model of an underwater multipath propagation.

Now, for developing a model for the speed of transmission we need to consider at least three effective interrelated variables, namely temperature, depth, and salinity.

In effect such layers in the oceans and deep-sea water temperatures are controlled by natural phenomena called *thermoclines*. Not far below the surface of the waters the temperature begins to drop, more rapidly in the tropical and warm terrestrial environments than in locations in cooler climates. The drop continues with the depths reaching a few degrees of centigrade. In frozen waters the temperature increases before declining again adjusting itself to the common deep waters' temperature. Figure 9.6 shows a simplified line indicating three main layers of underwater thermocline, called high latitude for very deep (below 1700m), mid-latitude (500–1700m), and the highly variable seasonal latitudes beneath the surface (above 500m) [6].

That is, we can use the thermocline-controlled temperature variation of the water to estimate both properties of the propagation speed, namely $Speed(T)$ and $Speed(S,T)$. One remaining significant property is the depth, where the density of the water increases improving the propagation speed. Using the summing up model of the components we have the following as our speed model,

$$Speed(T,D,S) = Speed(T) + Speed(D) + Speed(S,T) \quad (9.3)$$

where, $Speed(T)$ for its temperature component, $Speed(D)$ for the depth also called pressure component, and $Speed(S,T)$ for salinity component with the following values:

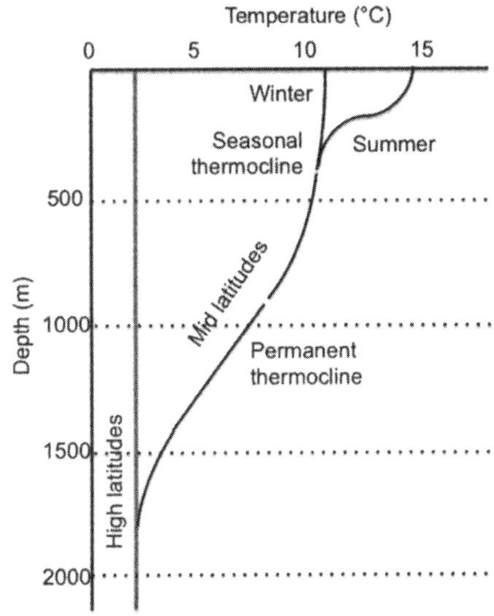

Figure 9.6 The temperature versus depth of water.

$$Speed(T) = 1449.05 + 4.57 \cdot T - 0.05 \cdot T^2 + 0.00023 \cdot T^3 \quad (9.4)$$

$$Speed(D) = 16.3 \cdot D + 0.18 \cdot D^2 \quad (9.5)$$

$$Speed(S, T) = (S - 35) \cdot (1.333 - 0.0126 \cdot T + 0.00009 \cdot T^2) \quad (9.6)$$

Where, speeds of the waves are in km/s, T is water temperature in degrees of Centigrade, D is the depth of point of measurement in km, and S represents water's salinity weight in *part per thousand* (ppt) [7].

In Figure 9.7 we have three areas of the acoustic rays propagation controlled by the gradient of the speed profile on the left: (a) the set of three rays leaving the upper transmitter located in negative gradient region continuously bend their paths further towards the middle (clockwise); (b) the single ray in the middle is located in the minimum speed region and is trapped traveling up and down in a small area dictated by the original angle of transmission, which is also called SOFAR channel; (c) the set of three rays leaving the lower transmitter located in positive gradient region continuously bend their paths further towards the middle (counterclockwise) [8].

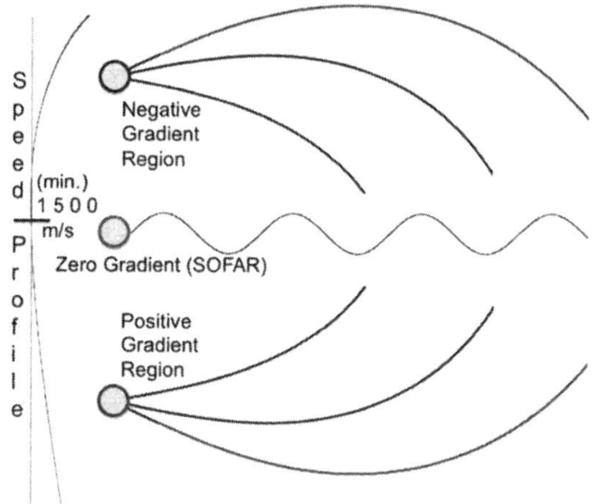

Figure 9.7 Three main areas of curved propagation rays: upper negative gradient region makes the rays continuously change their direction (clockwise), middle SOFAR ray, and lower positive gradient region makes the rays continuously change their direction (counterclockwise).

As shown in Figure 9.7 the minimum velocity oceanic water layer, also called the SOFAR channel (SOund Fixing And Ranging) and scientifically known as isothermal, is used for carrying special signals for straight long distances that can be used for special signals such as using the comprehensive nuclear-test-ban treaty organization (CTBTO) signals using international monitoring system (IMS) frequency signals. Being connected to a global network of seismic, infrasound, radionuclide and hydro-acoustic sensors with the hydro-acoustic part of the IMS using hydrophones deployed below the buoys in the SOFAR channel connected to shore stations by cables providing energy and data connectivity. Usual communications are organized for 1000 km carrying data at 100 bits/s [9].

For estimating the path loss of acoustic rays there are some commonly used models. It is naturally frequency dependent. Figure 9.8 shows the absorption rate in dB/km against any carrier frequency, f, as a good approximation for narrowband signaling. Without going into detail we present the graphical values of $a(f)$, as the depth related underwater *absorption rate*, where the upper envelope of all curves should be used for the design purposes [10].

Figure 9.8 shows six absorption curves. The main base line for the absorbed energy is seismic natured *shear viscosity* line. Then, an additional 4.67dB/km shift is added to the theoretical shear viscosity to get to what we normally observe in practice with the new parallel line of *structural absorption rate.*

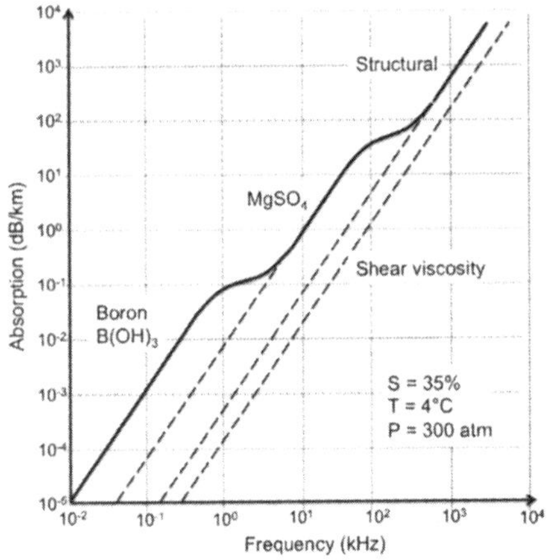

Figure 9.8 Changing absorption loss rate against the carrier signal frequency.

As this absorption drops sharply for lower frequencies two new components appear to become nonnegligible. For lower frequencies, a particular natural property of the deep waters, namely *magnesium sulfate* and *boric acid*, due to molecular relaxation phenomena become effective adding extra absorption loss of 10 dB/km each, to acoustic signals at all frequencies below 1 kHz and 60 kHz, respectively. An empirical curve for the envelope can be given by:

$$a(f) = \frac{0.1 \cdot f^2}{1+f^2} + \frac{40 \cdot f^2}{400+f^2} + 2.75 \cdot 10^{-4} f^2 + 0.003 \tag{9.7}$$

Finally, in order to ease up the design we add a linear approximated absorption to the family that can be used for quick guesses and in-the-field engineering works.

Now, for developing a model for the propagation loss through the water we consider the three effective variables of absorption, distance, and transmitting media. The effects of the transmission media can be included using two additional variables K and A for the geometrical condition and additional losses (noise and interferences), respectively, as,

$$Loss(d, f) = d \cdot a(f) + K \cdot \log(d) + A \tag{9.8}$$

The first factor K is associated with the transmission path, a factor representing the way signal is being geometrically propagated while A represents additional losses in the field due to surrounding random noises in the media (e.g., thermal and shot noises injected by devices, systems, and circuits), impulsive noises (e.g., manmade, biological, and other media related). For example, due to three-dimensional freedom in the middle of oceans' deep waters a higher dependency can be allocated using a value of 20 for this factor, while in the shallow waters two-dimensional propagation a smaller value of 10 would be more adequate.

In order to evaluate the bit error ratio (BER) of an acoustic propagation link we mainly need to consider the coupling alignment of transmitter-receiver, the loss, and unwanted interfering signals. For the loss we need to include the water absorption and geometrical disturbing signals mentioned earlier (9.3), rewritten for the attenuation and then normalized to the transmitting power density, $P_t/\Delta f = 1$, $\Delta f = B$ for narrowband signaling bandwidth, as[10]

$$Att(d, f) = d \cdot a(f) + K \cdot \log(d) + N(f), \tag{9.9}$$

where, K is the transmission media factor between10 to 20. The signal-to-noise-ratio is normally written in linear scale of powers that can be written as,

$$SNR = \frac{P_r}{P_n} = \frac{P_t/Att(d,f)}{N(f)\cdot B} = \frac{1}{Att(d,f)\cdot N(f)}. \tag{9.10}$$

where, Δf indicates the narrowband channel's bandwidth and $N(f)$ indicates the total effective acoustic noise density at the receiver. The overall noise power effective at the detecting point of the receiver then, in dB, becomes the sum of all effective noise densities as,

$$N(f) = N_t(f) + N_s(f) + N_w(f) + N_{th}(f) \tag{9.11}$$

Here, we have included typical noise components of: (a) turbulence noise at $10 \cdot log(N_t(f)) = 17 - 30 \cdot \log(f)$; (b) shipping noise at $10 \cdot log(N_s(f)) = 40 + 20(S - 0.5) + 26 \cdot \log(f) - 60 \cdot log(f + 0.03)$ with S being the shipping factor between **0** and **1**; (c) wind driven wave noise at $10 \cdot log(N_w(f)) = 50 + 7.5 \cdot \sqrt{w}) + 20 \cdot \log(f) - 40 \cdot$

[10] The power for acoustic and sound signals and associated noise elements use pressure, Pascal (re Pa).

$\log(f + 0.4)$, w is the speed of surface wind in m/s; (d) and thermal noise at $N_{th}(f) = 20 \cdot \log(f) - 15$ [11,12].

Figure 9.9 shows some narrowband SNR plots for acoustic channels using (9.10) for K=15, no wave noise, and S=0.5, reflecting a moderate shipping activities.

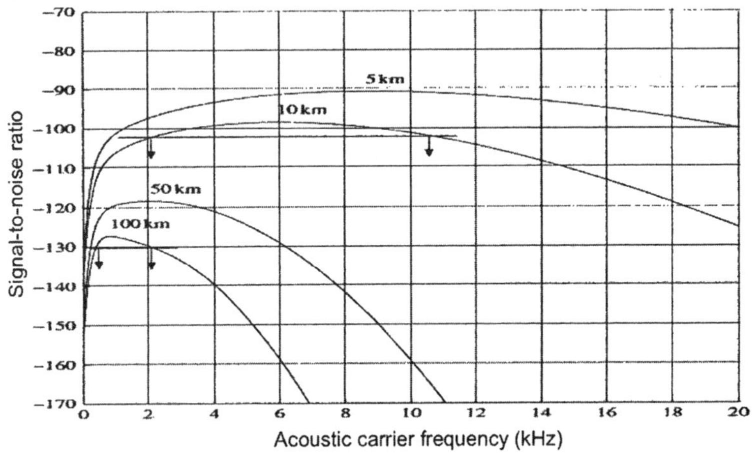

Figure 9.9 Typical narrowband SNR for acoustic propagation rays. For 10 km, e.g., for 5 kHz carrier we have practical bandwidth for a receiver with –102 dB sensitivity [36].

As for signal processing and other functions such as coding, modulation, filtering, and equalization most of the terrestrial proven methods can be effectively used when adjusted for lower frequencies of acoustic, with rich literature already used for sound, sonar, and similar systems such as hydrophones. For antenna design, beamforming, and array pattern efficiency refer to Chapter 6.

Two main devices traditionally used for UWSN systems are *hydrophones* (acoustic microphones) and *sonar* (acoustic scanners).

Hydrophones in the field function as an acoustic modem and are used for transmitting (often) powerful acoustic signals and receiving them with some local data and signal processing capabilities. Due to the long wavelengths of the acoustic carrier frequency all components such as antenna, amplifier, filter, and detector are quite large requiring a large body as a unit to cover them. The structure of their circuit-equivalent signal processing functions modeled very similarly to terrestrial wireless systems, using similar and often simpler models for modulation, filtering, and beamforming.

Sonar (SOund Navigation And Raging) is very similar to classic radar operating in water using acoustic frequency signals for detection. Due to its search feature sonar needs to be able to make use of antenna directivity in the form of scanning. In traditional systems the scanning directivity provided by mechanical movements of the antenna but better use of computing and low power VLSI has enabled dynamic beamforming. We have two types of sonar products. Passive sonar scans by just listening to the receiving signals, constructing images, and identifying the objects.

This type is particularly useful if operated free from heavy acoustic noises in quiet and remote places. It also features zero-interfering and low detectability if required. However, it suffers from weak reception and low range of operation especially if objectives do not generate or reflect sufficient signals or if detection and localization are for devices and vehicles that do not under generate any suitable signals.

Active sonars on the other hand due to their use of reflected signals have three clear features: (a) refection of sonar controlled signals makes the detection process much more efficient, (b) receiving own generated signals provides further efficiency and naturally comes with estimating the distance range (for this reason they are also called echo ranging sonars), and (c) better accuracy of the speed estimation (doppler shift). Active sonars are disadvantaged due to (a) dependency on reflect-ability of the target surface, (b) injecting unwanted interfering signals in the media for other devices, and (c) being easily detectable [13].

Another important problem with using high power acoustic communication systems for underwater applications is its impact and direct interference with the marine life. As Figure 9.10 shows our mid-frequency and high frequency range signals fall in natural communication bands for marine species, and as a results we are not only polluting their living environments we are actually deafening them.

For most marine mammals, sound is their main source of information and communication with their living environment including echolocation listening. Due to some mechanical coupling between their swim muscles and hearing organs many species have developed a good sense of hearing capability with optimum sensitivity between several kHz up to one 100 kHz, which makes them vulnerable to artificial acoustic modem signals.

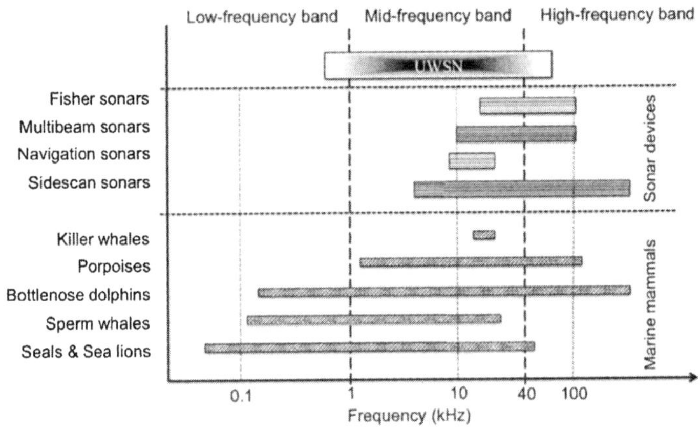

Figure 9.10 Frequency range of acoustic signals that fall in natural communication bands for marine species, drawn based on [14].

The main problems have been associated with sonar systems operating in 1–10 kHz. For example, many studies on this frequency have demonstrated that such signals can affect marine mammals by changing their normal behaviors [16]. National marine fisheries services (NMFS) has been investigating these effects and has reported that 160 dB (re μPa) creates noticeable behavioral disturbances and that 180 dB can cause injuries in their auditory systems usually, expressed as temporary threshold shifts (TTSs) and permanent threshold shifts (PTSs) [14].

9.2.2 Other Potential Propagation Methods

As discussed in Section 9.2.1 our commonly used traditional acoustic propagation technique as the main choice for UWSN applications shows weaknesses due to some shortcomings in some practical aspects including: (a) very high propagation energy loss; (b) limited bandwidth due to the low carrier frequency nature; (c) very long propagation delays; and (d) possible harmful radiations for sealife. Therefore, before we pay more for this underwater propagation choice of technology we should invest in exploring other known (RF, MI, OP) and unknown methods. The known methods such as RF and OP have been tried but not explored rigorously. Some fundamental research and investigation for these two methods is urgently required. The MI method is new, very different in nature and requires much more work to release its possible potentials. The following is just a simple discussion for their potentials.

RF

Using terrestrial WSN (TWSN) style radio frequency (RF) or electromagnetic (EM) propagation networking of today shows many limitations for UWSN applications. This, however, is not as serious barrier as it sounds if we are able to adopt it to a specifically designed architecture (e.g., accompanied with some chains of short distance relaying system).

As always various economical and quality factors force us to extend the TWSN connectivity as far as feasible into the waters. If GPS is used then for a mix technology solution, some islands enhanced with communication platforms plus a few mobile AUVs could enable the required connectivity through a synchronized relayed channel; this is a practical, good solution. In order to be able to support the new 5G for long distances and remote access, such as the control or localization of some objects or vehicles under water, we need to use a complex scenario; it would be a serious challenge to seek an object in the middle of a deep ocean. The use of platforms, for example, could help with the access and deployment of a complicated task, where a complex configuration of radio/air and acoustic/water relayed through various interconnection stages makes sense but cannot be counted as a good design solution [15].

For RF horizon we do not see any signs for radical innovations but we can try incremental ones, which in turn may guide us towards bigger steps. A few examples are discussed in the following.

Work from Uribe and Grote, for example, provides a unified simple analytical model that can be used as a design tool to explain the near field conflicting issues measured, analyzed, and simulated it for a new model, where we can see they can be used for long distance EM transmission up to 100m under various conditions and how the loss-rate, after the huge loss of over 120 dB in the first 20 meters, began to ease up. With around 130 dBm we seem to still have some residual of 5 dB over the low steady light noise to push data through (i.e., in many cases performing well for various underwater low data rate applications [16]). Another interesting work investigates the practical possibilities of low bit rate monitoring systems for *n* new 5G standards using a 500 MHz carrier; we can easily get data rates close to 500 kbit/s for a distance of 100m in the water [17]. Also, for under underwater examination, inspection, and repair AUVs are becoming the best option, where using short distance RF radars would help with such applications using a frequency of 2.4 GHz for low at this frequency salinity impact and low-cost product and components. Figure 9.11 shows a block diagram of the system with its radar frontend, an underwater antenna, an 8 dBm directional coupler, and a six–port interferometer [18].

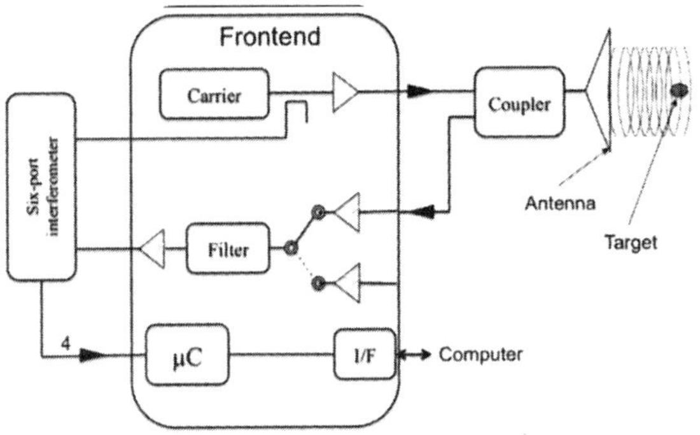

Figure 9.11 Block diagram of a robotic underwater radar [18].

Generally speaking we can see that the main bottleneck of RF propagation in underwater environment comes from the water interface design. We therefore recommend that much greater effort be allocated to this research area, as an alternative to the out-of-date ideas of acoustic, as we are desperately trying to save energy, protect sea life, and trigger new applications for the underwater environment.

MI

Another alternative propagation is the magnetic-field induction (MI) technique. Using MI for moderate distance underground communication is discussed in Chapter 10 (Section 10.4.2), which can give us a good idea of its usefulness for the UWSN applications as well. The main differences in UWSN technology are (1) longer distances, (2) easier mobility, (3) node localization, and (4) easier supply of energy. However, the same degree of deployment of extensive potentials is to be discovered. The following brief description of some published cases may help with planning to deploy this technology, as in most cases some basic and applied research efforts may be required.

Coming with the promising advantages of low propagation delay and less susceptibility to the transmission environments MI channels have been studied for a decade for shallow waters with omnidirectional (1D) antennas then being extended to tri-directional systems (3D) using MIMO with a more rigorous analysis of transmitting data over field to perform under the working scenario of 3D coils, complex path of reflections, and heavy absorption in water. Figure 9.12 shows the generation of 3D propagation. Performance figures for the study have been calculated based on 10 dBm transmission power, noise level –120 dBm; for a distance-to-coil ratio of 400 the system can provide some 1 kbps data for up to 20m distance in a conductive salty water. Under better conditions the distance could be increased to over 100m [19].

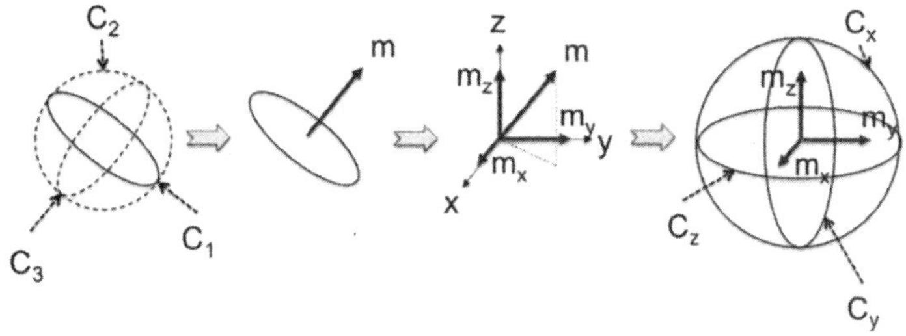

Figure 9.12 Conversion of 1D dipole to 3D propagation [19].

With the similarity of wet clay and seawater from the propagation point of view with a coil radius to 0.34m and current up to 2A, using the error rate of data using a simplified case of BSPK stays acceptable as the number of turns in the coil increases continuously. For example 240 turns on the coil can provide a communication distance of 80m. In the similar way as size of the coils in the waters are much easier than ground and clay the improvement for UWSN would be far easier to claim in practice [20]. A similar analysis comes from Silva and Moghaddam, where in their search for low power optimum carrier frequency selection they find a set of working figures including 26 kHz for 40% VWC (volumetric water content) operating at 30m [21].

An interesting review for underwater magnetic channels is from Akyildiz et al. showing the potential of MI comparable with optical at around 1 Mbit/s with ranges up to 100m [22].

Optical

Optical propagation technology, normally in the form of FSO (free space optic) ray, has been showing a promising performance in various environments; however, due to its limited applications so far it has not been adopted widely for WSN technologies. Optical communication through volumes of water signifies its own distinct evolutionary life and unique properties based on its sequence of natural developments point out that we should seek a way to include it in our search for our new UWSN design solutions.

With a promising bandwidth and the highest traveling speed with optical technology we should explore the possibilities further. For example, for low cost low power we can use LEDs and photodiodes for moderate data rates of several Mb/s for short distances (up to 100m) and even better if laser is employed. Practically, the light beam intensity loss of the signal traveling from one point to another in water varies with the wavelength of the light λ as given by the attenuation A expressed as,

$$A = e^{-k\cdot(x_2-x_1)} \cdot \left(\frac{x_2}{x_1}\right)^2 \tag{9.12}$$

where k is a wavelength dependent as $k = a(\lambda) + b(\lambda)$ with $a(\lambda)$ being the absorption coefficient and $b(\lambda)$ the loss factor associated with the geometry between x_1 and x_2 including the losses due to the dynamics of the water such as turbidity. The light absorption coefficient in water varies some with the content but extensively with λ with a minimum at λ=420 nm (UV) of say around $5x10^{-5}$ $5 \cdot 10^{-5}\ cm^{-1}$ for increasing linearly through the colors to IR (800 nm) to just over $0.01\ cm^{-1}$ [23].

9.3 REMOTE SENSING

Let's start with a very fundamental concept of using WSN technologies for underwater environments, which leads us to "generic UWSN" without thinking of the underwater signal transmission part. There we have an inexperienced researcher, who has been excited about *smart homes* and is now being confronted with a design case for a UWSN application scenario. If unguided the result could be random, and considering the confusion surrounding this researcher's mind we should not allow this to happen. Due to the long distances involved generic networking does not apply to UW. For designing an application the only practical forms of UWSN are either well spread wide area networking style UASN or remotely activated clusters at a very special small place. Obviously, the use of autonomy and basic artificial intelligence technologies, employing some widely available computing technologies such as agents, pervasive, dynamic, and opportunistic computing in conjunction with low power acoustics and other promising new underwater transmission technologies, should be used to build new technological applications to develop new breakthrough systems for these untouchable waters.

That is why instead of starting with complex classic networking UWSN we begin with remote access. This may help engineers and educators to reach directly towards finding solutions for designing required applications for smart sensors, instead of cracking the complex classic networking of generic UWSN.

Therefore in this section we divert our fundamental search from finding viable UWSN style solutions to accessibility and remote sensing, to penetrating bundles of sensors into the earth waters. That in real terms means the potentially over two thirds of the surface of the earth is being wasted by being covered by water, leaving humanity unable to access these vast, untouched natural resources and to enhance our knowledge of the evolutionary life on earth.

Accessing and controlling underwater sensors and associated devices via the internet does not belong to the dream world any more. Anyone should be able to maneuver an AUV in the remote spots of the ocean, monitoring and scanning suspicious locations. More useful applications such as locating and cleaning up debris, toxics, metals, and plastics could become new hobbies for many people [24].

Remote sensing using fixed platforms features reliability and low maintenance costs, and is easy to operate; however, due to some practical factors it may not be as useful for UWSN applications including: (a) location suitability as not all platforms can be placed in the right location for a UWSN design: (b) cost of running could be high due to business aspects of the platform; (c) logistics, where due to superseding priority of the main operation of the platform UWSN performance could be compromised. High priority cases that require sensor rich applications such as disaster and security become exceptional.

In general, fixed platforms come in a range of sizes from very large to smaller ones. Larger ones normally appear in docks, ports, and special coastal establishments. They can be used as an access point for UWSN applications. Smaller ones may appear in special places near seafloor mineral resources or special lengthy projects.

Mobile platforms, however, in floating ships or submarines could become particularly useful if modernized sufficiently to operate for the purpose. That is, classic AUV systems are regarded as a mobile platform to appear in the underwater technological arena in two distinct forms: (1) lightweight, agile, mobile extensions of robotics to form a platform or part of it where its tasks are defined and specialized to carry a dedicated sensing activity and (2) extensions the reach of fixed platforms for access and investigation. In practice its mobility and specialized design purpose is to convey tasks normally in very harsh and difficult locations for some operations.

Further classification of *AUV* can be as: manned underwater vehicles (MUVs); classic AUVs (due to control and drifting problems they can work with or as annexes to ships, fixed, or larger mobile platforms); and remotely operated vehicles (ROVs), which are more sophisticated systems with sufficient capabilities to ensure they can be safely controlled at a distance.

Their general applications include: oceanography (sampling, floor oceanographic mapping, geothermal), environment (various pollution and hazards, hydrocarbon spills, radiation leakage, environmental monitoring, inspection of pipelines and other underwater structures), ocean mining and oil (survey and resource management, maintenance and undersea construction), and other applications (mine search and disposal, underwater fishery ship and nuclear power plant inspections). Their robotic functions are regarded as prime requirements, which make them particularly useful for UWSN remote sensing applications.

As for the size, nowadays AUVs can be produced in much smaller sizes than the common ones used in special storage ponds such as nuclear storage. As demonstrated by Watson and Green, a small UWSN consisting of a few spherically shaped nodes with a target diameter between 130 mm and 200 mm of their μAUV can map a pond the size of an Olympic swimming pool moving at a speed under 1 m/s energized by a rechargeable Li-Ion battery pack with a capacity under 100 kJ [25].

Autonomous underwater gliders are lightweight and practical members of the unmanned AUV. As an emerging field of underwater vehicles they glide through the waters at low speeds using their buoyancy to move up and down and using the wings to keep up and push the vehicle forward. Their low cost design features long-duration oceanographic sensing and associated scientific missions for collecting and delivering data to the land or satellite stations. Their common design features include: (1) an aerodynamic shape with fixed wings and tail measuring around 50 kg 2 and meters in length; (2) capability for lengthy oceanographic observations and sensing missions; (3) control of their pitch and roll by internal mass movements; (4) balancing through pumping oil between external bladder and internal reservoir; and (5) drop weight capability for emergency cases [26, 27]. Figure 9.13 shows the main movements of a glider.

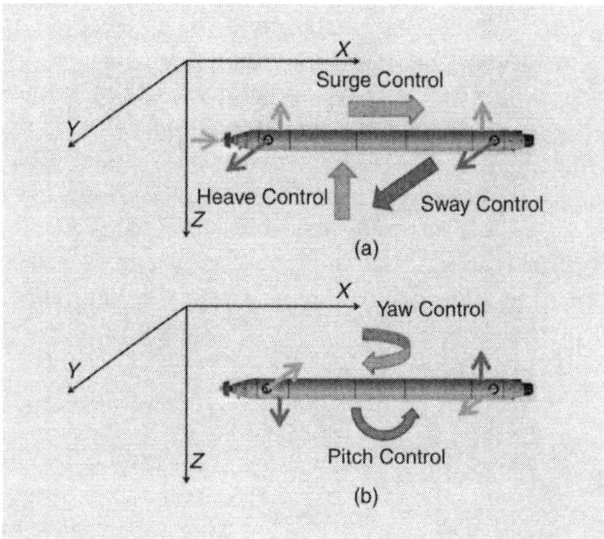

Figure 9.13 Directional thrust arrows for controlling the glider: surge (x), sway (y), heave (z), pitch, and yaw [27].

9.4 ENERGY

Due to underwater natural cooling condition for UWSN, sensors and other devices, their work in an underwater environment makes it seem as though they should come with some form of self-sufficiency for their energy. This is not true but is an assumption due to their traditionally being used as an integral part of underwater vehicles. However, under today's shrinking trends devices and systems should cover their own energy requirements. Reasons includes: (a) low cost UWSN systems and products need energy to function; (b) poor underwater access makes fueling and recharging costly and often impossible; (c) batteries have a very short life in underwater environment; (c) energy dependency is a weakness for wireless devices working in remote places; (d) autonomous systems and smart devices require far more energy than passive ones; and (e) the global market today requires "low energy" solutions, meaning that without energy considerations new UWSN design should fail in seeking a global market position. We mention one special case below but direct interested readers to the energy part of Chapter 11 where industrial applications of WSN are discussed.

Buoys, for example, originally used in navigation and industrial applications, now exist in a wide range of shapes and sizes all over the water and can be used as a basic platform for useful applications. Apart from their possible uses for UWSN applications, if equipped with self-energized smart sensors they could play significant roles in the future. Their unique positions of floating on the surface provides the possibility of making use of "wave energy" to enable significant access gateway for UWSNs. One interesting solution is from Symonds et al. who bundle three or more buoys for a basic design of generating around 50W of power to cope with high-power acoustic or EM communications down to the UWSN environment [28].

9.5 UWSN APPLICATIONS

We understand and make use of our knowledge to gain control over our technological developments through a responsive set of objectives to maintain our overall quality of life on earth. Considering that water covers 71% of the earth's surface, its volume of 1.386×10^9 km^3, and its average density of 1.025 g/cc, water dominates our living environment. Here, we discuss the importance of using advanced smart sensors in an underwater environment and its associated water industries[11] including rivers, shallow waters, deep underwater, and the floors of seas and oceans.

In order to cover our smart sensor needs, we can divide our applications into three main geographical categories of shallow waters, deep waters, and seabeds with special attention to the needs for underwater farming. The following points plus two following subsections, each can be divided further into three levels of urgency in the form of *immediate, short-term, and long-term industrial developments,* answering the question of "*why we need UWSN*?"

We have an ever-growing list of reasons to believe that further development of UWSN technologies in some areas are urgently needed. One is the rapid habitual change in the waters due to continuously rising temperature. Combined indicators are: (a) ongoing global warming; (b) surface water acidification; (c) rising water levels; (d) habitat losses; and (e) spread of invasive species due to immigration to foreign regions in search of colder climates [29].

[11] Sensors and UWSNs for their economical operations almost always compliment other industries to serve.

Another ongoing area in the use of sensors and sonar devices in is the *bathymetry* and associated scientific explorations, which are extremely important for oceans, seabed and storing projects. As the list of applications is too long to discuss here we look at a family-tree style classification as shown in Figure 9.14.

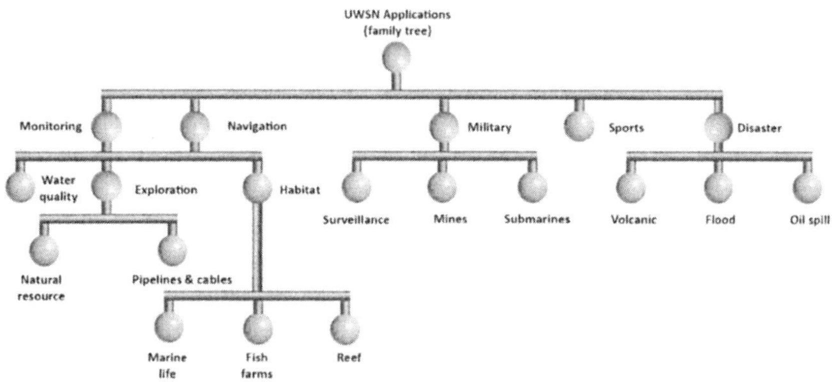

Figure 9.14 Main classification of smart sensor applications of UWSN [30].

9.5.1 Shallow Water and Coastal Floods

Some people relate depth to distance from the shore for rough guesses but estimates should not be used due to their element of risks. For industrial activities we take water levels up to 50m deep as our nominal level as shallow waters. It corresponds to 1 km with a very easy slope of 2.867 degrees. Under such shallow-water demarcation this category includes many areas such as major rivers, small lakes, water sport complexes, fishing farms, and water reservoirs on land plus all shores of the many seas and great oceans. This part of the environment to industry means bringing in unique start-up industrial opportunities for making the most of our emerging smart sensing technologies [31].

Shallow water safety concerns have been growing in recent years to become one of major concerns of today and therefore they require urgent attention. With so many lives at risk one can analyze the possibility of saving lives and associated economies using UWSN technologies due to: (a) vulnerability of ever expanding buildings for dwelling and industrial in the coastal and riverside areas; (b) pollution of rivers and shallow seas; (c) daily increase of risks on human lives in densely populated areas; (d) exhaustion of marine fisheries; (e) man-made water disasters (some dams built to control stream waters are now poorly managed and create flooding and man-made local draughts); and (f) unpreparedness against increasing recurrent of natural hazards (such as tidal waves, coastal winds, typhoons, and tsunamis). For example, tsunamis, addressed in Section 9.1.3 do not only involve the waters but also shores and coastal land areas, and thry pose great danger to life. This is particularly significant due to the fact that traditionally, riverbanks and seashores were selected as a starting point for large cities enabling sea trading as well as a natural supply of seafood. Then, as populations have been growing steadily, even exponentially under industrial activities, today many of them are being seriously threatened by life and health including dangers such as toxins, global warming, flood, and hurricanes. Figures show coastal areas accommodate about two-thirds of the world population (including those living within 60 km of the shores) [32].

As discuss in Chapter 10 the source of most large-scale fatalities is unpredictable underwater volcanic eruptions due to tectonic plate movements. In order to capture the most immediate uses of UWSN eruption oriented floods we should look further into the fatality scales of water flooding due to rising seas, winds, and storms.

Floods are short-term water overflows onto the seaside dry land that also can occur to lakes, rivers, streams, land fish farms, and canals. For example in 2015 flooding accounted for 40 percent of disasters in Asia-Pacific regions, costing the nations some $11.5 billion USD (i.e., 25 percent of their total economies).

Water flooding, the largest (over 90%) fertility contributor in the 1950s (see Figure 7.11) has not been repeated so far mostly due to better housing, infrastructure, and preparedness but this is by no means the end of it. That is, we still have very little control over its source. It is therefore very disturbing to our normal living conditions and impacts the financial side of quality-of-life, and industrial independence has not been significantly reduced. Supporting our earlier problems of high populations living near the shores losses increase further. As indicated in the "BP12" conference [33], we have too many cities and towns lingering on the edge of the shores, mostly expanding rapidly due to depletion of small villages and growing seaside industries. The OECD predicts due to climate and socioeconomical changes by the 2070s around 150 million people will be exposed to natural disasters caused by coastal flooding. The predication lists the five top nations as China, India, Bangladesh, Vietnam, and the

United States, where the first three will have around 10 million exposures each due to the climate change. The first four will have measures of around 10 million exposures each due to socio-economical disasters. There are 136 port cities around the world with over one million inhabitants. Most of these largest port cities are found in Asia (38%) where many (27%) are located in deltaic conditions subject to higher coastal flood risk and land subsidence [34].

We have a *tropical cyclone* as a rapidly rotating windstorm, low-pressure at the center circulating at the surface with spiral movements and thunderstorms and it produces heavy rains. Dependent on the location and strength it can be referred to by different names, for example, *a hurricane* in the Atlantic Ocean and northeastern Pacific Ocean but *a typhoon* in the northwestern Pacific Ocean while being called *tropical cyclone* and *severe cyclonic storm* in the south Pacific and Indian Ocean. *Cyclone* indicates the circular movement, clockwise in the Southern Hemisphere but counterclockwise in the North, typically between 100 and 2,000 km in diameter. Due to being generated from ocean's warm water the coastal regions are particularly vulnerable to tropical cyclones. As discussed in a United Nations' report we have a total of 86 tropical cyclones in a year with 48 per cent of them in Asia and the Pacific, the coastlines affecting nine million people and causing $11.8 billion USD of economic damage [35]. The amount of lifesaving and comfort needed in all the above places is so significant that, as mentioned earlier, with many technological UWSN solutions we should encourage and help the local industries to save the day.

9.5.2 Deep-Water Applications

It is interesting to see most existing UWSN applications fall into similar configurations as we have for TWSN cases. This is understandable as all industries are trying to extend what they already see as a successful technology extending off land over onto the water. This is particularly visible for shallow waters, with most industries only take on low risk penetration into water projects missing many exciting, potential, and long-term viable UWSN applications including valuable minerals and significant scientific and geological explorations opportunities. This factual truth becomes especially important and sometimes critical for advanced industries with limited national shores preventing them from starting new projects and missing out on low cost opportunities. Missing opportunities for UWSN applications are too many as deep waters and underwater floors provide the most valuable industrial opportunities. Examples are too many to count ranging from environmental observations and monitoring geological processes to monitoring sea lives including fishes, microorganisms, and mammals. Other deep-water applications including ocean-floor mineral extractions, and monitoring underwater establishments are exciting projects not to be missed [36].

In general underwater monitoring makes up well over 90% of UWSN projects, which can be categorized into four areas of toxics, hazards, leakage, and radiations. For example, the carelessness of industrial toxic problems can be indicated by the garbage issues that Evan-Pughe claims is because "our throw away culture" of the last six decades needs urgent attention. For her case she looks at the Pacific Ocean as the *Great Pacific Garbage Patch* between California and Hawaii, as it contains 140,000 tons of floating plastic being washed into the sea from rivers and sewers. Cleaning up such a huge amount of scattered garbage and continuous dishabitation of the waters sounds like the tip of the iceberg. That means we have a great task waiting for the USWN to search, locate, and destroy or pick up such debris before facing the near future's new residuals of man-made disasters [37].

9.6 SPECIAL CASE: OFFSHORE OIL AND GAS IINDUSTRY

Due to the global mismanagement of coordinating forces towards a greener and safer world we have entered a cycle of crossfire political upheavals for some 40 years, during which signs of global warming appeared. Since then not only has the use of nonrenewable energy did increased but with the rising global price of oil, by employing new technologies, such as satellite imaging, the oil and gas industries have been able to explore new wells on the oil rich floors of the waters. Now with the aid of smart sensors the oil and gas crises are being turned into the biggest legal and illegal business opportunities ever seen in the history of mankind.

The UWSN technological solutions as part of integrated monitoring and management systems are best achieved in the form of remote access under medium size networking. Availability of a wide range of smart sensors and associated devices and equipment makes such integrated solutions viable and effective. The integration is achieved through gateways and communication channels that can share the bandwidths for reducing the cost and making use of legacy, new, and highly reliable delay tolerant networks (DTNs).

The thriving key investment area across the oil and gas supply chain industries includes the production plants, refineries, petrochemical plants, exploration systems, pipelines, and transportation mostly at the verge of a new expansions. Such expansions are possible due to the provision of trustworthy, secure, and reliable wireless communications being used at the core of UWSN systems, enabling a new range of cost effective automation and control solutions, which would not have been possible before using traditional, wired systems.

These new multidisciplinary technological developments should encourage closer collaboration between users, application experts, hardware designers, and software developers providing more efficient systems to support more financially viable and secure UWSN industries.

Figure 9.15 shows a typical configuration to support fundamental operations for using the UWSN, mostly equipped with autonomy to control all the required technological and geographical functions from basic exploration to the final stage of production.

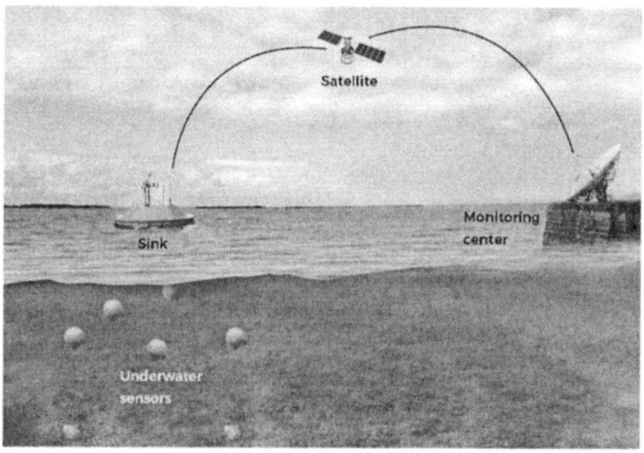

Figure 9.15 Typical configuration setup for using UWSN for oil and gas operation – from exploration to production.

9.6.1 Offshore Remote Production

Due to ease of accessibility and the cost effectiveness of extendable coastal operations at all required stages, a near to the shore oil and gas mining industry helps to maintain low cost production. As mentioned earlier, however, most rich and valuable fields are located further away in remote, and often deep spots that are not just inhabitable but are hard for humans to visit making operations difficult, and often very expensive.

As mentioned in Section 9.3 the remote sensing capability of UWSN upon specially designed platforms can help extensively with many underwater projects. In such cases most of the operations makes significant use of embedded autonomous functions, as a set of routine tasks, enabling remotely controllable simplified instructions. For simple industrial operations these can be translated into monitoring, complemented by some selective choices due to specific circumstances such as environmental changes in reaction to hazardous conditions, or extreme cases such as very high temperatures exceeding a given threshold or declining below another threshold. Equally, due to the changing demand speeding up or slowing down the process mainly by monitoring relevant subprocesses.

That is, a remotely enabled UWSN can provide a wide range of capabilities for oil and gas companies to meet their challenging new marketing demands, financial conditions, regulatory aspects, and production levels. Their typical technological capabilities as in remote monitoring oil and gas production include: (a) pipeline integrity monitoring; (b) tank level monitoring equipment; (c) condition based monitoring (CBM); (d) pipeline pressure relief valve monitoring (PRV); (e) refineries pressure relief valve monitoring (PRV); (f) wellhead automation; and (g) monitoring effects of offshore marine environment. The last set of functions is related to corrosions requiring intensive maintenance, and is therefore regarded as a group of tedious and labor-intensive tasks.

Without the use of remote control technologies due to the very high maintenance cost and expensive parts all traditional oil and gas production platforms are under pressure to maintain a high rate of production and they expect to operate continuously for maximum productivity. Where, with real-time data acquisition and monitoring using UWSN systems, we can enable the working platforms to: (a) cut costs; (b) comply with regulations; (c) significantly cut down times for maintenance; and (d) eliminate unexpected failures of the tools and machineries.

9.6.2 Seabed Operations

The oil and gas industrial underwater floors operational monitoring tasks can be divided into two stages. One is for exploration, drilling, and installation stages where usually a temporary dedicated platform enhanced with some mobile AUV systems is set up to coordinate the initial work. The other one is for the production and transportation stage where usually a lengthy network of transportation pipelines is established for a smooth and lengthy production.

The efficient movement of natural gas from production locations to consumption points requires extensive and elaborate transportation systems. In many cases, natural gas produced from a particular well will have to travel great distances to reach its point of use. Therefore transportation systems for natural gas usually consist of a complex network of pipelines, see Figure 9.16 [38].

One serious problem with natural gas is leaks. If manually operated, some gas detecting equipment regularly needs to be examined by experts checking for leaks.

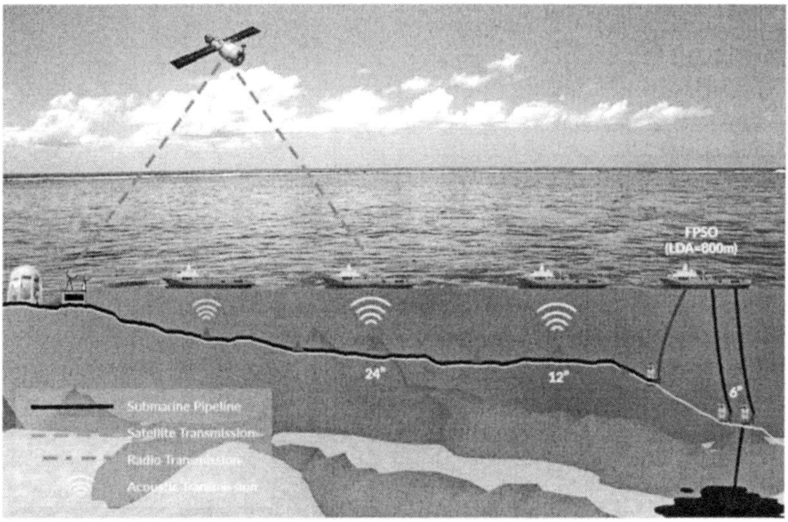

Figure 9.16 A typical deep-water monitoring system using logistic-support vessels.

This task is particularly important to ensure the product (gas) is not odorized. Under the existing well-matured acoustic enabled remote sensing technologies (UASN) such regular monitoring can be easily operated using modern AUV systems.

Figure 9.16 shows a deep-water monitoring system using logistic-support vessels equipped with a suitable set of underwater sensor networks.

Brazil Oil

One interesting business case is related to a new national oil company. In the past few decades some oil producing countries were keen to pursue their self-sufficiency in oil production, which would not succeed without recent technological advances in oil exploration and operations in deep waters.

Brazil has come up with numerous discoveries in the oceanic continental shelf, in an area as large as 70 km off shore, with water depths ranging from 120 to 2800m and is now generating a national production of oil and gas of up to 2.376 million barrels daily. The exploration and operation in the deeper parts of the ocean (e.g., water depths below 2000m), however, has been very challenging and is demanding the use of new technologies.

Therefore, new communication techniques should be used to obtain information on the subsea infrastructure. The activities of this industry are complex and dangerous, requiring a robust and reliable infrastructure to withstand the harsh and inhospitable conditions.

At somc stages the work went far beyond their strengths, limiting the operation in high seas with extreme slopes and exposed submerged structures presenting great instabilities to the operations.

This demanded constant monitoring as the traditional methods limited the observation points and excluded most of subsea equipment. UWSN technologies enabled monitoring of the full extent of the subsea infrastructure, and particularly in subsea pipelines, providing an operational control increase through constant equipment conditions and position checking, monitoring, and verification of submarine displacements that could damage structures in the marine environment.

Communication in these circumstances is subject to several limitations, including losses in the transmission channel, low bit rates, and large transmission delays. However, further developments of a new architecture based on underwater delay/disruption tolerant networks resolved the problem.

The underwater monitoring system consists of acoustic sensors, platforms, and logistic support vessels. The acoustic sensors are responsible for calculating the position and storing and transmitting the information obtained from subsea equipment. The sensors are distributed over the submarine infrastructure and subsea pipelines. Logistic-support vessels of oil exploration are responsible for the collection of messages generated by these sensors and subsequent routing to the control center.

Underwater acoustic sensors can be installed over the submarine pipelines for monitoring pressure, temperature, and flow and positioning control, see Figure 9.17. Vessels are equipped with a GPS system and radio communication, and in many cases, a satellite link and their routes match the subsea pipelines, becoming an appropriate option to capture sensed data. The information is generated and stored in the sensors until some vessel is available for message reception, as shown in Figure 9.18 [39].

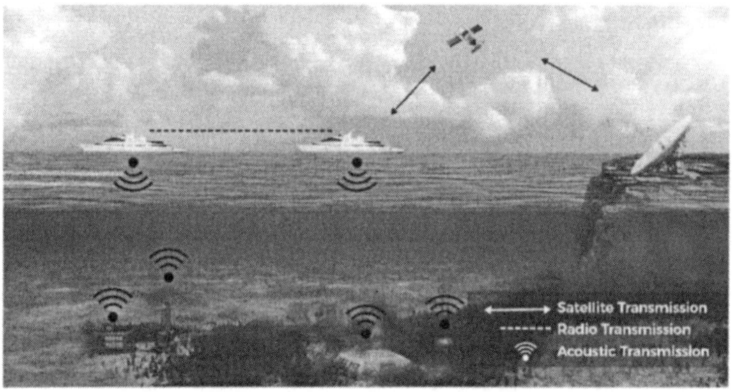

Figure 9.17 A typical deep-water monitoring system for laying down new pipelines.

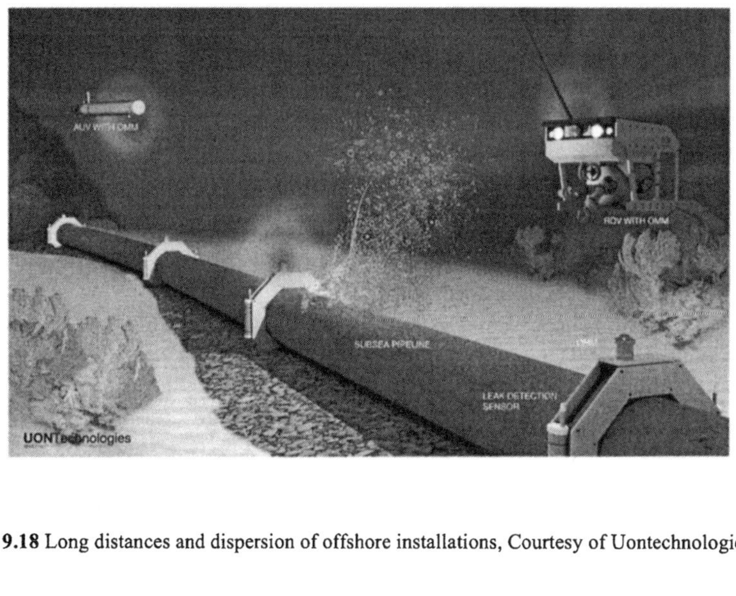

Figure 9.18 Long distances and dispersion of offshore installations, Courtesy of Uontechnologies [40].

Another interesting case is related to imported crude oil to the United States. Domestic production declined in the 1970s and the crude oil imported from Mexico, Venezuela, and the Middle East is mostly denser with higher sulfurs, so refiners invested in further processes.

Then recent Canadian production changed supplies and posed tremendous implications for the US national pipelines (transmission pipeline of 55,000 miles, 90% national) so that many of them had to change their flow directions reversed to transport Canadian crude oil, which is much denser for transportation and has to be diluted (bitumen). This made pipes vulnerable to release. Figure 9.19 shows statistics of the failure cause [41, 42].

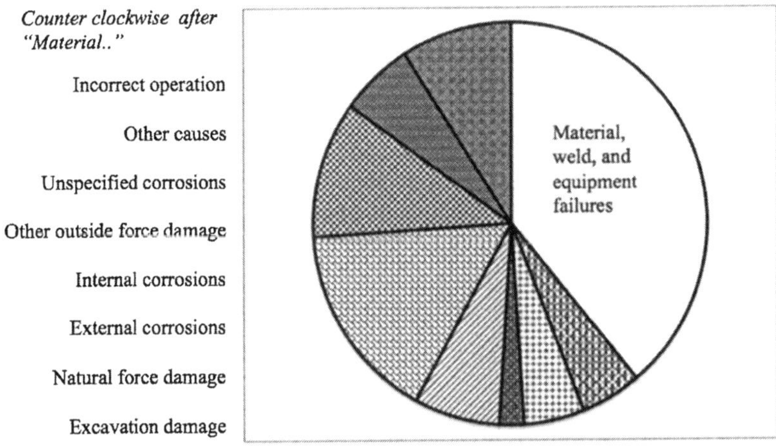

Figure 9.19 Failure cause contribution of diluted bitumen for transmission pipelines, drawn based on [41].

9.7 CONCLUDING REMARKS

Our critical view of the progress to date accompanies some sample solutions for the development of new smart sensor technologies for this extreme environment. We have observed and commented on a wide range of technological advancements and recent achievements for using them in the form of lightweight UWSN applications. Unfortunately the developments have not yet made sufficient progress for significant paradigms, which means that a large number of opportunities are missed. Those being overlooked are (a) a proper use of new computing and propagation technologies and (b) a full adoption of technological innovation as a process. Some persistent business bottlenecks are also present in many key development areas including poor public awareness and a lack of basic scientific breakthroughs. A lack of understanding of their uses for saving life in water related disasters, and better uses of the underwater environment are also matters of concern.

Although we have covered many details of using acoustic propagation of data through water, we believe many more alternative methods should be given attention to release their full potential.

References

[1] Jonkman, S. N., Loss of life estimation in flood risk assessment, Ph.D. thesis, Delft University, The Netherlands, 2007.

[2] Abt S. R., R. J. Wittler, A. Taylor, and D.J. Love, Human stability in a high flood hazard zone, *Water Resources Bulletin*, Vol. 25 No.4, pp. 881–890, 1989.

[3] National Academy of Engineering, NAP Report 2016 on "Frontiers of Engineering: Reports on Leading-Edge Engineering from the 2015 Symposium," http://www.nap.edu/21825.

[4] The Japan Society of Mechanical Engineers, Report of the JSME Research Committee on the Great East Japan Earthquake Disaster, "Lessons Learned from the Great East Japan Earthquake Disaster," 2014.

[5] United Nations Development Programme (UNDP), "Sustaining Human Progress: Reducing Vulnerabilities and Building Resilience", Human Development Report 2014.

[6] Thermocline – Wikipedia.

[7] Urick, R. J., "Sound propagation in the sea," Defense Advanced Research Projects Agency (DARPA) 1979, ADA319320.

[8] Cruz, N. A.(Ed), "Autonomous Underwater Vehicles", *Intech* 2011, ISBN 978-953-307-432-0.

[9] Papadakis, J. S., Underwater Acoustics, *2nd Underwater Acoustics, Conference and Exhibition (UA2014),* 22nd to 27th June 2014, Rhodes. Greece. ISBN: 978-618-80725-1-0.

[10] Dyer, I., "subject notes for 1385: Fundamentals of Underwater Sound Applications," Dept. Ocean Eng., Massachusetts Institute of Technology, Cambridge, MA, 1993.

[11] Lobo, V. J., PhD thesis, 'ship noise classification', Lisbon 2002.

[12] Santos, A. B., 'Inferring ocean temperature variations from shipping noise,' 2nd Underwater Acoustics, Conference and Exhibition (UA2014).

[13] Bartberger, C. H., 'Technology of underwater sound received noise.' Lecture Notes on Underwater Acoustics, 1965.

[14] Yang, Q. L., et al., "EFPC: An Environmentally Friendly Power Control Scheme for Underwater Sensor Networks," Sensors, 2015, www.mdpi.com/journal/sensors.

[15] Janssen, T., 'Localization in Low Power Wide Area Networks Using Wi-Fi Finge,' 2017.

[16] Uribe, C., and W. Grote, "Radio communication model for underwater wsn," *New Technologies, Mobility and Security (NTMS), 2009 3rd International Conference on*, December 2009, pp. 1–5.

[17] Loret, J., et al., "Underwater Wireless Sensor Communications in the 2.4 GHz ISM Frequency Band", Sensors 2012, 12, 4237-4264; doi: 10.3390/s120404237, www.mdpi.com/journal/sensors.

[18] Sporer, M., et al., "Underwater Interferometric Radar Sensor for Distance and Vibration Measurement," IEEE, WiSNet 2015.

[19] Guo, H., et al., "Channel Modeling of MI Underwater Communication using Tri-directional Coil Antenna." IEEE, 2015.

[20] Huang, Q., et al., "BER Analysis of BPSK based Magneto-Inductive Communication System in Clay Channel," IEEE Computer, UIC-ATC-SalC-CBDCom-IoP, SWC, 2015.

[21] Silva, A. R., and M. Moghadam, "Operating Frequency Selection for Low-Power Magnetic Induction-Based Wireless Underground Sensor Networks," IEEE, 2015.

[22] Akyildiz, I. F., et al., "Realizing Underwater Communication through Magnetic Induction," *IEEE Communications Magazine*, November 2015.

[23] Anguita, D., et al., "Optical Wireless Communication for Underwater Wireless Sensor Networks: Hardware Modules and Circuits Design and Implementation," IEEE, 2010.

[24] Goldberg, K. 'Beyond Webcams, Online Robots (Remote sensing),' MIT 2002.

[25] Watson, S., and P. N. Green, "Design Considerations for Micro-Autonomous Underwater Vehicles (μAUVs)," IEEE, 2010.

[26] Skibski, C. E., "Design of a Autonomous Underwater Glider focusing on External Wing Control Surfaces and Sensor Integration," Master of Science, Florida Institute of Technology, Melbourne, Florida, December, 2011.

[27] Spear, A., 'Icefin' 2017. 10.1109/MRA.2016.2578858: A. Spears et al., "Under Ice in Antarctica: The Icefin Unmanned Underwater Vehicle Development and Deployment," in *IEEE Robotics & Automation Magazine*, Vol. 23, No. 4, pp. 30–41, December 2016.

[28] Symonds, D., et al., "Low-Power Autonomous Wave Energy Capture Device for Remote Sensing and Communications Applications," IEEE, 2010, 05617902.

[29] Demertzis, K., L. Iliadis, and V-D, Anezakis, "A Deep Spiking Machine-Hearing System for the case of Invasive Fish Species," 978-1-5090-5795-5/17, European Union, 2017.

[30] Felemban, E., et al., ' Underwater Sensor Network Applications: A Comprehensive Survey,' 2015,896832, http://dx.doi.org/10.1155/2015/896832.

[31] Chizhik, D., A. P. Rosenberg, and Q. Zhang, "Cohérent and differential acoustic communication in shallow water using transmitter and receiver arrays," *OCEANS'10 IEEE SYDNEY*, Sydney, NSW, 2010, pp. 1–5.

[32] Livi-Bacci, M., 'A Concise History of World Population,' 5th Edition, Wiley-Blackwell 2012, ISBN: 9780470673201.

[33] Bally, Ph. (Ed), "The International Forum on Satellite EO and Geohazards", the Santorini Conference, Santorini, Greece, 21–23 May 2012.

[34] Pickering, M. D., 'The Impact of Future Sea-Level Rise on the Tides,' PhD Thesis,, National Oceanography Centre Southampton, 2014.

[35] UN, 'Specific Hazards: Handbook on Geospatial Decision Support in ASEAN Countries,' ESCAP, 2017.

[36] Heidemann, J., 'Underwater sensor networks: applications, advances and challenges,' Phil. Trans. R. Soc. A (2012) 370, 158–175.

[37] Evans-Pughe, C., 'Can we engineer our way towards cleaner oceans?' Engineering & Technology February 2017, www.EandTmagazine.com.

[38] Ribeiro, F. L., A. C. P. Pedroza, and L. H. M. K. Costa,'Deepwater Monitoring System Using Logistic-Support Vessels in Underwater Sensor Networks,' in Proceedings 21st International Offshore (Ocean) and Polar Engineering Conference--ISOPE-2011, Maui, Hawaii (Vol. 2, 327–333).

[39] Ribeiro, F. J. L., A. C. P. Pedroza, and L. H. M. K. Costa, 'Underwater monitoring system for oil exploration using acoustic sensor networks,' *Telecommunications Systems*, Vol. 58, Is. 1, January 2015, pp. 91–106.

[40] Uontechnologies, http://www.uontechnologies.com/about.php, retrieved online December 2017.

[41] Transportation Research Board and National Research Council. 2013. TRB Special Report 311: Effects of Diluted Bitumen on Crude Oil Transmission Pipelines. Washington, DC: The National Academies Press. https://doi.org/10.17226/18381.

[42] National Academies of Sciences, Engineering, and Medicine. 2016. Application of Remote Real-Time Monitoring to Offshore Oil and Gas Operations. Washington, DC: The National Academies Press. https://doi.org/10.17226/23499.

Chapter 10

Underground Applications of Wireless Sensor Systems

This chapter presents the requirements and recent developments of underground smart sensing systems and their applications, with particular attention to the need to establish new industries for adopting theses future advanced solutions. Potential applications are plenty with a wide range of opportunities varying from (a) better understanding of the crust, its behavior and properties safeguarding our lives from natural disasters; (b) better understanding of the soil and near surface properties of the crust as the foundation of dwellings, industrial activities, and for valuable resources.

10.1 INTRODUCTION

In order to enable new technologies to thrive in their quest to convert the existing dirty, dark, inaccessible, dangerous, and highly complex underground (UG) environment into a somewhat smart, and partly visible environment, we need to accomplish two complementary technological breakthroughs in the three critical areas of sensing, computing and communications, at the same time staying committed to operating under the forces of the global open market rules of the near future.

The need to enhance our underground systems with some agile computing capability requires the use of lightweight, low power, and agile processors in the nodes, cluster-heads, and other parts of the *wireless underground sensor networks* (WUSNs). In most cases the speed is not as critical as other features such as energy-efficiency, autonomy, and some selective touches of "intelligence" to ensure they perform to their basic requirements. Another urgently required development area is in signal propagation. So far we have three main candidates: (1) the traditional electromagnetic (EM) waves with some serious weaknesses against many materials including water, (2) the new magnetic field signaling platform in the form of magnetic induction (MI), and (3) enhanced seismic wave technology. Developing specialized smart sensors requires new kinds of sensors to provide us with (1) new underground visibility, (2) penetrating capability, and (3) enabling some controls to safeguard us from the moving crust.

Opportunities are plenty including some to address our burning need for survival, some to prepare us for ever-threatening natural events such as earthquakes and some seeking to develop more effective methods for preventing their effects. Under these new WUSN enabled systems we should be able to advance our agricultural processes to enable the production of better food for a better quality of life. We hope our new underground smart sensors industries will fulfill their fundamental role and demonstrate these three basic capabilities: (1) communication as required for penetrating, clustering, and interconnecting devices and systems for the required coverage of the media, (2) self-sufficiency and basic capabilities for managing a device's resources such as energy and interfacing, and finally (3) a sufficient degree of artificial intelligence, as required for taking smart decisions.

The rest of this chapter is organized as follows. Section 10.2 looks into the requirements to justify developments of small, medium, and large scale WUSN technologies. Section 10.3 looks into studying the crust for understanding, analyzing its behavior, and predicting earthquakes. Section 10.4 looks into WUSN key technologies. Section 10.5 scans through potential underground service solutions. Finally, Section 10.6 provides a brief insight into the monitoring of mines.

10.2 UNDERGROUND SENSING

It is our natural desire to understand and make sense of our knowledge about the earth, and to gain control of our technological developments on earth to give us a responsible set of objectives for maintaining the quality of life on it. As a responsible, intelligent species of evolutionary life on earth, we need to be able to take on responsibility for life on earth. Therefore, to secure our extension of life on earth, and for our children and future generations, we need a better understanding of the crust and its limits in sustainability, maintenance, and usability features. In order to ensure the effectiveness of our advanced dwelling and technological developments we need to agree on a set of unified objectives for "advancing humanity," which could begin with gaining full control over our *underground technological developments.*

For example, we need to know about the weak points of the crust, the quality of the soil, and its characteristics related to land sliding, water-holding, and seismic behaviors. The strength and conditions of the surface should be analyzed as to their suitability for different developments and uses such as foundations for buildings, agriculture and the construction of new mines, roads, tunnels, underground storage, and foundations for heavy structures, bridges, and high-towers. In brief, the three most desirable areas of development are: (1) watching the crust; (2) watching the ground and promoting industrial activities; and (3) promoting global agriculture and farming.

10.2.1 Watching the Crust

We are sure the question of *where we live* has crossed everyone's mind? The simple answer of "*on the surface of the crust*" can escape the imagination of one trapped in the comfort zone of a cozy office or walking on a crowded street. The earth's crust, the layer that we call it home, is thick and strong to our scales but very thin when compared with other geological aspects of the earth. The crust varies extensively from 5 km in the oceans to 55 km in the high mountains (e.g., in the earthquake prone plateau of Iran the crust thickness varies from 30 to 55 km). With a simple calculation for a 20 km average crust thickness and earth radius of 6378 km we get a *volumetric ratio* percentage of $vr = 0.94$. That is, a crust containing less than 1% of the earth's volume is safeguarding us from the roaring high-temperature magma governing the inner earth. This amazingly thin skin must be sufficiently strong to support us in life [1].

Physically, crust is three times heavier than water, divided into two sublayers of *upper crust* and *lower crust*, composed of different materials with two densities of 2.715 and 3.125 g/cm^3, respectively. This thick and solid layer is strong enough to make us feel secure, but at the same time it is only a relatively thin skin with some weaker points.

As far as the safety of the crust is concerned we can do very little at this stage but watch and listen using our sensors, and be prepared. Building a multidimensional (3D+) map of the earth complete with the history of the movements of all the critical plates would enable us to make the right decisions in an emergency, such as the impact of rising atmospheric temperature on the crust.

In order to understand the crust and its behavior better, we have already some tools in place for monitoring its activities at the surface and just beneath the top layer. These tools, however, are insufficient to enhance our knowledge for significant impacts. The crust is vastly unknown and can in no way be counted as a smart media to satisfy our *smart sensing* objectives. Our behavioral studies for further down below the surface are few, and mostly sporadic. Studying the seismic waves to measure the movements of underground natural forces and their impacts on the crust has been progressing rather too slowly.

10.2.1.1 Enhanced Satellite Observation

The existing satellite systems (as a long-term investment) are able to collect for observatory purposes high quality images, which can be analyzed, and used in measuring the surface of the crust. One natural use is the daily recording of events and watching the changes in their footprints. *Observation* for monitoring the earth is what satellites do best, and we can easily share them with all earth observation systems (EOS) and services to enable them to watch and analyze the changes. EO images are usually sufficient for detecting slow movement and visualizing any surface deformations of the crust.

Integrating EOS, with special global positioning stations, can also provide more accurate information. Such systems can be further enhanced using extended complementary ground and underground measurement using various sensors. For example, the EOS of NASA comprises a series of satellite missions and scientific instruments in the Earth orbits, designed for long-term global observations of the surface. Enhanced using powerful imaging devices and empowered by local and remote supercomputers these can detect every minor change and predict any significant forthcomings. This program (EOS/NASA) started about 20 years ago, and others such as the Committee on Earth Observation Satellites (CEOS) join a wide range of industrial and humanitarian applications including old mines and quantifying the soil [2, 3].

10.2.1.2 Global Warming

So far we have clear deductions regarding air pollution and global warming issues. Monitoring the steady rise of average atmospheric temperature from –5 degree Celsius in 1978 to 55 in 2008, indicates a significant annual rise of 2 degrees a year over thirty years. A simplified plot for the above measurement, Figure 10.1, prompts a significant message that not only raises concerns about upcoming global warming, but also possible side effects as a whole on the surface, or deeper parts of the crust [4, 5][12].

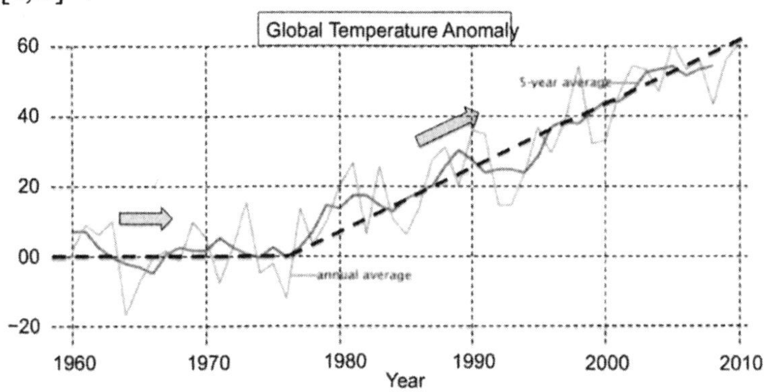

Figure 10.1 Global-warming trend for the period 1960–2010, drawn based on [4].

That said, all agree with the temperature rise but none have yet considered the impact of increasing temperatures on the crust. That is, we know very little about the crust's behavior and its reaction to persistent high atmospheric temperature. There are many unanswered questions to answer. Would the high temperature soften or crack the crust on very dry lands? Should a cracked or soften crust lose its strengths (e.g., in the dried out deserts or at the thinner spots of the oceans)? We therefore need some massive WUSN setups to help us to identify any developing significant changes.

10.2.2 Watching the Ground and Promoting Industrial Activities

Watching, observing, and monitoring the ground to make sure our industrial activities do not interfere with the natural behavior of the ground, also helps with designing successful products and services (see Section 10.5) and developing a feasible future WUSN paradigm.

---

[12] NASA figures do not include recent movements towards a toxic free world such as the rapid replacement of traditional coal burners with solar energy in the USA, Europe, and other places. However, the TAC 2017 report of 'doubling down on climate progress' indicates they won't significantly change the trend.

This can be divided into three areas of long-term observation, safety, and industrial sustainability. This may allow us to address all of our essential underground monitoring needs including watching the old mines, sustainability of new mines, and seismic-aware construction of mines, tunnels, and other confined spaces. It is important to monitor important natural and man-made places such as historical caves, and underground reservoirs, in addition to watching the further release of radioactive materials, toxic gases, wastes, pollutions, hazardous explosives, and other unhealthy or dangerous activities. For example, there have been many industrial incidents at a global level representing the bitter truth of growing cumulative injuries and long-term illnesses. This highlights the future impact of improper management.

10.2.3 Promoting Agriculture and Farming

Recent advancements in sensor technologies and animal care, and the decreasing cost of automated machineries in conjunction with the surprising rise of food prices at the international levels, has initiated a trend towards a new agricultural paradigm for self-sufficiency at national and local levels. The SME style agriculture and farming industries under *back to the nature, back to farm* may revive once again abandoned farms and this will work towards stimulating life in small villages.

Due to two critically important factors we would like to discuss the importance of local and independent agricultural needs: (1) the importance of interactive processes and dynamic functions that WUSN can offer new infrastructures (i.e., agricultural applications) and (2) the availability of abandoned and vastly reuseable data.

The need for the development of smart agriculture for better living counts as an essential requirement[13]. Astonishingly, despite so much agricultural technology being available today, we hear of isolated unrest caused by poverty[14] and shortages in overgrown cities for new migrants from abandoned villages. That is, despite some statistical figures showing that *the overall global production exceeds global consumption*, there are many populations who still taste the bitterness of hunger and dependency.

Nowadays low cost sensors and a progressive internet are offering new viable superior solutions for developing and reviving virtually all small

[13] Many of us may have already forgotten historical starvation periods caused by natural disasters and man-made troubles such as devastating wars.

[14] We may need to refine term "poverty" as relative and comparative.

and medium sized agricultural units as a way of life free from being trapped in the smoky, over priced large and over-populated cities. However, traditional self-sufficiency and hobby based farming and agriculture has no room for a professional approach to compete with the growing multinational organizational food supply chain industries. We therefore need to introduce a few new definitions to help with a better understanding of the above reform.

Smart agriculture and farming (SAF) can be defined as a new trend, and as a core process to make better use of smart sensors and associated technologies, as a primary requirement of reinventing new farms enhanced with lightweight networked computing to enable the small agricultural farmers a high ROI under low financial obligations.

In practice, we have the possibility of extensions to SAF (see Table 10.1) by adopting more specific scenarios such as centralized agriculture and farming (CAF), distributed agriculture and farming (DAF), and cooperative agriculture and farming (KAF), and by setting up new small and medium size agricultural and farming (SMAF) enterprises. Table 10.1 highlights the basic features of WUSN enabled choices for reviving SMAF industries. Further services and technical details supporting the concept can be found in Chapter 17 of [6].

Table 10.1

Comparable Specifications for Five Members of SAF Family

Technology	Availability	Technology complexity	Maintenance req.	Five year paradigm	Long term paradigm
SAF	Abundant AG sensing	Very low	Low	Yes (>5k)	Yes (>2m)
CAF	Advancing UG sensing	Maturing	Progressive	Yes (>1k)	Yes (>0.5m)
DAF	Various UG communication	Complex	Various (Scenario)	No	Yes (>0.25m)
KAF	Computing (SW) abundance	Mostly established	Fully automated	No	Yes (>10k)
SMAF	System's cost Product distribution High	Low	Moderate	Yes (>3k)	Yes (>1.25m)

10.3 MODELING THE EARTH

Following our earlier discussions on acquiring a better understanding of the earth, mysteries of the "moving crust" as the main source of earthquakes, and associated natural disasters, we require some global models to help our knowledge base so we therefore begin with the Pangaea concept.

As the scale of the work we are dealing with is a long-term ongoing project, it enables us to build a tomographic image of the Earth using natural seismic waves called the *global seismic-wave tomographic model* (GSTM)[15]. GSTM requires ongoing lengthy detailed measurements and an assessment of the Earth involving seismic wave detectors, accelerometers, and sensors in a massive scale distributed all over the earth.

10.3.1 Pangaea, the Moving Crust

Under our existing technological capabilities we can do very little to manipulate, control, or impose any meaningful impacts to the crust but in order to be prepared to safeguard our lives, we need to expand our knowledge and understanding of it. So far our technological capabilities have been observation of the surface, and with some images, of the inner earth. Deductions include the discovery of major changes in a 3D earth image, one dimension for the radius and two for the stratum surface or layers. These collected data are vital for explaining changes on the ground and inner earth such as landslides, coastal subsidence, and (other) tectonic processes. Figure 10.2 shows earthquake casualties for 1900 to 2012. For example, the International Forum on Satellite Earth Observation and Geohazards Conference identifies in their 35 earthquakes recorded during 1900-2012 as many as 1m casualties. The 1,500 severe volcanic eruptions in the last 12,000 years is called *Holocene Era* [7, 8].

Looking at the structure of the earth we have the mantle as the layer directly below the crust where the magma, a hot melted rock, leaks towards the surface through weak points of the crust, assuming the crust is made up of large plates, causing earthquakes, volcanic eruptions, tsunamis and other unpredictable natural changes on earth, then known as *Pangaea concept*.

[15] GSTM is named as a futuristic cooperative network encouraging all seismology databases to be integrated for the good of humanity.

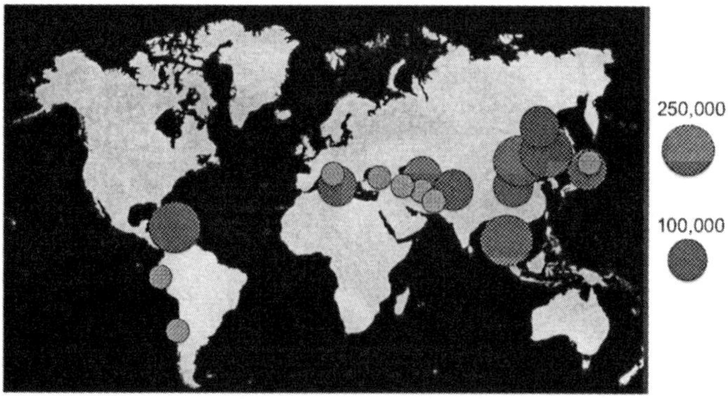

Figure 10.2 Earthquake human casualties for period 1900–2012, drawn based on[7].

Pangaea is an old theory claiming that some 200 million years ago the earth had only one very large bulky continent but due to some interactive forces (e.g., core magnetism) it gradually started to break up into pieces, which are moving away from each other. Then, in the course of time, the earth's continents reached their existing locations (see Figure 10.3).

It is easy to imagine that this theory has helped our geoscientists to discover *today's continental plates*, and to observe the issues where their broken edges do not join together [9].

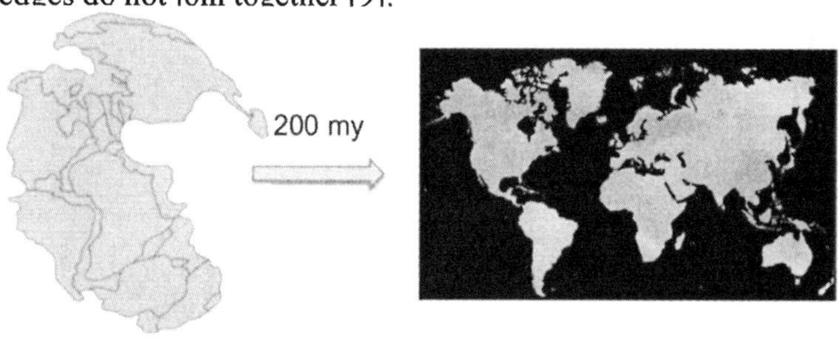

Figure 10.3 Pangaea theory indicates that our main continents have been moving over last 200 million years, Drawn based on [9].

As the above theory is gaining ground, and our sensors and devices are deployed measuring more accurate movements of the continents, we discover more.

10.3.1.1 Measuring the Move

The developing models are complex but we may get a better idea of the process using a simplified example.

Let's assume that these plates have been moving with the same speed for over 200 million years. We can then estimate the average speed of movement today, based on having the Africa and Euro-Asia plates stuck together while each one of other three plates, namely North America, South America, and Australia, have moved about a quarter of the Earth's circumference, *D*, in 200 million years in time, *t*, then a simplified calculation for an average movement of the plates gives us 2.5 cm per annum[16]. This figure may not look very significant, but due to its persistent and widely distributed nature counts as the *main* source of the earthquakes on earth, which initiated the rediscovery of the plates and extended works leading to our futuristic advanced crust monitoring models, generically called GSTM.

10.3.1.2 Reference Points

In a geodetic monitoring system of the crust, a set of points is established covering a region of interest. The movement of these points is measured using the relative distance to another point, called a reference point. So, in order to obtain the real state of crust movements or its displacement in the monitored area we should select some reference points with stabilized measurements throughout time (i.e., a measure of sensitivity).

10.3.2 Global Seismic Tomographic Model

Tomography by definition means visualizing the inner construction of an object, normally three-dimensional, using information carried by waves traveling through or inside the object. For construction of GSTM we use *seismic waves carried information for collecting and reconstruction of the Earth.*

What is a *seismic wave*? The wave part of the term means propagation of a detectable signal energy, which we will discuss here and in further detail under WUSN technologies in Section 10.4. The word seismic traditionally means earthquake. We, however, use it in a more classic sense as earthly signals. These natural signals, unique to the earth enable

[16] We have used: https://solarsystem.nasa.gov/planets/earth/facts [39].

us in two ways: (a) they are the primary supplier of our ever-growing database for GSTM. Every little signal could provide us with some information from the deep and out of reach inner Earth, which can be analyzed, processed, and stored to help with our understanding of what is going on in various layers and (b) secondly, they introduce us to a new technique for propagation of information through the ground. This is a new method for communicating through the ground. This important technological advancement is growing fast, and is expected to enable us to use the underground as a media in the future. They are also called earthly signals or earthly energy.

These waves in nature are generated from collision of the plates or other solid objects. They are naturally made of mechanical energy to appear in some form of low frequency vibration and local movement power that can easily travel through the solid layers of Earth and are detected by seismic detectors and suitable low-frequency sensors. Due to the low frequency nature of them they can use acoustic systems. While traveling underground, they can easily be conducted, reflected, and refracted on the way.

In general the earthquakes, normally caused by some volcanic eruptions, inner activities, and as a result of various collisions, then generate a large number of seismic wave signals propagated in many directions. As shown in Figure 10.4(a) we have two main components, *P-wave*, the penetrating (primary, pressure, compressional) wave in the direction of propagation, followed by *S-wave*, the shear (secondary) wave spreading over the surface of the force plane and vertical to the direction of seismic propagation.

When traveling through inner layers they are called *body waves* but when they reach the surface the P component vanishes and the surviving S spreads over the surface before passing its devastating energy to the objects on the surface. Due to the nature of these two basic waves their velocities vary similarly but P-waves travel much faster than S-waves. The difference, as shown in Figure 10.4(b), is more significant in the core, especially when traveling through the "outer core", where the S-wave is normally blocked. Figure 10.4(c) shows a typical traveling pattern of a seismic wave through the inner earth.

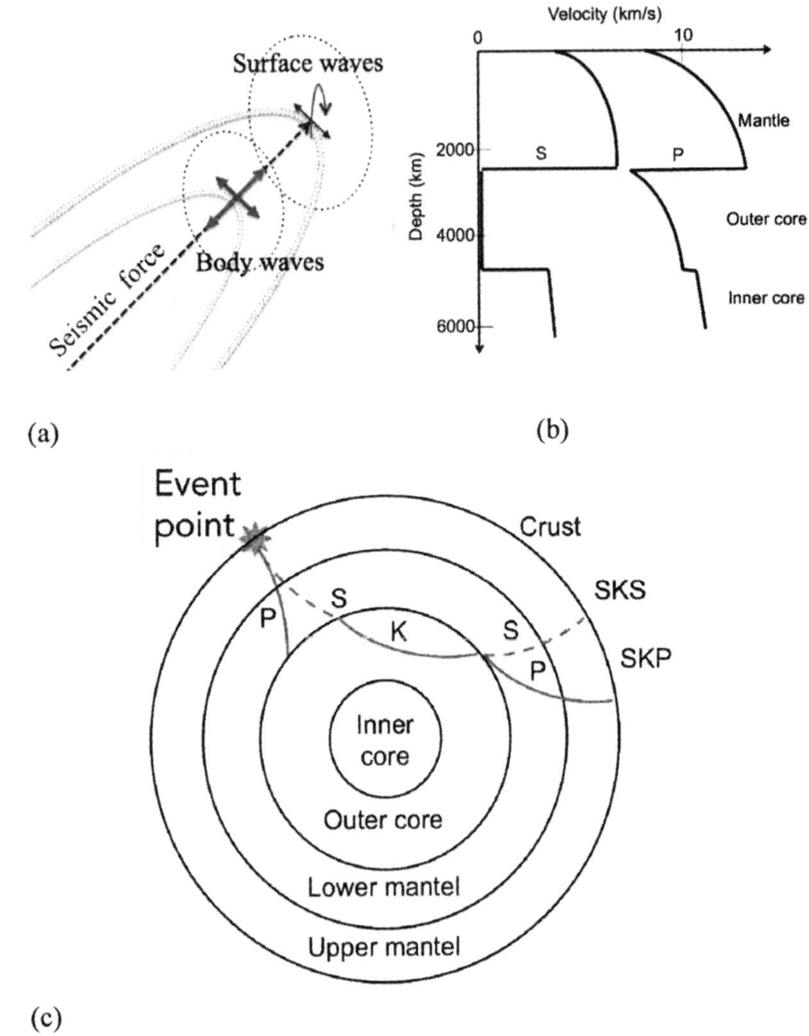

Figure 10.4 Natural seismic waves: (a) vector representation of the waves, (b) velocity of the seismic waves for various depths from the surface of the ground, and (c) various patterns of seismic waves for traveling through the layers. Adapted from Wikipedia [10].

It is worth mentioning that we have a few other seismic waves that normally arrive with the earthquake waves. These are normally slower than P and S waves and appear on or just below the surface of the Earth. They have been named after their investigators as Rayleigh, Love, and Stoneley waves. Figure 10.5 shows the sequence of signals appearing on the surface of a sample recorded in Hawaii in 2006 with the following details. P-waves arrive first, very strong but plotted with much reduced amplitude due to their termination at the surface and being fully spread and converted into other waves, which diminish before the S-waves arrive. We then notice the surface waves (Love, Rayleigh, etc.) with much stronger, physically devastating waves, which create the earthquake impacts [11].

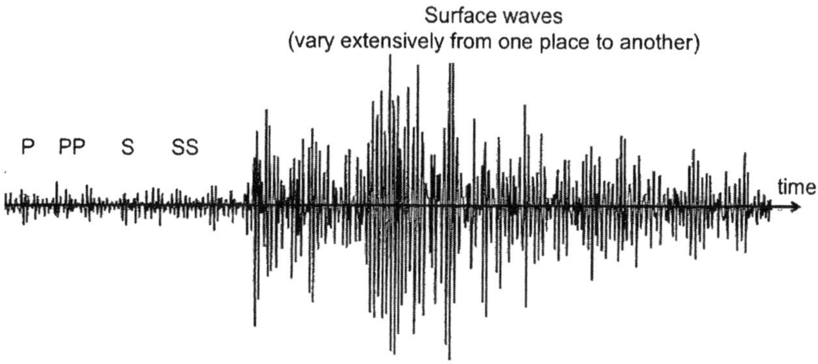

Figure 10.5 Earth seismic wave recorded, a typical example drawn based on [11].

A GSTM works in a large networked style compromising of three groups of devices to provide a fully operational service, namely: (a) geophones, sensors, accelerators, and detectors; (b) specialized seismic recorders and local signal processing; and (c) a network of information processors and controllers in the form of a distributed database system.

In order to create a GSTM as a useful knowledge source of prediction to generate early warning alarms, we need to make use of the very basic *theory of plate tectonics*, introduced over a century ago by Alfred Wegener, that the Earth's crust is not one piece but broken into smaller plates.

The GSTM can facilitate with relating the earthly incidents to the plates and can help to clarify the plates and their boundaries. The models for analyzing the observation must be able to predict travel time tomography cases using common techniques such as the *inverse predictive process* (IPP) [12]. A contribution from Koulakov et al. [13] uses such a system for gathering and analyzing the signal data at various local and international parts of the database. For example, Ivan Koulakov in his extensive work of a tomographic model analyzes the inconsistencies between P and S velocity models to refine various characteristics of the system, where he makes use of IPP techniques examining 32 regional seismic areas along with the database for the upper mantle beneath Asia [14].

10.4 KEY WUSN TECHNOLOGIES

The WUSN, commonly known as a underground extension of WSN, is in effect not very accurate. Due to the EE nature of underground it embraces the complicated deployment and service scenarios of WUSN and reminds one of a classic WSN. For these reasons we avoid using the generic WSN topologies for underground smart sensing design. So far, devices used in WUSN are traditional and can vary extensively from one application to another.

In order to make WUSN more comprehensive for those seeking the connection, and who would like to see it as a new extension of WSN, we divide WUSN upon its application into two groups of (1) near surface underground services (NSUS) and (2) global scale systems. The latter case is similar to those discussed in a previous section (e.g., GSTM) but for NSUS cases the applications in effect require a highly variable wireless sensor system (WSS) style deployment, where each application needs to be optimally designed and tuned under its own working environment. See Section 10.5 for further discussion.

So far, all of our natural propagation candidates such as electromagnetic, magnetic induction, and seismic technologies show weaknesses in providing an ideal link for communicating through the ground. Similar problems should be mentioned for energy and uses of intelligence. Most of the existing applications, for example, make use of nodes using mote style technologies, most of which are optimized for terrestrial WSN (TWSN[17]) scenarios and become inadequate.

As we know for an adequate development to make sense of underground *smart sensors*[18] and to make them work for us, we need to address the three main network properties of communications, energy, and local intelligence.

---

[17] Term TWSN in this Chapter (Book) has been used for 'terrestrial WSN' only, which should not be confused with TWSN as 'tunnel WSN' occasionally seen in the literature.

[18] Term 'sensor' is used in a generic format so includes all similar devices.

10.4.1 Electromagnetic Propagation

As we see, apart from special applications such as GPR, electromagnetic is not generally a preferred propagation choice for most WUSN applications. It, however, shows some desirable features in specific underground cases including: (a) short distance superiority for dense and heavily populated networks, (b) nearness to the surface and so able to use above the ground terrestrial links, and (c) low cost networking using TWSN electromagnetic parts for their well established off the shelf availability.

It is also interesting to see the effect of frequency of carrier and the soil content of the path especially the water content in the soil imposes significant loss to the signal. For example, an earlier work from Akyildiz and Stuntebeck [15] shows some 70 dB/m path loss for 1.4 GHz frequency signal at the first meter of the transmitter in soil sample with a 5% wetness. Then, 1m signal loss increases by 30 dBs for 25% VWC showing the influence of water. For the carrier frequency dependent losses doubling it to 2.8 GHz will show, in dBs an increase of 13, 38, and 60 for 5%, 10%, and 20%, respectively. Further updated results from Akyildiz et al. [16] for longer distances and slightly lower frequencies show more interesting results. At 300 MHz a loss of 35 dBs at 0.6 m with 5% VWC increases only 20 dBs per meter and for 25% VWC it jumps up only further by 7 dBs per meter. This extra jump reduces to 3 dBs for 900 MHz. The official name for the soil's wetness is the measured *volumetric water content* (VWC)[19].

Two main limiting factors for signal propagations: (a) heavy losses for practical distances and (b) highly variability of loss due to uncontrollable moisture and content of the soil such as stone, sand, silt, and clay.

For the radio wave propagation of the soil measured upon the statistical properties of the soil. Three main properties are permittivity (ε) for dielectric behavior, permeability (μ) for magnetic behavior, and electrical conductivity (σ). The complex loss constant ($\gamma = \alpha + j\beta$) includes phase distortion and amplitude attenuation, $e^{-2\alpha d}/d^2$, where, α is the media's attenuation constant and d is the propagation distance. For a system design we need an error free reception, which requires a measure of two basic power components: (a) useful signal power S at the detecting end of the link and (b) cumulative noise power as the total noise N. [17] Then, the SNR determines the signal quality of the link given in decibels

---

[19] Dampness or wetness of the soil, have been shown in the literature using both terms of *volumetric water content* (VWC) and *volumetric moisture content* (VMC), we use the first.

as $10 \cdot log(S/N)$. Where, S = signal power and N = overall effective noise power. For a complex underground path link design, we expressed in decibels as $S_d = S_t + G_t + G_r + G_a - L_f - L_m$ and $N_d = N_t + N_r + N_a + N_m$. Then, S_d, N_d are the detector received signal power and its noise; S_t, N_t are transmitter signal and noise; G_t, G_r are transmitter and receiver antenna gains; N_r Receiver noise excluding the LNA[20]; G_a, N_a are LNA's gain and noise; $L_f = 20 \log 4\pi d/\lambda$, is free space signal loss; λ effective signal wavelength in the media; *d* overall distance of the path; L_{em}, N_{em} are media losses of the signal and noise[21] [17].

To continue with the mathematical modeling of underground electromagnetic propagation we look into the path loss (L_{em}) for the specific underground media. For this we use the commonly acceptable estimated loss formula of 0.3–1.3 GHz frequency range [16],

$$L_{em} = 6.4 + 20\,log(d) + 8.69 \propto d + 20log(\beta) \tag{10.1}$$

The above formula is for a direct underground channel with constant magnetic and dielectric properties of permeability, μ, and permittivity, ε. For nonmagnetic materials $\mu = \mu_0 = 4\pi 10^{-7}$ H/m. However, permittivity, ε, is for the dielectric characteristics of the media changes upon the carrier frequency, materials, and their wetness. Its complex behavior of loss therefore depends on two complex components of $\varepsilon = \varepsilon' - j\varepsilon''$. To help designers with a practical approach we make use of the *Peplinski principle* and their experimental results [18]. Drawing as an example of the loss due to permittivity components we plot ε' and ε'' for their end frequencies of 0.3 GHz and 1.3 GHz using their *Field 1* (50% sand, 35% silt, and 15% clay) as the media for five VWC points of 5%, 10%, 15%, 20%, and 25%. Figure 10.6 shows a plot of two ε component constants. Then, Figure 10.7 shows the path loss L_{em} using (10.1).

The pair set of results estimated in Figures 10.6 and 10.7 can be adopted to design the connection links between the nodes of networked smart sensor systems for any practical range and carrier frequencies shown in the Figure 10.7. The water content is less effective for lower frequencies. However, due to normal fixed-distance deployment of WUSN

---

[20] Front-end amplifier also called *low noise amplifier* (LNA), critical functioning part of the receiver.

[21] If underground media are non-uniform we usually have some multipath effects caused by reflections and refractions of the layers; therefore some adjustments for both signal and noise components of the path are recommended.

nodes (e.g., in an array) one can estimate towards a working space between two nodes to work for the range upon its shortest in the range corresponding to the highest wetness of the ground.

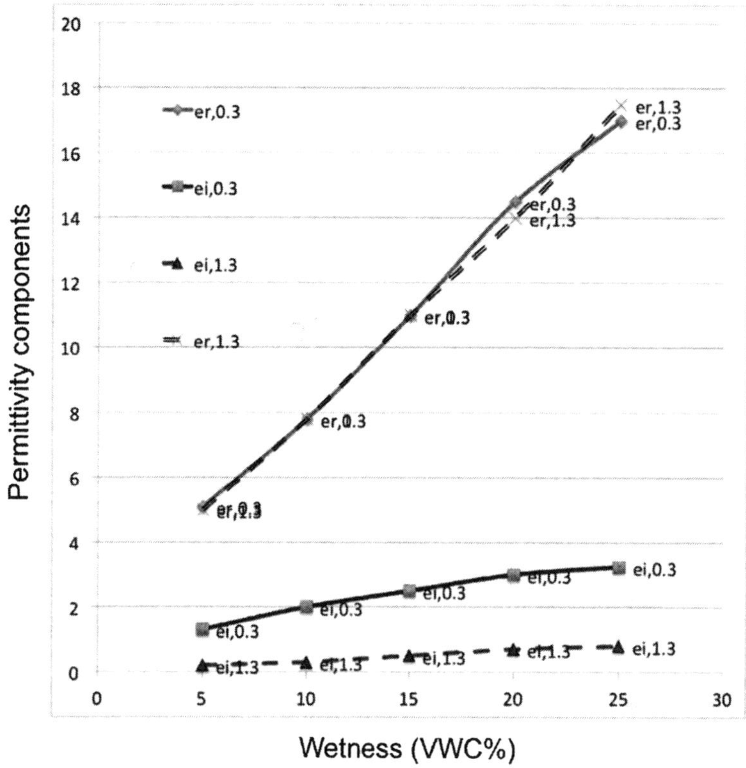

Figure 10.6 Redrawn of two ε (permittivity) components for Peplinski Field-1 mixture [18].

10.4.2 Far-Field Magnetic Propagation

Unlike a traditional electromagnetic wave, wireless communications propagation using a magnetic field layer in form of magnetic induction resolves many shortcomings in underground communications design problems.

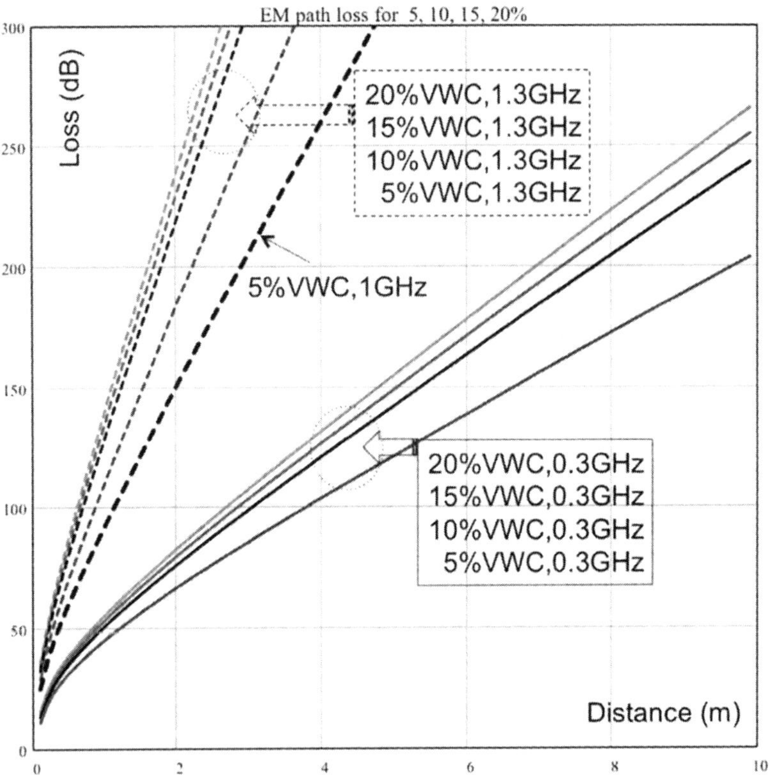

Figure 10.7 Path loss for some electromagnetic propagation cases: 5%, 10%, 15%, and 20% VMC.

Historically we already have many sensing devices in operation for very short distance applications called near-field communication (NFC) technology for uses in radio-frequency identification (RFID) and medical applications working at 13.56 MHz, which are naturally electromagnetic products are not suitable for practical WUSN. Their magnetic induction versions however have shown the potential to cover longer ranges and can therefore support WUSN technological developments.

The magnetic version of NFC, called near-field magnetic induction (NFMI) communication technology has shown some uses. However, its longer distance version has shown potential for WUSN and *wireless power transmission* (WPT) over relatively longer distances; we therefore prefer to call them *far-field magnetic induction* (FFMI)[22] or simply magnetic induction.[23]

The original work uses classic magnetic principles to override the signal's sensitivity against the medium, where two conventional coils are used, one as the transmitter and one as the receiver antenna. Due to the conventional spread of the energy in a magnetic field the transmitted signal faces the natural 3D loss effects (i.e., the rapid roll-off creating a loss of power of sixth order for the distance). A simple magnetic induction linked coil is shown in Figure 10.8(a). The link loss power has a mathematical model as, $\left[64R_0\left(R_0+\frac{1}{4}\mu j\omega w_t\right)\right]/[(\mu^2\omega^2 w_t\, w_r a_t^3 a_r^3)\cdot d^6]$ [16]. Here, w_t and w_r are the transmitter and the receiver coils' number of winding cycles, a_t and a_r loss are the transmitter and the receiver coils radii, R_0 the coil wire's resistance per unit length, and d is the linking distance. In practice, however, we normally use similar coils on the both sides transmitter and the receiver (i.e., $a=a_t=a_r$ and $w=w_t=w_r$), we can reduce it to $4\frac{1+jQ}{Q^2}\delta^2$. This model then has only two independent design variables, *linking quality factor*, $Q=\frac{\mu\omega w}{4R_0}$, and *relative volumetric distance*, $\delta=\frac{d^3}{a^3}$, that we can use for compacting the signal power loss to,

$$L_{mi}=6.02+5log(Q^2+1)-20log(Q)+20log(\delta) \tag{10.2}$$

Above loss power can be designed for any operational frequency, f via $Q(f)$ and d in $\delta(d)$.

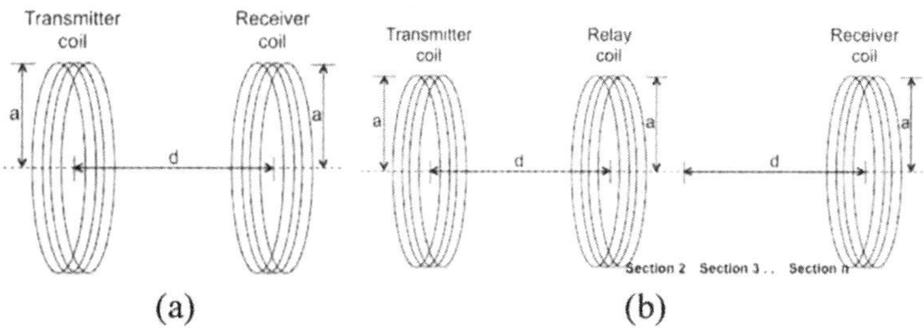

Figure 10.8 The schematic of magnetic induction linking: (a) simple link connecting two coils and (b) equally spaced n-section relay connection.

[22] Our use of 'far' in FFMI under practical state of technology in WUSN today is relative to the NFC distances, therefore much shorter than TWSN.

[23] NB: Due to terminological similarity between 'magnetic induction' and 'electromagnetic induction' there is considerable confusion between electromagnetic and magnetic induction in the industry.

As shown in Figure 10.9, it is interesting to see that path loss starts with an order of 6 for short distances and increases with the distance rapidly to over 100 dB. Such low signals are unacceptable for common receivers.

In order to extend the distance one solution is to make use of repeater style relaying guided coils as shown in Figure 10.8(b). That is, we simply add (n-1) relay sections to the same coil-link shown in (a). The power loss calculation of the signals then becomes a combination of an ordinary link followed by any number of relaying sections each with,

$$\text{eq} = section\_power\_ratio = \left(\frac{\delta}{Q} + j\delta + \frac{1}{\delta} + \frac{Q}{1+jQ}\right)^2 \tag{10.3}$$

Then the overall loss for an uncompensated n-section-link-relayed connection becomes $\left(4\frac{1+jQ}{Q^2}\delta^2\right)\left(\frac{\delta}{Q} + j\delta + \frac{1}{\delta} + \frac{Q}{1+jQ}\right)^{2(n-1)}$. In order to achieve a better performance we can create a matching condition for each relaying section we use a compensating capacitor, $C = \frac{\mu}{2\pi a R_0 Q^2}$. Using C then simplifies the relaying section loss formula to $\left(\frac{\delta}{Q} + \frac{Q}{\delta}\right)^2$ for a total n-section-link loss in dB of,

$$L_{total} = L_{mi} + 20(n-1)log\left(\frac{\delta}{Q} + \frac{Q}{\delta}\right) \tag{10.4}$$

Figure 10.9 shows some samples of signal power loss for three cases of simple link (first section, L_{mi}) using (10.2), uncompensated relaying, and compensated relaying loss using (10.3). The seven plots in Figure 10.9 from to the top (higher losses) to the bottom (lower losses) are explained as follows.

The solid curve shows the loss for an uncompensated relaying section using a 1 GHz carrier signal. It starts with a 60-dB loss at 1m increasing to 140 dB as it approaches 20m of distance. We, then have two link L_{mi} curves for 100 MHz and 1 GHz carriers. The first one (100 MHz) comes with three lines attached showing the slopes of the curve at its three points. The first slope line is touching the curve at about 1m of distance with 40 dB loss showing the highest slope of 18 dB per octave (short distance slope), then the middle slope line touches the curve at about 6m at about

80 dB showing a slope of 12 dB per octave (mid distance slope), and the third at 10m a slope of 6 dB per octave (long distance slope). Next the link L_{mi} curve of 1 GHz shows an improvement of 15 dB over 100 MHz. We then have four performance curve, for compensated section cases with carrier frequencies of 100 MHz, 300 MHz, 1 GHz, and 2 GHz as they go down in loss, respectively. For example the effect of compensating capacitor around 80 dB cannot be overlooked.

The link loss results estimated in Figure 10.9 for magnetic channels show some significant performance factors of this magnetic induction configuration. The carrier frequency plays an important role but other design factors are also becoming useful. An increase of carrier frequency in effect reduces the loss. Also due to initial loss (near-field effect) using more sensitive receivers for longer single link distance is far more effective than making relayed connections.

Then, our equally spaced n-section-link design becomes an easy task of selecting the best n for a minimum total loss given by (10.4) as in Figure 10.10. Carrier frequency for all plots in here is 1 GHz. As shown in Figure 10.10, the loss of an n-section-link increases with n. That is, cascading may become an important design parameter that can be easily put in a design table.

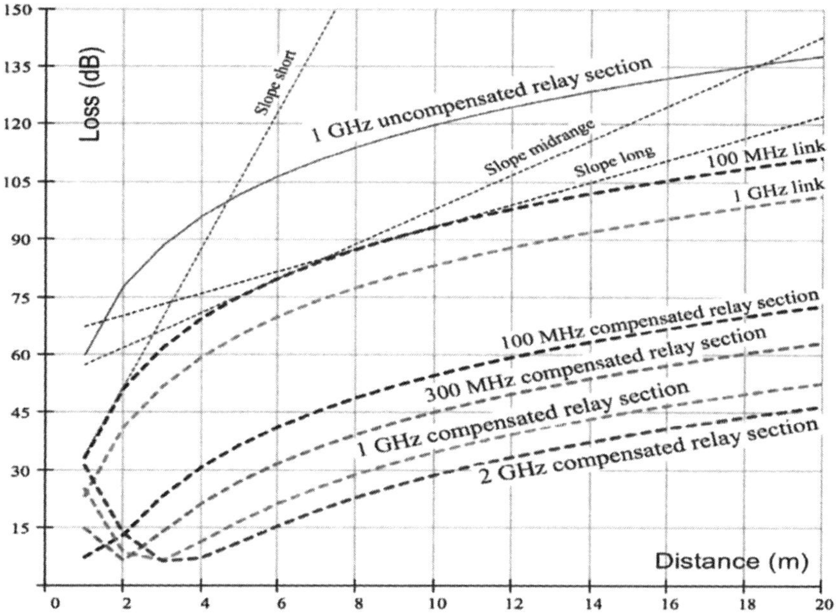

Figure 10.9 The magnetic induction link loss and section loss plots for various frequencies and compensated conditions.

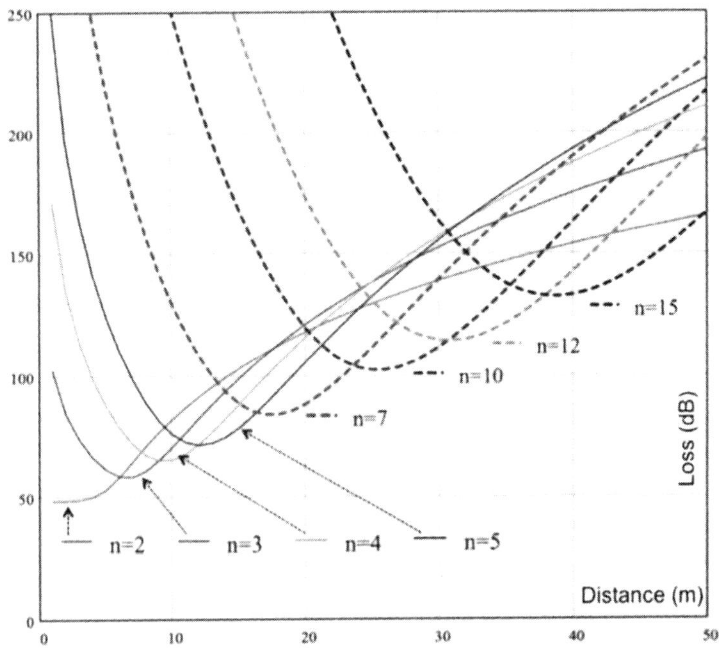

Figure 10.10 The equally spaced *n*-section-link magnetic induction relayed overall loss design.

10.4.3 Seismic Wave Propagation

If we look into the evolutionary nature of human ears, which are structured for a mechanical style hearing the author is surprised that we have not adopted this naturally advanced technology for making use of seismic waves for underground communications. The evolutionary paths of many animals such as moles and farm rats (Middle Eastern blind mole-rats) use mimicking the vibrational cues through their natural receptors (e.g., vibrational stimuli are transmitted through mechanoreceptors, corpuscles, and nerve endings to locating the source). Therefore, another potential signaling technology now emerging to become a promising base for WUSN development is to use *seismic waves for underground communications*.

Mentioned earlier in Sections 10.2 and 10.3.2, seismic waves have been around and used extensively in geophysical projects building a leading base for predicting many natural disasters and saving lives.

In nature seismic communication exists in the form of vibrations as a process of conveying information through mechanical (seismic) vibrations between many creatures such as mammals, reptiles, birds, and insects. Scientifically speaking, we take it in much broader aspects of technological and device communications, where "seismic waves" stands a better position as a new phenomenon, a three dimensional mechanical waveform that suits the nature of ground media to penetrate through the tough layers of the ground enabling us to use them to communicate through the ground. This rich global scale technology can now enable us to climb to a new plateau of WUSN technology, entering a new stage with stronger hopes to help to build a new underground industrial paradigm.

Some of the existing earthquake seismic waves cannot be counted as broadband, due to their scattered range of frequencies making them quite hard to analyze; however, due to their limited spectrum those adopted for industrial applications in earthquake prone areas can be both low cost and far more effective for their existing uses. Having these for specific applications we can see a wide range of products for seismic monitoring systems, as shown in Figure 10.11, working in the frequency ranges from 0.01 Hz to few MHz.

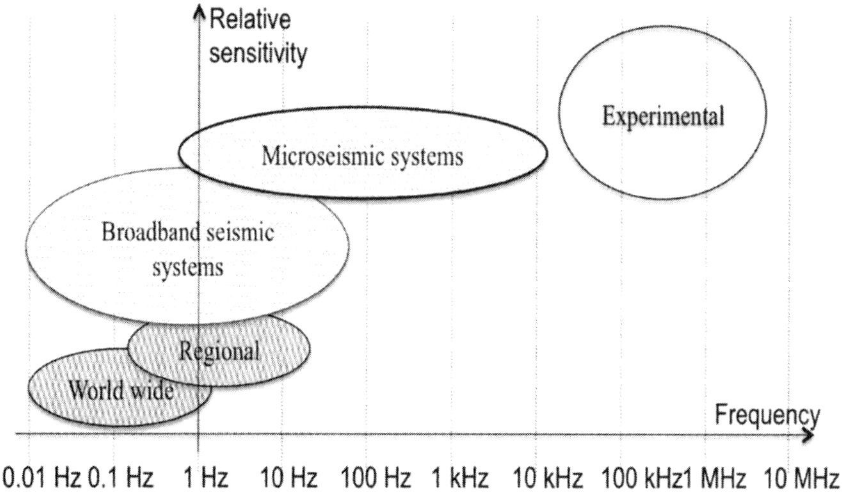

Figure 10.11 Today's seismic waves working spectrums [19].

As shown in Figure 10.11 seismic wave spectrum can be divided into five frequency waveband spectrums from the natural earthquake signals transferring the bursts and cracks info in the form of mechanical vibrations through the ground. From a WUSN deployment viewpoint such *vibration signals* can be taken as a potential technology for carrying data between sensors. For this we need to know the loss of energy, which obviously depends on the path. For this we look into the *intrinsic attenuation* [20] and use a dimensionless quantity variable. The Q, also called the *seismic attenuation factor* comes with a general definition, $Q = 2\pi E/\Delta E$, where, E is energy of the wave and ΔE its one cycle average change. The value of Q varies with the content in the ground layer (medium) from as low as 5 for soft and pure soil to around 100 for rocks increasing to very high for metals such as 10,000 for steel. Therefore, it is useful in practice to measure the Q of a medium before any designs, where we use a single wave, also called 1-D, for our measurements, which normally could produce the best result if the receiver is located at the surface. For this case the Q equation can be simplified to, $Q = -\pi A/\Delta A$, where, the *effective amplitude* of the wave, $A = \sqrt{E}$ and the negative sign shows that the amplitude ΔA is reducing. This represents an exponential loss formula that can be expressed in terms of distance, x, and time, t, based on the following expressions, $Q = -\frac{\pi A}{\lambda \frac{dA}{dx}} = -\frac{\pi A}{\frac{2\pi v}{\omega}}\left(\frac{A}{\frac{dA}{dx}}\right) = -\frac{\omega}{2v}\left(\frac{A}{\frac{dA}{dx}}\right)$, where wavelength $\lambda = \frac{2\pi v}{\omega}$, $\omega = 2\pi f$is the frequency, and v is wave's propagation velocity. It then can be shown in its exponential derivative form as, $\left(\frac{\frac{dA}{dx}}{A}\right) = -\frac{\omega}{2vQ}$. This results in the following decay functions of the seismic waves for any distance at any time written for any effective amplitude and energy as, $A = A_0 e^{-\frac{\omega}{2vQ}x} = A_0 e^{-\frac{\omega}{2Q}t}$ $and\ E = E_0 e^{-\frac{\omega}{vQ}x} = E_0 e^{-\frac{\omega}{Q}t}$

From the above model we see that loss depends on the signal frequency and Q of the media that propagation takes place. In practice the inhomogeneous layers of the ground can change the velocity of the waves and with it the Q. Figure 10.12 shows the Q and P-wave propagation for the experimental results [21].

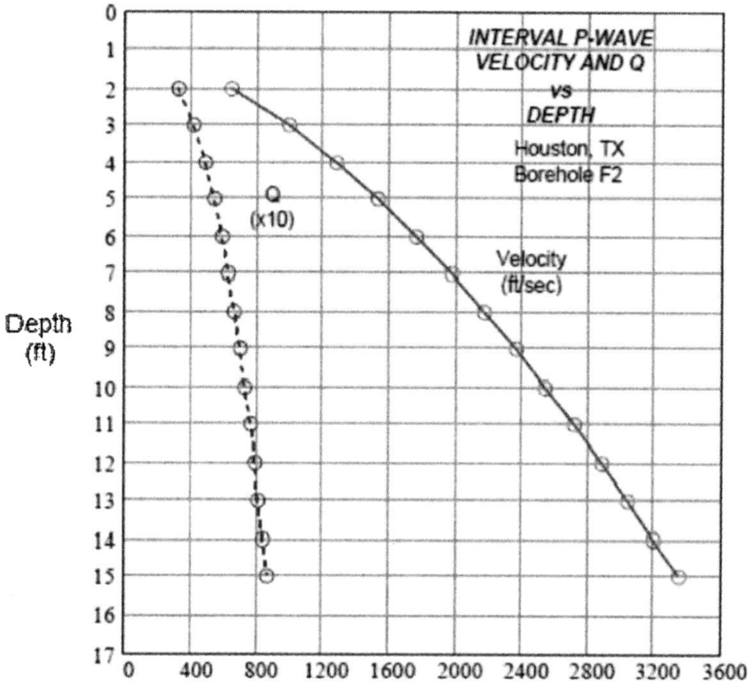

Figure 10.12 Plot of Q and velocity against the depth for some experimental P-waves drawn based on [21].

Should this single ray propagation go through a multilayer medium with n sections for distances of $d_i: i = 0,1, to\ n$, at the points in time of $t_i: i = 0,1, to\ n$ through n sections with $Q_i: i = 1,2, to\ n$ and $v_i: i = 1,2, to\ n$ the energy loss can be given as,

$$ln\left(\frac{E}{E_0}\right) = -\omega \sum_{1}^{n} \frac{x_i - x_{i-1}}{v_i Q_i} = -\ \omega \sum_{1}^{n} \frac{t_i - t_{i-1}}{Q_i}$$

It is important to mention that Geophone is one of very popular seismic wave devices used for various underground services.

10.5 NEAR SURFACE SERVICES

Under the changing world of technological developments we follow new global trends and therefore write this section upon the new approach of "*design for provision.*" Globalization impacts of the telecommunications and computer technologies are forming the future of the global information society evolving further from product paradigm into the service paradigm led by the smart sensors. That is, in the case of WUSN the technology should be taking on a holistic business trend evolving from product design into service design, a *fully automated and self-maintained deliverable system at the user's doorstep.*

This is not an idea but a *change process* happening now. It started decades ago, noticeable through experiencing too many static, unstable, and failing industries, where WUSN is not an exception. On the positive side virtually all WUSN applications do not require digging very far into the ground. On the negative side it is easy to come with exponential figures of cost and accessibility.

For the above reasons we propose design of advanced lightweight WUSN, also called WUSS applications. To that effect we address our potential small WUSN systems under *near surface underground sensor services* (NSUSs).

As a successful new technological development, an innovative engineering community should posses some new capabilities to ensure it does not fall into any possible *discontinuity gaps*. Discontinuity gaps may not have been noticeable in the earlier stages of industries' first and second technological revolutions, but under new global market forces they could become critically important. We therefore look into the whole process of service deployment as a new design approach for any missing gaps in advance.

Due to the scale of development and its financial implications one may not have sufficient knowledge about some of the points. The risk can be taken but it is better if it is expected and possibly calculated or predicted than a surprise. For this chapter one should be aware how the WUSN technologies are developing very slowly but we still need to identify how recent developments can enable new capabilities, knowing the *state of the art* for technological availability.[24]

[24] Unfortunately these days much academic research aims at publication for promotion rather than production and humanity advancements.

In order to initiate new WUSN opportunities we address three groups for service deployment namely, *scanning as a service* (SAAS) and *localization as a service* (LAAS) as one group, *monitoring as a service* (MAAS) as the second group, and finally *networking as a service* (NAAS).

10.5.1 NSUS: Scanning and Localization

Scanning and localization are of *discovery in kind.* Both require signal penetration through the surface of the ground or direct access to the area of discovery. Scanning and localization services are not always independent. Many projects make use of both combined or are integrated with other complementary technologies such as RFID, imaging, and communication.

10.5.1.1 Scanning the Ground

Underground scanning normally enables us to measure properties of the ground such as the quality of a new soil in an agricultural land or generation of a specialized map for the ground and its near surface layers. It can be used to check the safety of old mines for structural risks. New mines, tunnels, and similar excavations all could benefit from some preliminary scan before setting up a strategy to ensure their lengthy risk-free project life, especially in natural disasters prime areas.

Lengthier scanning services are designed to measure dynamic behaviors of the ground or objects buried underground. Short-term scans are for those that require identification with some unknown features while long-term scans normally serve uncertainty or with risky conditions to facilitate an early warnings and further test. It is important to mention that the *ground penetrating radar* (GPR) is one of very popular devices used for scanning and localization services.

The Landmines

One of the humanitarian applications of GPR is the discovery of *remaining landmines* in the troubled lands and warzones, where too many forgotten and lost mines could remain active, creating a risk to life and public fears.

Although the task sounds simple and trivial, in practice it could be costly, inconvenient, and complex due to the facts that: (a) most land mines are in very poor areas, where people cannot afford daily living expenses and humanitarian budgets cannot be stretched very far, (b) very dangerous and risky operations are required to be set up, (c) uniquely designed specialized equipment and devices are required for the work, and (d) systems, machineries, and robots used for the operations should be shock-resistive therefore heavy. Some good examples for demining developments are described in Habib's book [22] claiming that some 100 million scattered mines pose significant hazards in more than 68 countries across the world. Landmines are cheap and easy to make and lay. A typical landmine consists of a firing mechanism, a detonator that sets off a booster charge, and an explosive charge, and the body of the mine, a plastic or metal casing, which is placed on or inside the ground. A number of fused activation mechanisms used including push, release, move, sound, magnetic field vibration, electronic, or remote command. There are some 2000 types of landmines around the world categorized into two groups, antipersonnel (AP) mines and antitank (AT) mines.

Scanning for Drilling

Drilling into the ground for various reasons such as tunneling, mining work, a pass way, or storage area is a service that requires scanning the ground beyond its visible surface. Technologies used for activating sensing and imaging technologies are electromagnetic, seismic, and magnetic. The presence of a seismic source generating energy from a tunnel-boring machine (TBM) can generate a survey.

In many cases GPR or tomography technologies could be sufficient but for hostile conditions of rocks and layers. An interesting work from Bellino on 'tunnel seismic while drilling' (TSWD) does not require a separate source as it uses seismic waves for correlating geophones with signal sources. For driving TBM they use accelerometers as seismic sources resulting in strong coherent events, corresponding to P and S signals. With the sensors distributed all over the tunnel walls then the seismic detectors (geophones and accelerometers) analyze them to control drilling and busting the rocks [23].

To name other drilling and scanning techniques we can mention "in-tunnel horizontal seismic profiling (HSP)" and "vertical seismic profiling (VSP)" to work in with the horizontal axis. Then, more advanced exploration systems for hard rock environments is the "tunnel seismic prediction (TSP)" system followed with more sophisticated software in "integrated seismic imaging system (ISIS)." On the tomographic side we have the tunnel reflector tracing (TRT), true reflection tomography (TRT), tunnel seismic tomography (TST), true reflection underground seismic technique (TRUST), and tunnel geological prediction (TGP) [24].

10.5.1.2 Localization

Underground localization is a short-term service that often follows a scanning project, where the locations of certain hidden but known objects or features are to be measured. Techniques effective for localization vary from the use of natural seismic waves to radio and magnetic induction technologies. In some special cases we may have to get help from and use combined technologies such as directional scanners and imaging systems.

Localization requirements of advanced sensors and WUSN projects are neither new nor fully developed. Three main areas of the need for advanced sensors for localization are: (1) subsurface scanning for the natural contents and new materials; (2) subsurface scanning for discovery of the objects or artifacts; and finally (3) upper crust monitoring and observation of seismic waves and their predictive behaviors to enable us to locate more stable ground areas to be used for heavy and large industrial structures such as, tall and heavy towers.

Utility Rediscovery and Mapping

There are many old utilities such as pipes (water, gas, etc.) still in use and due to their length and depth they should not be lost or forgotten. This happens in old towns with too many new roads, reconstructions, and logistic changes or due to various natural disasters and man-made troubles so that the maps and data of old utilities have been lost and cannot be located. The rediscovery has been a complicated, expensive, and sometimes dangerous process that many local governments cannot undertake. New sensor technology can help with redrawing the utility map. One good work for such localization is using seismic technology for locating the utility in such places. The Young report from NAP provides an interesting work on the development of using some multisensor platforms for locating the pipes and other utilities long buried under the ground [21] (see Figure 10.13).[25]

[25] Dielectric nature of the materials constituting the ground creates two components for each wave. For example, S-wave has two components of SV for vertical and SH component for horizontal components (plane, perpendicular to the direction)

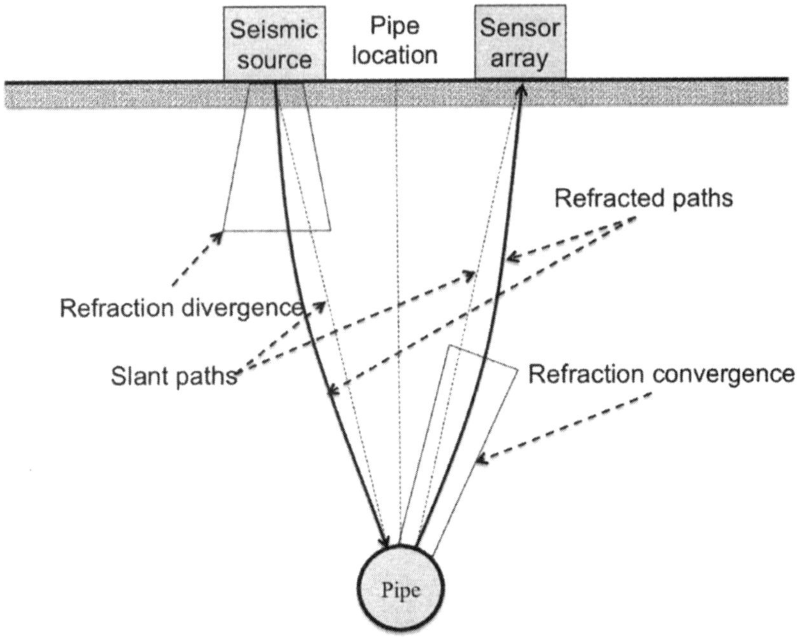

Figure 10.13 Utility locating process, using a simplified schematic for the experimental measurement [21].

Another basic but interesting utility discovery work from NAP is from Hammerschmidt [25] for location and identification of stacked and deep utilities. In his active acoustic method the injected sound pulsed into the pipe utility propagates though the medium in the pipe then leaks out towards the ground surface where will be detected. This method uses a very low frequency system and converts the *time of flight* (TOF) to estimate the location of the utility. Figure 10.14 shows the schematic of a two-sensor version of the active acoustic technique. Looking at this method it is easy to see a considerable potential in acoustic signal processing for improving performance of this device.

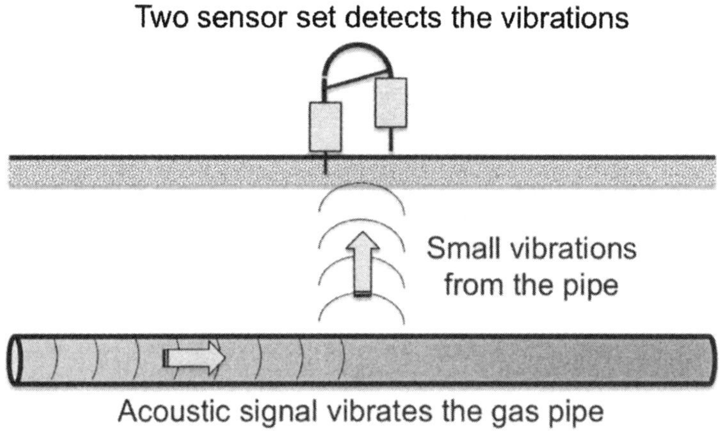

Figure 10.14 Using a two-sensor active acoustic for pipe localization [25].

10.5.2 NSUS: Monitoring as a Service

Monitoring applications of WUSN can be broadly divided into three *management service categories*: (1) earth management dealing with natural disasters, (2) underground industrial management, and (3) controlling underground industrial mismanagement.

10.5.2.1 Earth Management, Dealing with Natural Disasters

In order to be able to continue with our intelligent life on earth we need to maintain our monitoring of the planet and prevent it from any serious mishaps while saving ourselves from natural disasters. So far two major steps been taken that we should be proud of. One is monitoring to prevent any further increases in the earth's overall temperature. It is, as far as we know, due to the excessive CO_2 emissions caused by careless industrial development and mindless contributors to heavy traffic in large cities (see Section 10.2.1). The second is building up a knowledge base to understand the earth at a global scale.

Dealing with Excessive CO2

So far we have managed to come up with two acceptable solutions for reducing the excessive CO_2 in the atmosphere. One is to keep it out of the atmosphere stored away in the crust for safekeeping and future uses. The second is to convert it into new energy sources. Due to the other roles that

CO_2 can play storing CO_2 within the skin of the crust has been proposed, whereas drilling and production of oil wells still continues nonstop we can follow Benson's analysis of the design case in her two related works of 2005 and 2008 [26, 27]. The facts supporting the practicality of this idea are: (1) wells are already vacant due to the gases that have been taken out and replaced with water, (2) CO_2 as gas can easily be compressed when injected into lower layers of the crust (e.g., 30 times in volume for 1,000m depth), (3) natural storage being made available for the original gas makes for very low leakage (practically zero over hundreds years) for monitoring and safekeeping.

The second solution sounds more productive and less risky due to its industrially motivated process of regenerating hydrogen and other chemical processes as indicated by Luberti [28]. Although it could be more effective it requires long-term commitments and specific resources. Under the title of "design of a H_2 pressure swing adsorption process at an advanced plant for cogenerating hydrogen and power with CO_2 capture" in order to reduce the existing dependency on nonrenewable energy resources he proposes a new setup for new *carbon capture and hydrogen purification technologies* process of using an emerging clean coal technologies called integrated gasification combined cycles (IGCCs).

Monitoring Earth for Disasters

The WUSN services for dealing with *Earth disasters* are mainly concerned with understanding the volatile nature of the Earth, predicting earthquakes, and generating early warning signals. Services for this category have been started and discussed under GSTM in Section 10.3.2. Here, we will have large number of potential opportunities to expand and use WUSN to full extend for the operation.

10.5.2.2 Underground Industrial Management

Knowing the nature of Earth and its behavior near the surface we can manage our industries and make better use of our resources using our smart sensors technologies under the WUSN. For underground industries three significant areas are: (a) monitoring services in agricultural and farming industries, (b) monitoring services in mine industries, and (c) monitoring services in oil industries.[26]

---

[26] Due to the high level approach taken for this book detailed solutions are not included. Most of the design solutions are simple interconnections of sensors and devices.

Monitoring Agricultural and Farming Industries

Agricultural services can vastly benefit from adoption of WUSN and WUSS technologies. The industrial requirements and their importance have been discussed in Section 10.2.3. For the detailed technologies we direct interested readers to [6].

Mine Industries

Mines and minerals are the bloodlines of the heavy industries. Historically, these industries started with copper, bronze, and ore before entering the first industrial revolution for coal as a source of energy. Due to the unique WUSN service requirements for monitoring mines we discussed it as a *typical case* in Section 10.6.

Microseismic Technology for Monitoring Oil Wells

In parallel with the long history of seismic waves and their moves towards the global Earth model microseismic monitoring has shown its value as a tool for understanding and modeling in the oil and gas and mining industries. For example, Figure 10.15 shows a schematic drawing of a heavy-oil production facility in Canada using a vertical array of geophones installed for monitoring the steam injection through a series of underground wells into a layer of heavy oil [29].

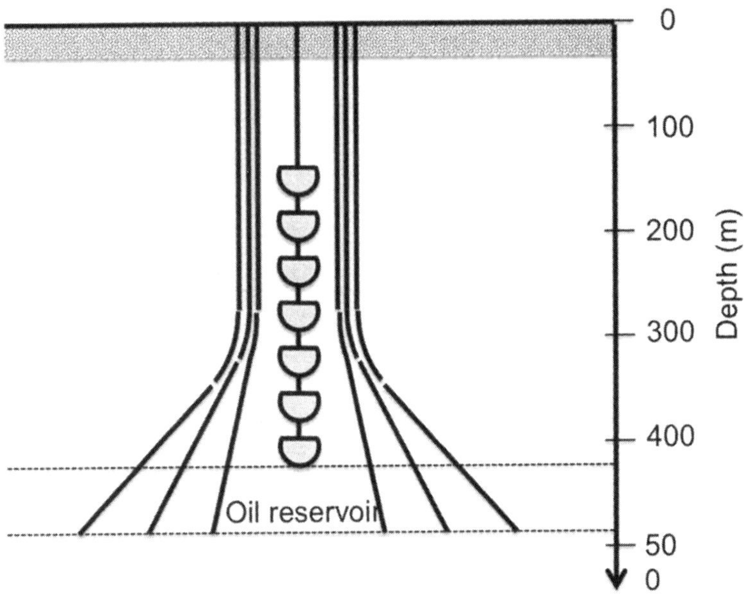

Figure 10.15 Schematic drawing of a heavy-oil production using an array of sensors [29].

10.5.2.3 Controlling Industrial Mismanagement

Due to their historical developments factories and heavy industries share the ground we inherited to live on. This sharing is like a double edge sword; it can bring prosperity and a better quality of life if managed wisely or threaten our lives with dust and poisonous technological side effects if badly managed, leaving us with man-made industrial hazardous, toxics, and ground pollution for generations to deal with. Here we address these issues briefly in two parts. One is under "toxics," finding ways to deal with them using sensors and specialized devices. The second part comes under poor design, poor management, and leakage.

Industrial Toxics

Toxic or toxicant means poisonous in making a substance deadly to a living creature. Therefore, toxicity represents its degree of effectiveness to endanger human health, a living creature, and the degree of contamination it contributes to the living environment. Toxics exist in the form of solid, liquid, gas, and in some cases radiation. They are most effective when they come in contact in the form of liquid or gas. The liquid forms are normally *chemical* or *biochemical.* Chemicals are normally natural including mercury and lead or *man-made chemical products* such as toxic acids used in chemical-war substances. Natural ones are easier to predict and locate. However, man-made ones are normally wastage from the old, primitive industrial activities, often chemicals. Biological toxics are much harder to identify and often need specialized sensors.

The biochemical toxics have very long list, which is growing due to their developments and uses for medications. However, with their use for underground cases we are limited to organic compounds such as methyl alcohol and poisons from plants and other living organisms. The biological toxicants produced by fungus, plants, and micro organs called toxins. Animal toxics are called venoms.

To this effect monitoring buried underground toxics of various kinds (e.g., chemical, nuclear, biological, lab rejects, leakages, corrosion,) threatening our lives are still around appearing unexpectedly. For example, Harris et al. report that since 1988, there have been 17 documented incidents in the United States and Canada in which carbon monoxide (CO) is suspected to have migrated through ground strata into occupied enclosed spaces. One group of dangerous trapped toxic gases contaminating the top layer of the ground are half burnt fuels such as carbon and nitrogen in various through natural processes or various man-made forms. The gaseous products of explosive detonations include many of the kind including CO, nitric oxide (NO), nitrogen dioxide (NO_2), methane (CH_4), hydrogen (H_2), and oxygen in the right places to be used again as dwelling places and industrial sites [30].

Dealing with toxins, gaseous ones in particular have been always very poorly managed, expensive, and tedious. Therefore, most poorly managed industries prefer to pay hefty court charges rather than cleaning their hazardous remains. As the development of sensors and better detectors are too slow we seek global efforts to activate new toxic sensors and better WUSN development possibilities to keep our precious lands clean. Tomography style monitoring is promising. A recent work from Neumann et al. sounds like a promising approach to deal with many gases using one system. The interesting part is generation of 2-D maps of the gas density distribution using the well-established method of computerized tomography (CT) [31].

Underground Industrial Leakage

Leakage is contamination, which means pollution plus contagion. This is the second part of our mismanagement of industrial hazards. Sensing for leakage control comes in two forms: (1) leakage out of a dumping ground and (2) leakage from existing in service systems. Monitoring hazardous objects buried in the soil needs to be known before being discovered (e.g., we still have many atomic disposals of old experiments and the residuals of the nuclear power stations, throughout the world). They are assumed safe and decaying slowly, but mostly are still unknown to local dwellers.

People have the right to know their locations, information about their safety, and any plan of action in case of a leak. These days they should be equipped with some advanced sensing (WUSN) and early warning systems. For the leakage of existing in-service systems the providing industry, being utilities or industrial pipelines, should adopt proper detecting and recovery systems.

Today's industrialized life could not be maintained without the use of pipes. Pipes come in various forms and shapes and for a wide range of services such as: (a) clean water distribution, essential; (b) domestic gas distribution, smart living; (c) international gas and oil distribution, trading; and (d) industrial site distribution using structural pipeline systems common for water, gas, and many other goods as alternative to classic transportation methods using tankers and other vehicles on land or in sea is often preferred due to low cost, high reliability, and security. They, however, depending on their structure and service are subject to corrosion, cracks, and other malfunctioning issues.

Leak detection normally adds a significant weight on saving time, efforts, and losses. The methods developed for detection and localization of the leak come in two kinds. The *out-of-pipe* method, though sound convenient and easier in effect, has been shown to be unreliable, expensive, and often complex. They mostly rely on the secondary effects of the leak (i.e. the content spreads significantly). Therefore, they suffer from two drawbacks: (1) too many expensive detectors and (2) environment depend sensitivity (e.g., soil), which leads to very poor and unreliable design. *In-pipe* leak-detection, however, though conceptually sounds complex and difficult but it provides a much better solution under an infrastructure. One interesting design works based on the simply change of pressure. As all pipes, gases or liquids, inherit a forced flow of material under pressure to be transported then an accidental leak creates differential pressure at the point-of-leak (POL) to be detected. This design uses an in-pipe method proposed by Chatzigeorgiou et al. [32, 33]. It uses a sensing vehicle consisting of a basic miniature robot-looking device designed to move freely along the pipe carrying a floating disk equipped with a small cylinder looking membrane attached to it at the outer edge. The detection works upon using the differential pressure between the media inside of the pipe and outside causes a small disorientation to the position of the disk, detected and location is reported to main controller. Figure 10.16 shows the basic setup for the remote control of the robotic system. The upper part of Figure 10.16 shows the general installation and leaking problem. The middle part of and Figure 10.16 demonstrate four stages of how the leakage detection and temporary cover works. The lower part of Figure 10.16 shows the laboratory set up for an experimental design.

10.5.3 NSUS: Networking as a Service

Following our earlier introduction to WUSS versus WUSN and its related discussion in Section 10.1, it is now time to analyze this issue further for the sake of viable opportunities for the NSUS market. For doing this we should consider all viable wireless smart sensor opportunities. With a five to ten year plan we estimate that contributions from all sectors of industry could examine the possibility of building a near future industrial paradigm. In order to estimate the possibilities one can rely on the future service capabilities, including the services we have already discussed in this section plus those to be added upon integrating our networking (NAAS) main key features, *cross layer design*, *energy*, and *autonomous level intelligence*. Before discussing these details, we should, however, examine our new opportunities arising from integrating deployment under WUSN and WUSS approaches.

It is common sense to benefit from integrating the markets for larger numbers, where mass production would benefit both users and industries. This can be demonstrated for NSUS applications in Figure 10.17. As Figure 10.17 shows, scanning and localization products operate with a small number of devices, but monitoring is best implemented using more integrated number devices. The seismic networks, however, growing continuously use very specialized devices in very large arrays. WUSS solution approach is very low cost as it does not use any particular networking but it still suits all three NSUS cases. The WUSN provides optimized networking solutions for all but due to the networking overheads its performance drops very rapidly for smaller application scenarios. The boundary varies expensively depending on the number of devices and the homogeneity of them. Therefore, it suits seismic networks with millions of similar devices best.

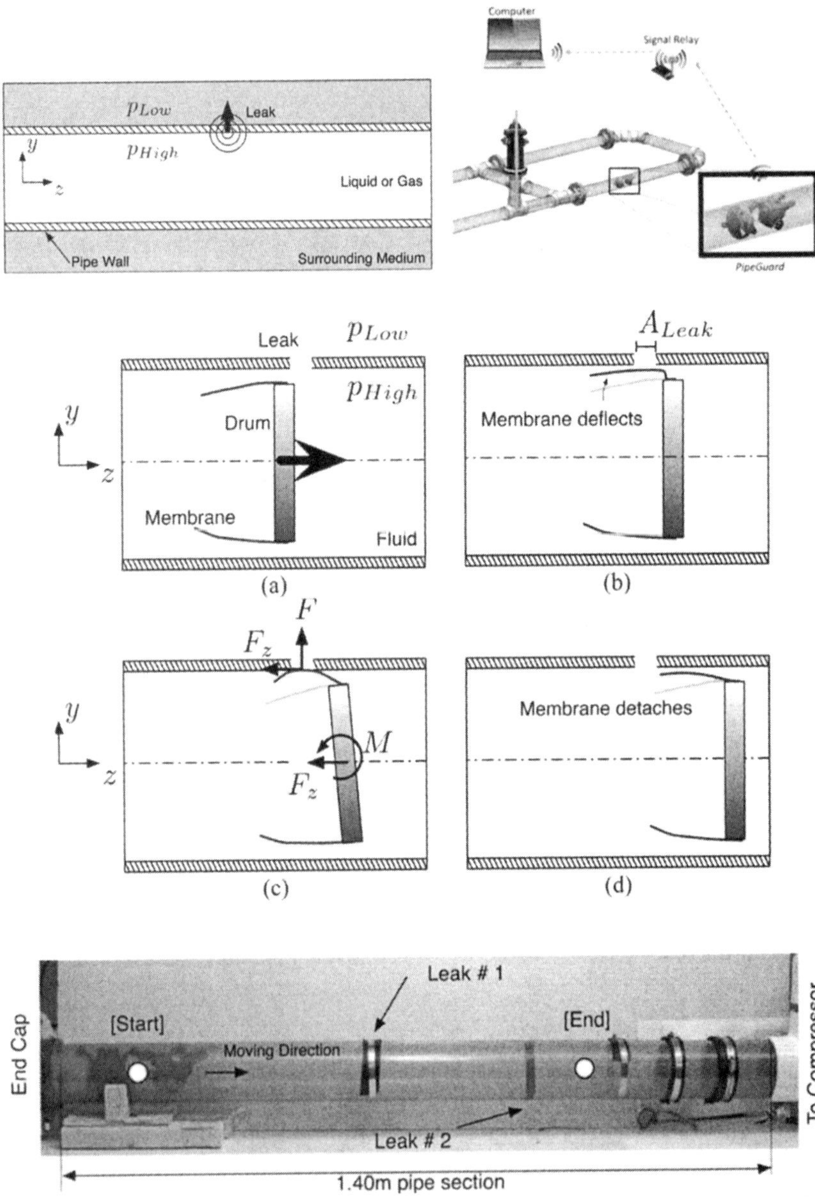

Figure 10.16 Leek detector: (a) the basic setup including a remote control of the robot, (b) schematic drawing of the functions parts, and (c) demonstration of leakage detection [33].

However, underground industries, mostly unaware of WUSS solutions, have not seen many suitable and low cost WUSN solutions and therefore they do not see new potentials of wireless smart senor technology. Therefore, scanners are bare and basic and monitoring solutions are still very expensive to adopt. Therefore, they miss their unique opportunity while the underground sensor industry remains heavily under-performed. This confusion has created a growing false market under *special projects* costing industries and therefore industrial nations enormously.

Now, we see a way to save the industries and promote a new underground smart sensing paradigm under a new breed of *medium range ad hoc wireless underground sensing* (WUSX). As a missing link this new lightweight networking solution is expected to break down large projects into smaller ones while promoting basic services such as SAAS, LAAS, and smaller MAAS to grow into light, low cost, and very highly performed systems. Such integration therefore should expand very rapidly around a paradigm focal point as shown in Figure 10.17.

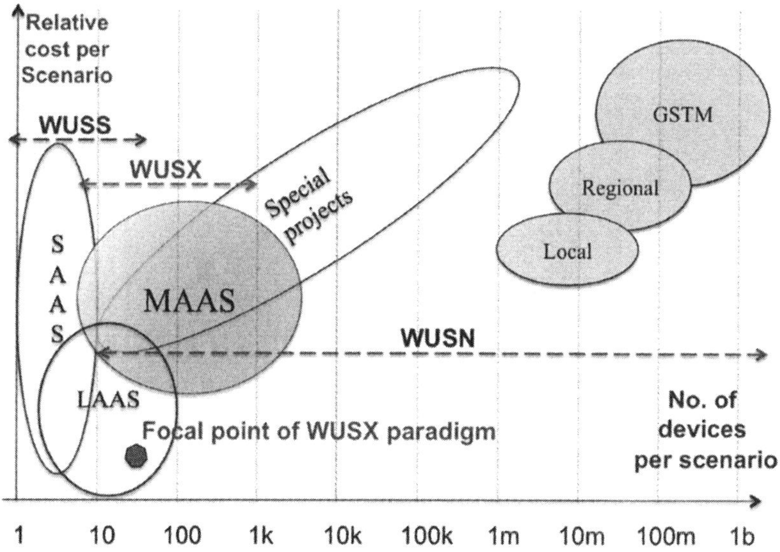

Figure 10.17 WUSX paradigm for future wireless sensor technologies.

10.5.3.1 Cross Layer Design

As discussed in all chapters of our earlier book [34] the traditional layering systems such as open systems interconnection (OSI) and its associated enhanced IP have been designed and optimized to bring acceptable performance, efficiency, and other specific benefits such as speed and reliability to the information. Earlier data networks demanded extra computing and signal powers that are no longer valid for new devices and systems and definitely unsuitable for most WUSN systems operating under extreme underground environments. However, in order to make our new systems compatible for interworking with those and other practical reasons we need to use them. Instead of redesigning another short-lived set of standards we have invented cross-layer technology.

Using cross-layer design (CLD) techniques can be suitably configured for saving energy through reduced protocol stacks, thereby bypassing most of nonessential traditional overheads and handshaking protocols, refining the packet handling schemes, and making error correction techniques less cumbersome. CLD can help with reducing waste of scarce resources such as energy and bandwidth in extreme environments. For example, Lin et al. propose a distributed cross-layer protocol design for WUSN using magnetic induction technology for significant gains in throughput and energy [35].

The CLD is one of the key techniques to help with deployment of the WUSX style for NSUS applications; however, if combined with using an intelligent overlay management system (see Chapter 7) then the simplified system design would enable provision of a highly agile and dynamic network. Figure 10.18 shows the general approach to a CLD-overlay smart sensor design.

Design for optimized deployment (DOD) is expected to provide a high-level design tool with a programmable map of the network. This is a flexible design process that makes use of selectable configurations to enable trade-offs between main deployment factors such as energy consumption, environmental parameters (here underground), and serviceability (quality of service, network maintenance, and service management).

10.5.3.2 Energy

Energy to keep a network alive and active means dependence and is costly to maintain. There are two options to extend the dynamic life of a network, *efficiency* and *sufficiency*. The first option for WUSN often means a limited source of supply and therefore consumption should be controlled and minimized in two levels of: (1) nodes and (2) network, both attempting to reduce the consumption to minimum. The second option is energy harvesting with various methods of using "battery-less" devices using transmitter energy, wireless power transfer (WPT), and energy harvesting.

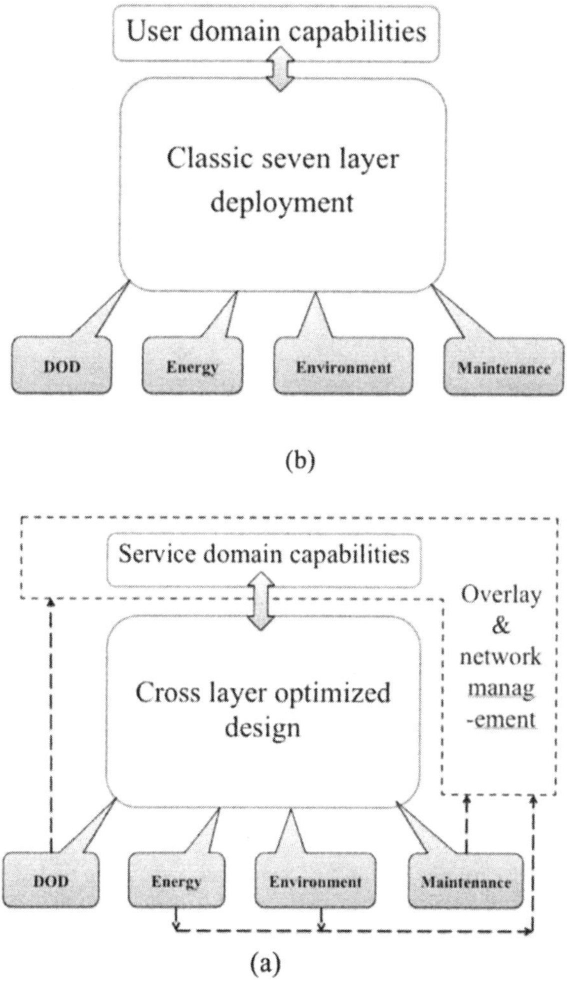

Figure 10.18 Basic structural difference between (a) classic networking deployment compared with (b) advanced networking deployment of smart sensors (WUSX) after adopting CLD for compressed networking layers, enabled overlay, and enhanced network management.

10.5.3.3 Intelligence and Autonomy

With the adoption of robotics the general fear of uncertainty of using well-known artificial intelligence (AI) is fading away. This is good news for making use of controlled intelligence for the adoption of WUSX. Then, using the well-established function of system autonomy can be taken in full. We brief two associated advancements in these areas under unstructured networking and look-up table (LUT) style controlled automation and adaptability.

Unstructured networking has been around for a sometimes but due to its general misunderstanding of *highly redundant networking and interactive hand-shaking dictated by structured topologies* very little attention has been given to it. However, advanced concepts like WUSX can make good use of the natural flexibility embedded in such networks. The main feature of this adoption is these networks' flexibility, where a dynamic system can make maximum use and create true ad hoc networking solutions for new levels of performance and capability.

Automation and adaptability of a network are new features that we can make use of at all levels of nodes, clusters, and networks to help with the lightweight design process of WUSX or WUSN. Being designed around a look-up-table (LUT) or simplified autonomous function such as waking up and shared monitoring for autonomous configurability approach to deployment of unstructured networking using distributed topology control algorithms.

For example if reduced overall energy consumption is a prime objective of networking (NAAS) then using low energy nodes, or ideally zero-power autonomous RFID node networking as suggested by Cook et al. would initiate a new approach to WUSX deployment, where energy sensitive nodes would have longer life for networking and therefore higher viable service periods [36].

10.6 TYPICAL APPLICATION: MONITORING MINES

Our earthly minerals, lying down on or just under the surface of the crust buried and waiting to be excavated, count as the fundamental sources of materials required for fueling thriving industrial nations. Initiated by technologically advanced multinational heavy industries mineralogy expanded rapidly some two and half centuries ago with the discovery of the steam engine to materialize earlier da Vinci's dream designs.

Technically speaking the mine industry could have been the leading force of smart sensor paradigm if a *true free open market* could have been available since the last half a century, but as we know until recently many miners were using a so-called bird in the cage (canary) technology as their early warning system. Now, with growing smart sensor awareness on the horizon and public awareness, the mining industries feel the pressure to adopt suitable technologies to address four critically important problem areas miners at work: (a) isolation; (b) health; (c) safety due to unpredictable rare events; and (d) systems failure.

Isolation

Rich with today's communication technologies working in isolation could become hard and cause serious psychological problems for sensitive workers. Sense of isolation could shorten the working period and trigger anxiety. Feel of accompany on the other hand would enhance time efficiency and performance, which are regarded as the main objectives of working in harsh environment. For this part of the problem the working miners should always have a minimum of one reliable communication link with the outside world.

Health

Whatever the minerals, mining is a *hard*, *injury prone*, and *illness prone* job. The workers should be physically fit and their working time should be shortened to be continuous without any long break periods. Wearing masks and helmets is essential but these and additional tools could easily make them hot, uncomfortable, or ill.

Mine-specific issues include dangers and uninvited hazards such as water and damp and toxic gases. A sudden rise in highly toxic carbon monoxide (CO), the inherent flammability of methane (CH_4), and insufficient oxygen are real. Blasting usually releases CO along with hydrogen sulphide (H_2S) and nitrogen dioxide (NO_2). Continuous inhalation of dust, carbon dioxide (CO_2), and nitrogen, can disturb the blood system resulting in chronic diseases [37].

Safety

The immunity and safety of miners is now coming into the attention of authorities throughout the world, but unfortunately some places are far behind. In South Africa for example, mines are still functioning as they were some 100 years ago with a cyclic drill-blast-clean process. Holes are drilled in with the *mining face explosives* placed in the holes for a mine wide detonation (entire mine cleared before the explosion). The fumes are then vented out 4 hours later, when reentered for clearing the blasted materials. This cycle then repeats on a daily basis [38].

Earthquakes and other natural disasters could also easily endanger the safety of miners and nearby communities. Therefore, the use of adequate preventing structures and technology should be adopted as the industry's prime objective.

Systems Failure

Nothing is worse than unreliable technology for incident prone mine working environments. Financial limitations and unawareness of new technologies due to poor management often means taking risks with miners' lives and the environment around the mines are merely responsible.

For the mining industry environmental hazards play a prime role for prosperity and are far too serious to be overlooked. Taking risks on the system's failure could result in disasters.

Wrong policies and the negligence of the local government could also endanger the safety. In the case of inactive mines (abandon due to low productivity or lack of resources) it can cause landslides and trigger other hazards with serious risks to people and the environment. For example, the 2012 report of ESA impacts of the landslides and hazardous risks caused by inactive and abandon mines discuss.

10.7 CONCLUDING REMARKS

This chapter discusses the important aspects of adopting wireless smart sensors for an underground environment for our urgent uses such as the discovery of the sources of earthquakes and industrial opportunities. On the quest to find the optimum solutions for propagation we have examined three main competitive technologies, electromagnetic, magnetic induction and seismic waves. By combining these three, we can support a large number of smart sensing opportunities with a wide range of prime services and applications. We have also revealed and characterized many potential markets to enable industries to move from small-scale production into large and global production so as not to miss the biggest underground sensor paradigms. This includes "smarting up" the crust for better visibility, and saving lives while enhancing the quality of life on earth through thriving underground industries. A promising WUSX future requires innovations in agile and fast reacting smart sensors, which can make use of any of the progressive propagation technologies.

References

[1] Alinaghi, A., et al., "Seismic tomographic imaging of P- and S-waves velocity perturbations in the upper mantle beneath Iran,",*Geophys. J. Int.*, Vol. 169, No. 3, 2007, pp. 1089–1102.

[2] https://en.wikipedia.org/wiki/Earth_Observing_System.

[3] Petiteville, I. (ed.), "Satellite Earth Observations in Support of Disaster Risk Reduction," *European Space Agency*, Special 2015 WCDRR Edition, www.eohandbook.com/eohb2015/files/CEOS_EOHB_2015_WCDRR.pdf.

[4] Riebeek, H., "The Carbon Cycle NASA, EarthObsevatory, 2011 https://earthobservatory.nasa.gov/Features/CarbonCycle.

[5] Madsen, H., "Doubling Down on Climate Progress," *Environment New Hampshire Research & Policy Center*, 2017, www.environmentnewhampshirecenter.org.

[6] Sheikhpour, S. et al, "Agricultural Applications of Underground Wireless Sensor Systems: A Technical Review," In *Wireless Sensor Systems for Extreme Environments: Space, Underwater, Underground and Industrial*, Ch.17, pp. 351–380, H. F. Rashvand and A. Abedi (eds.), John Wiley and Sons, The Atrium, England, 2017.

[7] Bally, Ph., "The International Forum on Satellite EO and Geohazards," *The Santorini Conference*, Santorini, Greece, May 21–23, 2012.

[8] Roser, M., and H. Ritchie, "Our world in Data: Natural Catastrophes", https://ourworldindata.org/natural-catastrophes/.

[9] https://en.wikipedia.org/wiki/Pangaea.

[10] https://en.wikipedia.org/wiki/Seismic_wave.

[11] Dziewonski, A., and B. Romanowicz, *Seismology and the Structure of the Earth*, Elsevier, 2014 (Vol.1, Treatise on Geophysics).

[12] Rawlinson, N., 1996, "Methods and codes used in seismic tomography," *Australian Geological Survey Organisation*, Record 1996121, Canberra, Australia, p. 49.

[13] Koulakov, I., et al., "Feeding volcanoes of the Kluchevskoy group from the results of local earth- quake tomography," *Geophysical Research Letters*, Vol. 38, L09305, 2011, doi:10.1029/2011GL046957.

[14] Koulakov I., "High-frequency P and S velocity anomalies in the upper mantle beneath Asia from inversion of worldwide traveltime data," *J. Geophys. Res.*, 116, B04301, 2011, pp. 1–22, doi:10.1029/2010JB007938.

[15] Akyildiz, I. F., and E. P. Stuntebeck, "Wireless underground sensor networks: Research challenges," *Ad Hoc Networks*, Vol. 4, 2006, pp. 669–686.

[16] Akyildiz, I. F., et al. "Signal propagation techniques for wireless underground communication networks," *Physical Communication Journal*, Vol. 2, No.3, 2009, pp. 167–183.

[17] Rashvand, H. F., Lecture notes, 2010 Series, Part II, *Signal Bandwidth and Noise*, MSc Course on Mobile Communication Systems, School of Engineering, University of Warwick.

[18] Peplinski, N., et al., "Dielectric properties of soils in the 0.3–1.3 GHz range," *IEEE Trans. Geosci. Remote Sens.*, Vol. 33, No. 3, 1995, pp. 803–807.

[19] Baig, A. and T. Urbancic, "Magnitude Determination, Event Detectability, and Assessing the Effectiveness of Microseismic Monitoring Programs in Petroleum Applications," *CSEG Recorder,* http://csegrecorder.com/articles/view/magnitude-determination-event-detectability-and-assessing-effectiveness.

[20] Hedlin, K. M., L., and G. Margrave, "Seismic Attenuation Problem: Annual International Meeting, Society of Exploration Geophysicists," *Proc. 8th PIMS-MITACS Industrial Problem Solving Workshop*, UBC, May 2004.

[21] Young_G. N., et al., *Utility-Locating Technology Development Using Multisensor Platforms*, The National Academies Press, Washington, D.C., 2015.

[22] Habib, M. K., "Humanitarian Demining: The Problem, Difficulties, Priorities, Demining 001 Technology and the Challenge for Robotics" In *Humanitarian Demining Innovative Solutions and the Challenges of Technology*, pp. 1–56, Habib, M.K.(ed.), InTech Europe, Croatia, 2008.

[23] Bellino, A., *Advanced methods for rock discontinuities estimation in tunneling*, for PhD degree, Mechanics, Politecnico di Torino, 2013.

[24] Nguyen, L.T. and T. Nestorović, "Advance Exploration by Stochastic Inversion of Tunnel Seismic Waves - A Numerical Study," *Proc. of the World Tunnel Congress* 2016, San Francisco, April 22–28, 2016.

[25] Hammerschmidt, A., et al., *Innovations to Locate Stacked or Deep Utilities*, The National Academies Press, Washington, D.C., 2014.

[26] Benson S.M., and P. Cook (Coordinating Lead Authors), "Underground Geological Storage," In *IPCC Special Report on Carbon Dioxide Capture and Storage*, pp. 195-276, Intergovernmental Panel on Climate Change, Cambridge University Press, Cambridge, U.K, 2005.

[27] Benson, S.M., and D.R. Cole, "CO2 Sequestration in Deep Sedimentary Formations", Elements, Vol. 4, 2008, pp. 325–331.

[28] Luberti, M., *Design of a H_2 Pressure Swing Adsorption process at an advanced IGCC plant for cogenerating hydrogen and power with CO_2 capture*, PhD thesis, School of Engineering, The University of Edinburgh, 2015.

[29] Vaezi, Y., "Applications of Seismic Interferometry" in *Microseismic Monitoring*, PhD thesis, Geophysics, University of Alberta, 2016.

[30] Harris, M.L., et al., "Toxic Fume Comparison of a Few Explosives Used in Trench Blasting", *Proc. 29th Annual Conference on Explosives and Blasting Technique*, Nashville, TN, International Society of Explosives Engineers; February 2–5, 2003, pp. 319–336.

[31] Neumann, P.P., et al., "Tomographic Reconstruction of Soil Gas Distribution From Multiple Gas Sources Based on Sparse Sampling," *IEEE Sensors Journal*, Vol. 16, No. 11, 2016, pp. 4501–4508.

[32] Chatzigeorgiou, D., et al., "Reliable Sensing of Leaks in Pipelines," *Proc. ASME 2013 Dynamic Systems and Control Conference (DSCC2013)*, Palo Alto, California, October 21–23, 2013 (doi:10.1115/DSCC2013-4009).

[33] Chatzigeorgiou, D., et al., "Design of a Novel In-Pipe Reliable Leak Detector," *IEEE/ASME Transactions on Mechatronics*, Vol. 20, No. 2, April 2015, pp. 824–833.

[34] Rashvand, H.F, and Y.S. Kavian, *Using Cross-Layer Techniques for Communication Systems*, IGI Global, 2012, (ISBN 1466609613, 9781466609617).

[35] Lin, S.C., et al., "Distributed Cross-Layer Protocol Design for Magnetic Induction Communication in Wireless Underground Sensor Networks," *IEEE Transactions on Wireless Communications*, Vol. 14, No. 7, 2015, pp. 4006–4019.

[36] Cook, B.S., et al., "RFID-Based Sensors for Zero-Power Autonomous Wireless Sensor Networks," *IEEE Sensors Journal*, Vol. 14, No. 8, 2014, pp. 2419–2431.

[37] Mathas, C., "Toxic Gas Sensors Replace Canaries and Mice in Mines," *Digi-Key Electronics*, 2015-04-23, https://www.digikey.co.uk/en/articles/techzone/2015/apr/toxic-gas-sensors-replace-canaries-and-mice-in-mines.

[38] Green, J., "Toxic Gas Sensors Replace Canaries and Mice in Mines," *Proc. 5th IEEE Robotics and Mechatronics Conference of South Africa*, November 26–27, 2012.

[39] NASA, Planets, *Earth: By the Numbers*, https://solarsystem.nasa.gov/planets/earth/facts.

Chapter 11

Wireless Sensor Systems for Industry and Transportation

"Many essential human needs can be met only through goods and services provided by industry... Industry has the power to enhance or degrade the environment; invariably does both"

World Commission on Environment and Development, 1987

Taking the opportunity to view, analyze, and discuss the industrial world of WSN through innovation angles this chapter provides a unique and intensive analysis of three interrelated processes: (1) requirement and uses of the innovation process for WSN enabled industrial production and transport systems; (2) the impact of adopting WSN technologies in the production process; and (3) the state of the art and progress in design and deployment of WSN services in harsh and extreme environments including underwater, underground, manufacturing, space, and transport systems.

11.1 INTRODUCTION

Intelligence: humans have been the beacons of our developing civilizations enabling honest industries to thrive upon bright ideas and serious deployment of new solutions, where our clever and artful designers are developing superior technologies to move the progressive wheels of industries. Under such a climate our mathematical skills were born long before our known history with their unique strengths and accuracy, to count as the mother of our computing and information technologies. This presents as one of our greatest tool for enabling the smart sensors as industrial devices and instruments to make our living concise and knowledgeable through bringing valuable information about ourselves and everything around us into the information world where we have the capability of processing, analyzing, and making decisions. Wireless sensors provide painless penetration to smart sensors allowing them to achieve their ultimate ability in utilizing the upmost agile distributed intelligence to master our new systems and offer further penetration into unknown extreme environments to help engineering in our industries and our lives with them.

Now, let's try to examine what lies behind the process. The basic model of innovation identifies need as the main force behind any successful product. This model comes from the fact that intelligent humans know what they want, enabling producers to identify main properties of the potential market to make a calculated set of strategic plans to supply the right product at the right time. These follow the innovation process staged to make this happen through various phases of R&D, tooling, production, and marketing. In order to identify the potential demand the necessity of the original need is measured upon the Maslow hierarchy. Figure 11.1 shows the original hierarchy, which can be used to predict the market strengths including its scale, size, and its duration. The two bottom levels of the hierarchy (physiological and security) usually go together indicating the most essential components of the need therefore; they are also called the basic needs group. That is, when one lacks any components of these two levels then all higher-level components become insignificant. Therefore, basic components or lack of them normally means a long lasting burning market opportunity. Devising a solution in this area leads to great innovations and deploying any solutions means successful innovation without any failure [1].

It is interesting to see that all of our smart sensor applications are centered in the basic need area. The earlier innovations in the basic levels have been around a long time but most of them are out of date (pre-WSN) and urgently need to be upgraded as soon as possible. Then as we consider the extreme environments (EEs) we realize that such harsh facts demand WSN solutions to enable industrial users to expand their operations in these areas in parallel with explorers and scientists to investigate such unknown areas.

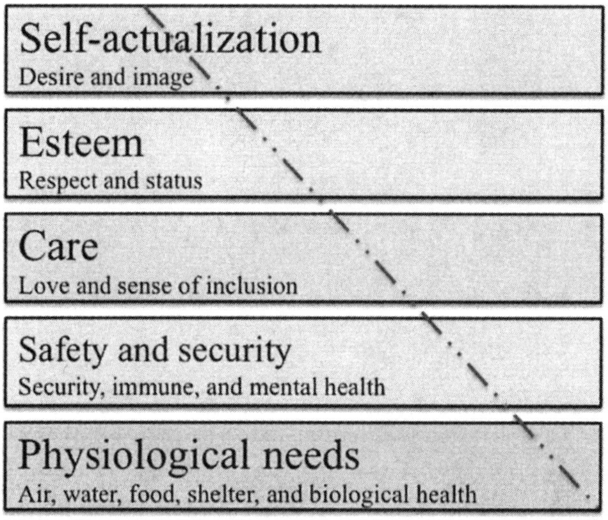

Figure 11.1 Maslow hierarchical model of the needs [1].

Throughout the last seven decades this model of need and the five associated generation models of innovation, have been challenged and often criticized as in some cases due to oversimplification of the process they did not work. We, however, believe the model is still valid in its own right and that any failures are merely due to the poor interpretation of two main factors of the innovation; the need and the idea that enables the innovation. Obviously the market forces and financial aspects of the business, not significant in a small-scale operation, can grow into major forces in a globally saturated market due to some political upheavals. That is, though this model is becoming hard to apply to new technologies and their markets its principles are still valid and should be updated to suit today's global technological forces. Some of our industries have been troubled with new world orders and may take a few more decades before they find their optimum positions. The free market is now replaced by free information. Highly regarded valued expert design engineers are replaced by low cost academic researchers, meaning that failure of innovative work is not due to the models of innovation but merely due to lack of dedicated engineers and engineering managers who seem to be vanishing from the scene. We therefore need to hold onto our principles and analyze new projects and judge them upon their suitability for a sustainable industrial innovation.

In our case of deploying wireless smart sensing applications the technology is still fresh; it is only two decades old. It comes with new brand capabilities and by looking at the need hierarchy it is easy to spot its significant potentials for many new technological paradigms for level 1 and level 2 indicating significant market value for any new smart sensing products enabling provision of WSN services. That is, sensors, transducers, and accelerators have been used in industrial automation and production for many decades but the dimensions of "wireless" and "smart" are new and significant in their own right generating potential vertical markets. Let's discuss the industrial impacts and also their service penetration onto the challenging environments addressed in this book.

Wireless in its generic format brings many significant advantages including: (a) removing communication and control cables and wires from the sites; (b) adding mobility and physical independence between devices and units; (c) enabling low cost updates and upgrades; (d) encouraging integration of various tasks and services such as surveillance and ad hoc monitoring with production lines; and (e) fostering new modular design technologies. The sensing specific features of wireless are even more promising for lifting up the barriers of obscurities in the process. For example, a simple thermometer that comes with the significant overheads of using wires, cost and complexity of installation can become totally invisible costing nothing. It can be moved about easily at zero cost at any time. Without the comfort of wireless most manufacturers would not have bothered enhancing valuable stores for production, services, and shared databases.

Smart sensing enables the sensors with some ICT functions. The low power lightweight processors such as Motes or equivalent hardware (circuits) help sensors to become somewhat autonomous and therefore an active node of the WSN able to select, filter, and preprocess the samples (data) for local decisions and required actions before sending them to the central processor or database. Smart also means being autonomous to choose to join in as a member of a cooperative group or cluster for a common cause. Examples include moving an unmanned autonomous vehicle (UAV) or enabling a robot to follow or track an object.

As the energy harvesting is maturing rapidly adding more autonomy to the sensors they gain more independence in being able to move around and become multifunctional sensing agents. The new downloadable software features of the middleware technology would help with having more agility and lightweight dynamic nodes for further smart sensing capabilities.

The rest of this chapter is organized as follows. Section 11.2 examines global aspects of the innovation process. Section 11.3 discusses the industrial production process for some extreme environments, while Section 11.4 discusses three main aspects of service deployments. Then Section 11.5 provides a brief discussion of evaluating the design and development of innovative production processes and service deployment of WSN for various extreme environments.

11.2 INNOVATION ASPECTS

The industrial requirement of the WSN comes with two interrelated dimensions of: (1) design, development, and production of WSN systems and (2) deployment and provision of the service. We should use our utmost design skills to take care of three main aspects of innovation: (a) superior quality and performance; (b) low and very competitive cost of production and therefore provision of the service; and finally (c) long lasting (i.e., sufficient life for the product to compensate higher infrastructure costs including the impact of any possible side effects).

Now, in order to be able to analyze our industrial projects for progress and impacts of producing and implementing WSN products we refine the above three aspects of innovation into three industrial WSN processing subgroups: (A) quality control production; (B); automation for speed and efficiency; and (C) controlled innovation side effects. It may be worth mentioning that subgroup A mainly considers the factory side of production quality but it should also include the outlet of the production. The same argument also applies to C to extend the side effects to include all aspects – production, products, and WSN services.

11.2.1 Quality Controlled Production

Quality is normally a subjective feature of a process, being a production or a service. However, in the production line we define a measurable feature to control the quality in either case. The role of sensors in a quality control task is important. They can automate the process for minimizing waste and reducing production time. They can also become an integral part of advanced robotics to maintain quality for any set of practical specifications.

Sensors can measure the product's specifications and inform the production line to take action if levels begin to drop. Some sensing devices such as chemical sensors may not be as fast as mechanical and physical features, where the quality of raw materials and subproducts are required to be checked before we end up with many unwanted or low quality products in the process. We also use sensors for controlling the product line for quality. It becomes more interesting to have sensors controlling the production of "sensors for quality." In this particular case we have two sets of sensor systems to consider. One set is for the production of sensors, where sensors of the second set, as products in this process are examined for quality and performance.

11.2.2 Automation in the Production Line

In the light of rapid prototyping and programmable devices virtually any sizeable production line can benefit from process automation. The main objectives normally are speed and efficiency. The efficiency could be compromised by a few other parameters such as raw material efficiency, energy consumption, and requirement for human supervision.

The automation could be easily expanded to minimize the use of manual workers for the very clear reasons that robots and automated machines not only bring down the cost for a very competitive market price but they also guarantee the required quality if designed properly. Robots themselves now mostly have embedded programmable sensors or very reliable interfaces for them. Programming is also modular and automated to fit various production lines at a minimum of effort and cost. Most important of all robots and automated systems are far more reliable than humans and they suit long hours of operation with far fewer mistakes and errors.

11.2.3 Controlling Innovation Side Effects

Every time we achieve a new technological breakthrough we may need to generate some new products, which are usually called radical innovation[27]. This in a competitive market means joining the bandwagon of technological superiority to dominate a highly competitive market. Introduction of this innovative new product to the market, however, injects sudden changes to the market and all of its related environments. If the

[27] Radical innovations change the marketing landscape. A large-scale production for the global market has its own unique rewards but also due to many unexpected incompatibilities (habitual, demographic, and cultural) and limited useful/valid/working life they can leave behind the discomfort of side effects.

infected environments are sensitive to the impact then such a sudden change leaves some undesirable scars behind called *innovation side effects*.

In some cases the introduction of a new idea could be managed gradually making the changes digestible and therefore controllable leaving no scars. This is called incremental innovation. For the case of using WSN in extreme environments, however, we have different impacts. The first is that an extreme environment is impeccable in the sense that it has no sensitivity to any change. The second is that any radical innovation is welcomed and regarded as a positive penetration into an unknown environment. We should, however, bear in mind that some spots of an EE could be already discovered and therefore sensitive to a radical change and also any new penetration could be regarded as radical due to the nature on most extreme environments.

In general, a new innovative project with the side effects in mind should consider all of the following five stages in the operation: (1) dis-invention of, and old or remaining side effects[28]; (2) taking on the planned production process; (3) testing the relevant devices and components; (4) design the required service; and (5) evaluating possible side effects of the new service.

It is interesting that WSN itself can bring the required capabilities (while identifying and removing) any possible side effects. That is, we can devise new WSN-enabled systems to use WSN as the key technology for measuring the effectiveness of a new project as well as dealing with the old side effects: (a) devising and measuring the remaining old side effects; (b) devising and measuring the side effects, which could be injected due to deployment of the new project; and (c) setting up a required database to demonstrate the time span for all significant aspects of the side effects assigned to a project.

11.3 INDUSTRIAL PRODUCTION

The process of an industrial wireless sensor network (IWSN) is new and can be defined as an *innovative adoption of wireless smart sensor technology into industries* creating three main dimensions of innovative production (process), innovative products (sensors), and provision of innovative services enabling new industrial WSN paradigms led by the

---

[28] In many advanced societies side effects of any poorly managed previous innovative services may leave significant scars, which may cause serious problems for new services to remedy.

extreme environments. As these developments are potentially environment-dependent, then all above-mentioned aspects of productivity varies significantly from one environment to another. Therefore, such an environmentally dependent opportunity can easily distort many experts' vision. This invisibility could be regarded as the main reason why so many marvelous paradigms have not been utilized. Therefore, these unique business opportunities may never be identified unless a clear set of definitions is set and some estimated costs for innovation components useable in the extreme environments are established.

In order to measure and analyze the industrial aspects of WSN we consider a generic classification for extreme environments industrial production as shown in Figure 11.2.

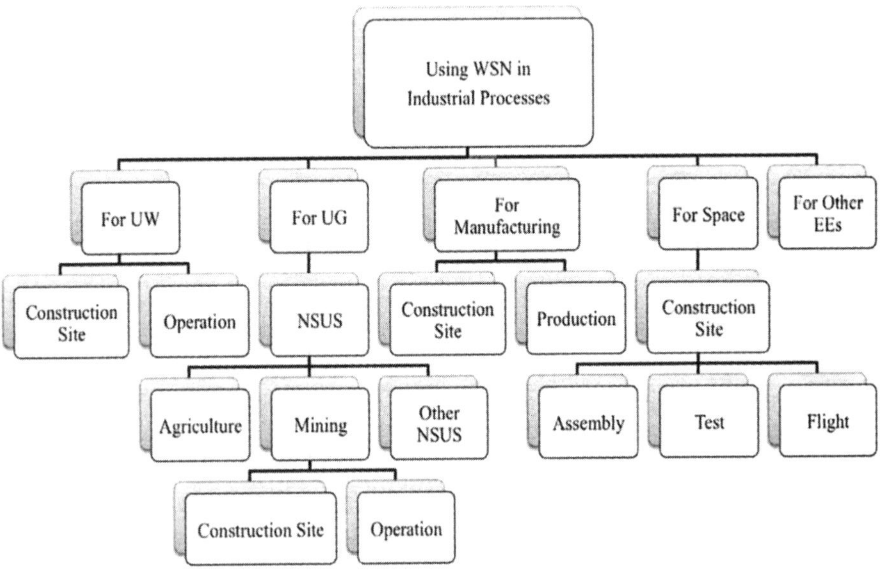

Figure 11.2 Family tree classification adopting WSN technologies to production in various extreme environments.

11.3.1 WSN for the Underwater Industry

In order to deploy basic shallow water's underwater wireless sensor network (UWSN) exploration and monitoring service we need to establish a loosely connected network placed at a location with some moving and some floating sensors[29] towards their underwater area that requires the service. Using today's acoustic enabled narrowband services compromises basic measurements (sensing) of some physical, chemical, and biological properties of the water-contents plus a small gateway, normally sitting on a buoy or a platform enabling a service usually controlled from another location. An ad hoc style networking ensures connectivity of the floating sensor nodes, which usually are relayed upward maneuvering around without losing their connection. Typical sensing data may feature: temperature, conductivity, turbidity, pH, oxygen content, seismic, and bathymetry[30] features.

In general the sensor nodes are designed to be smart enough to maintain the three basic tasks of communications (signal propagation using acoustic, radio, seismic, and magnetic technologies), collecting data, and maintenance. The last task includes, localization (maintaining themselves within an acceptable area), energy management, and basic signal and data processing. A more complex service requires some cluster heads to coordinate multisensory operations. Such a simple operation is called remote sensing, where most of the data processing and control functions are centralized at a large platform or in the database center somewhere within an establishment connected (e.g., via the internet or satellite).

The production of the parts for such basic applications is normally supplied through some commercial-off-the-shelf (COTS) system and deployment is straightforward usually automated using embedded and programmable devices. The design of such networks is simple and deployable by the nodes, either being thrown in the water one after another or placed around an expected location then adjusted automatically under the control of a program downloadable through the constructing network [2].

Toxics and Underwater Garbage

Another interesting area of development for UWSN application is due to the growing outcry of eutrophication. This is happening due to growing offshore industries and recent expansions of leisure boating and water sport activities. As an example, we have our growing toxics and garbage

---

[29] A fixed interconnected set of sensor nodes sitting on the seafloor would be an easy UWSN option but it may not be always possible.

[30] Bathymetry in here is for basic distance measurement. For geological and isobaths applications we may require dedicated sensors.

moving into the oceans. Following the "Pacific Ocean as Great Pacific Garbage Patch" in Chapter 9, we see much more going on with polluting of lands and waters. Mining companies are dumping more than 180 million tones of hazardous mine waste each year into waterways and seas throughout the world. They are affecting the water's natural properties at places with uncommon heavy metals and chemicals affecting marine life. For example, the Canadian shallow marine dumping of Island Copper and Kitsault mines have traveled long distances of up to 35 km. Not to forget various incidents with the underwater pipelines [3].

Now, many nations are facing the problem of identifying, locating, and cleaning, removing, and de-activating the effects from such growing wastes. The task is not simple (by any means) and now water wastes are becoming new serious problems that can be dealt with practically using smart sensors. For managing and removing such waste and cleaning up its effects we need to design a whole new range of fast acting sensors, a large number of UWSN systems enhanced with waterproof powerful smart sensors to monitor, detect and collect such widely ranged items. Here, we may have to call in our user friendly remotely operated autonomous underwater vehicles (AUVs).

AUV (Underwater)

Also known as mobile platforms, the AUVs, with their basic and new capabilities such as mobility, communication, and extensive sensors, are able to deal with such a large-scale problem. Their key features include linking up and controlled remotely with self-sufficiency features (energy harvesting, local autonomy, local data processing functions). They are now able to communicate using greater bandwidths and longer distances using acoustic technology.

Their floating and moving managed under a well-designed movement control system seems to suit many water and underwater applications. They are also equipped with flexible floating sensors and clustered sensors to enter smaller places for their observations. Due to their size and weight they can accommodate further attachments such as accelerometers and actuators and are able to manage mechanical interactions when required.

Due to their similarities to the automobile and boat industries the production of UAV systems can be regarded as a common practice and up to date with new computerized mechanical engineering. Also, due to their relatively small size and wireless technology the integration of smart sensors and UWSN technologies is also straightforward, cost effective, and flexible. These vehicles now have been produced in a wide range of sizes with a variety of new functions. In fact, using new WSN capabilities enhanced due to extra sensing attachments and devices can be designed upon request at no extra cost [4].

11.3.2 WSN for the Underground Industry

Until recent advances in smart agriculture and heavy machines, most underground industries were crude and primitive. Mining has been the number one source of GNP in many industrial nations and therefore continued to stay strong at the cost of many large-scale casualties. With the rise of sensor technology and robotics, however, the two prime underground industries, namely agriculture and mining, are due to change. Sensors can help the agricultural and farming industries to improve greatly through improved sensing devices and machinery. That is, the production of sensors and small devices has been or should become abundant in time, while machines remain to suit large-scale operations only. The deployment of the required wireless underground sensor networks (WUSNs) should improve the quality but not quantity (see Chapter 10). Therefore, we see limited potential for new paradigms for the agricultural and farming industries.

On the other hand, mining industries can benefit greatly from UWSN technologies leading to dedicated paradigms through new safety (against fire, collapse, earthquake, and early warning systems) and reliability measures coming from use of the smart sensors and robotics. All three industrial stages of the innovation infrastructure for the provision of service (production, the sensors, and UWSN deployment) are now set to rise towards the global market following the innovation rules mentioned in Section 11.2. However, due to the nature of mining the initial preparations and level of investment depends on the depth, size, and required speed [5, 6].

11.3.3 WSN for Manufacturing

Use of WSN at manufacturing sites is moving ahead in industrial nations as they are planning to upgrade their old production sites and establish new ones. We see that integration of smart sensors along with robotics and automation is becoming the common practice.

In order to support the innovative requirements mentioned in Section 11.2 new or upgraded manufacturing sites for producing WSN products requires considering a good use of basic tools compromising (a) robotics and automation; (b) lightweight intelligence in the form of agility, agents, and augmented reality (AR); (c) tagging RFID rich material control and using the industrial Internet of Things (I^2oT); and (d) optimized dedicated smart sensing for the production line (IWSN). We discuss items (a–c) here, but address IWSN technologies in Section 11.4 for more details.

Robotics

The use of robotics in the production line is well established in principle and being used throughout the world for large-scale productions replacing experienced workers for many features including quality, accuracy, speed, and reduced human error productivity; as we move into global markets any of these features could become necessary for businesses to survive. Robotics as a process is regarded as an extension of automation showing superior features when used in harsh and extreme environments enabling a wide range of WSN applications. They are particularly useful for dangerous services such as rescuing survivors of disaster areas, and able to go through fire, radiation, and explosives or submergence in water. We therefore count all of complex, dangerous, and harsh manufacturing industrial sites as extreme environments.

Our manufacturing sites producing sensors, devices, and special components for the extreme environments often require unique accuracy in the product, which also requires a basic accuracy in the production line. Assembly lines for special production of MEMS and nanotechnology sensors not only require accuracy but they need to follow certain lengthy instructions making the tasks too complex for manual workers.

In many complicated production line cases we need some sensor rich robots to monitor and produce the required products so that they can work in a very complex environment, handle many sensor rich piles of complex data, and follow given instructions for end products matching the required specifications.

Robots can show their utmost usefulness on the speed in production and assembly lines for all tasks; being routine or complicated. In a well-designed production process one robot can handle many tasks at the same time and be able to move from one production line onto another for assembly and so on without any training.

In many cases in order to speed up a very complex production process the use of multirobot assembly to deal with different stages of the production brings more flexibility into the process. Such a complex process becomes feasible through the high speed networking of them through a central controller upon an optimizing program such as using a multitemplate system to make better use of the resources (e.g., a mechatronic system) [7, 8].

I2oT

With the abundant use of RFID devices helping production lines to produce WSN systems we often need to recover our production system from a specific ambiguity for which we need access to our databases and ultimately get help through the internet. The impact of the *Internet of the Things* (IoT) and its standards on manufacturing and industrial environments has been quite significant.

Although the IoT itself has not been progressed as originally expected its industrial version seems to be catching up much faster. The industrial IoT (I^2oT) seems to be the most viable development, where through the internet global industries can cooperate for their fair share of the growing global markets. In other words, I^2oT is already in existence for two reasons: (1) standards are developing under G5; and (2) the only useful parts of the IoT are for industrial, business, and marketing applications that are already under development, and those most relevant to the quality and speed of production make extensive use of sensors and measuring devices. The network control and management of a manufacturing site and its process control unit can be greatly enhanced with new industrial applications and adoption of smart manufacturing, (i.e., a clever use of the potential market demands, and control of the production upon its available supply machinery).

Upon the cyber-physical systems (CPSs) model of the core technology we have a data interface between users and the cyber world. Figure 11.3 shows its 5C (connection, conversion, cyber, cognition, configuration) architecture. It also shows the significant role of WSN on both ends of the production process and its applications. That is, though its manufacturing side can be clearly integrated its applications in extreme environments have a way to become available to the internet users. In some cases experts suggest that the factory, as we know it under the so-called smart-factory, initiative is due to be reduced to a workshop production [9, 10].

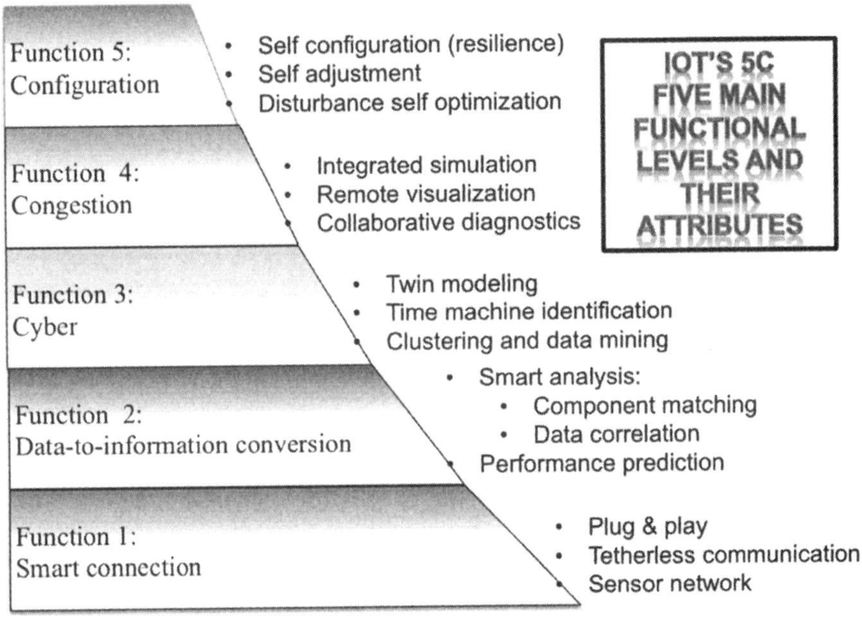

Figure 11.3 5C architecture of the industrial IoT-enabled manufacturing system [9].

As without a proper support from industrial communities no significant technological developments can happen then it is essential to convince the industrial forces to act promptly so that they do not miss any opportunities. For this let's examine the commonly used IoT[31] technology roadmap, shown in Figure 11.4. Figure 11.4 disassociates the use of smart sensors from the IoT's first four phases of development while for certain we well know smart sensors roles in the second phase, obtained from the vertical market applications. That is, none of the objectives mentioned on the list have been successfully deployed due to distractions injected from the third phase, ubiquitous positioning. The third phase objective of 'locating people and every objects' is too ambitious and ambiguous to the industrial communities. The fourth phase leading to 2025, teleoperation and telepresence, comes with the new expansion of WSN into extreme environments. This big challenge of sensors as great potential development can only be managed and deployed through the concept of WSN and associated smart sensor technologies.

[31] Although this road map mentions the generic name of IoT it in effect addresses the industrial aspects of the process and thus should be called I^2oT.

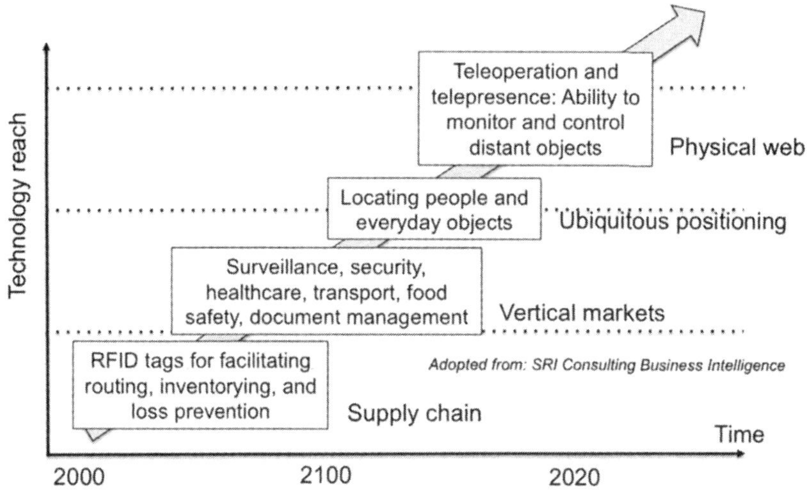

Figure 11.4 The IoT technology roadmap [9].

11.3.4 WSN for the Aviation & Space Industries

Due to closely controlled industrial production process of aviation and space we almost always consider novelty and innovation for this environment. Although this part of industrial production process in general includes any vehicles above the ground due to their short flight courses airplanes, helicopters, and short flying AUVs belong to the land (aviation) and so are not considered for the extreme environments vehicles. Therefore, only high altitude platform stations (HAPS), satellites, and space mission vehicles count in our space related extreme environments.

However, the requirements for long lasting missions such as Mars are much tougher than the satellites and those visiting the moon. Due to their different survival sensitivities each group have their own unique production and service resulting in the use of very specific sensors and WSN technologies for production of these vehicles.

In this book we have some other chapters discussing various aspects of space and associated developments. For space as an extreme environment we have Chapter 2 discussing the networking aspects of WSN, Chapter 3 for the resilience and longevity design, Chapter 4 for the electronics and hardware challenges, Chapter 5 for material aspects the design, and Chapter 12 for the use of WSN in space applications covering critical design and developments of space as a extreme environments.

11.3.5 WSN for the Transport System Industries

Here we look into the case of enhancing transport systems via a better use of smart sensors and associated WSN technologies. Due to similar smart sensor enhancement the future system is called the smart transport system (STS). There is, however, another closely related approach for upgrading the transport system upon making better use of the massive information being abundantly made available over the internet, called the intelligent transport system (ITS), which is usually achieved through adopting the IoT.

Now, let's analyze this to see what the transport system is made of and what areas of it require either smart sensors or the internet, if not both. In every community there are two ongoing types of service provision with a ratio varying extensively from one to another upon the choice, affordability, and mentality of the users and the local government. The ratio and associated features are out of the scope of this chapter. However, the structure of the transport system and its components are the main subject of our discussion.

We have four distinctly different transportation methods for moving on earth: (a) roads, a historical invention and the most natural way of traveling from one point on land to another being developed from on foot all the way to use of motored vehicles; (b) the invention of railroads sometimes counted as a way of traveling on the road but we would like to keep them separate for their various economic and technical differences; (c) boats, another historical invention to complement the road, crossing waterways being developed into motorized ships and underwater transporting vehicles; and finally (d) flight for the speed and associated convenience. There are also a few other possible components of transportation, which due to their low usage we will not discuss here. Let's discuss the potential of these four components for being significantly enhanced by adopting the WSN.

Out of the above methods, we observe that for local and national transportation using a vehicle on the road remains the most popular mainly due to its door-to-door features.

Unfortunately the saturating volume of traffic forces many city travelers to choose trains and for long distances and traveling overseas other alternatives such as flight, train or boat may also be required to complement the need. There is also the problem of global warming due to the CO_2 of the existing fueling system, which still persists, and we need to deal with it. To add to the problems, despite significant improvements in the manufacturing of the vehicles due to human errors and carelessness casualties on the roads are still quite high. Finally road maintenance has always been high causing inconveniences and long delays, which often increased due to earthquakes, flooding, damaged bridges, tunnel collapse, and land subsides.

Using smart sensors and WSN technologies would help extensively in most cases. However, they are usually costly and very time consuming if the works require closure of the road. Many unaware authorities may not be experts or unconvinced of the benefits and look for alternatives solutions. It could be the reason that many have been attracted to the ITS and better use of information made available through the IoT.

As expected one of the potential domains of progress for the IoT is transportation. Enabling access to the vast amount of global information over internet on the move can help us as a traveler considerably in various aspects of life and safety and of course also help the industry and transportation system and its infrastructure. Typical examples of industrial aspects include: (a) interactive infrastructural developments and maintenance using structural health monitoring (SHM) sensors and associated WSN technologies; (b) interactive traffic management, monitoring, control, and intervention; (c) emergency traffic management for saving lives and liabilities due to natural and man-made disasters; (d) smooth and energy efficient operations of public transportation systems (synchronizing schedules for bus, train, airline services); (e) transportation business management for time sensitive goods and cargo distribution; and (f) overall monitoring, safeguarding, warnings, and alerts for road safety and convenience of public road travelers upon breakdowns, accidents, delays, thefts and road blockage [11].

The alternative solutions means improving existing transport systems. One is trying to reduce the production of CO_2 caused by transportation in a combination of two sides of the problem: (1) at the source–change the fuel source from natural gas and diesel to biofuels and other renewables; (2) encourage users to support the move and make more use of public transport. Figures 11.5 and 11.6 show the potential for enabling reduction of CO_2 and similar gases, measured in g/km.

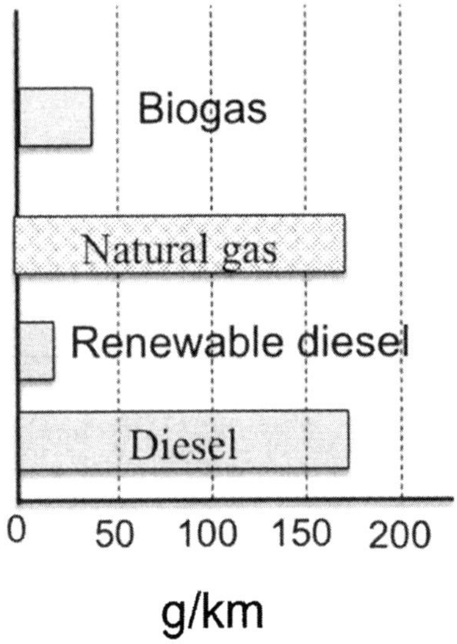

Figure 11.5 GHG (CO2 and similar gases in g/km) produced for 1 liter of four fuel types against almost zero emission electric vehicles, drawn based on [12].

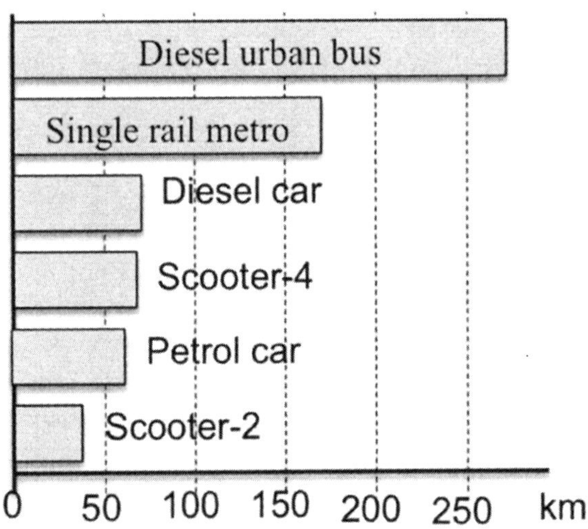

Figure 11.6 The distance of traveling for 1 liter of fuel, drawn based on [12].

The above method deployed by informing the drivers at various points on the road using specially designed displays (e.g., vehicle to roadside infrastructure connection (V2I) or ad hoc style connection through other vehicles as vehicle-to-vehicle (V2V) connections[32]). Figure 11.7 shows the new ITS style messaging against the earlier designed display for Scotland. Abbreviations are: variable message sign (VMS), closed circuit television (CCTV).

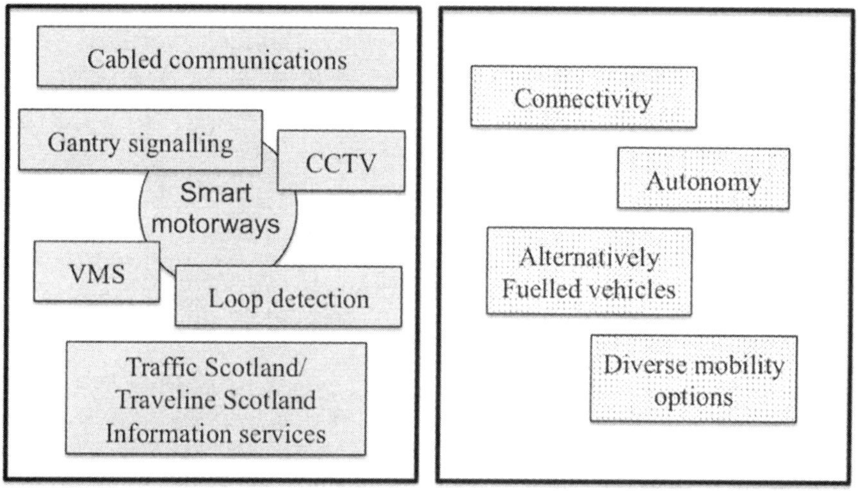

Figure 11.7 Displaying messages templates for Scotland. The complex traditional ITS on the left versus the new style ITS, drawn based on [13].

On the driving side WSN can be used for generating self-driving vehicles. Five levels of automation have been proposed to enable any embedded level of automation from driver assistance to full automation, as shown in Table 11.1.

11.4 IWSN SERVICE DEPLOYMENT

In this section we discuss two groups of WSN service deployments. One group is design and deployment of WSN specific application, where the WSN itself is accountable as a service for gathering information from the environment, sharing it, and interacting with other systems. The other group is being used as an integral part of another system, the host. In the

[32] A nearby vehicle can offer the opportunity for safety reasons to help.

case of serving a host, we need a larger system that requires the WSN services to perform. In most cases these two services are different unless there is no further interest or use for the collected sensing information apart from the host.

Table 11.1

Five Levels of Vehicle Driving Automation [14]

Driving Level	*Steering, Acceleration, and Deceleration*	*Monitoring of Driving*	*Fallback Performance of Dynamic Driving Task*	*System Capability (Driving Modes)*
Driver assistance	Human driver	Human driver	Human driver	A couple driving modes
Partial automation	System driver	Human driver	Human driver	A few driving modes
Conditional automation	System driver	System driver	Human driver	Some driving modes
High automation	System driver	System driver	System driver	Many driving modes
Full automation	System driver	System driver	System driver	All driving modes

Due to the fact that a host based deployment varies heavily with the application and the nature of its objectives we cannot get much involved in any detailed design aspects but instead we can address the generic design aspects of WSN systems for some specific aspects of performance and requirements. Here we look at three main topics of WSN deployment, namely: energy, platforms, and networking.

11.4.1 Energy Aspects of WSN

Energy for a WSN means survival throughout the service, which is also called its lifetime as the source of energy runs out. Due to their nature, wireless sensors originally used batteries to operate. It is essential for smart sensors to have the required energy (usually electrical) to operate as a node of the WSN. In general the communication part of operation takes some 50% of the power with 25% for computing and the remaining 25% for sensing and housekeeping tasks. In extreme environment cases the communication tasks could easily increase to 75% and over.

In the design process we have a wide range of methods to reduce consumption. Energy requirements for occasional communications can be reduced using ad hoc networking. Energy consumption for transmission is normally twice as much as for receiving signals; therefore in good energy design networks the nodes try to listen more than calling out randomly. The WSN low duty nodes normally take long sleeps until they are needed to operate. Therefore, apart from their data transmission periods the nodes have only their receivers active. Beamforming at the cluster level is particularly helpful (see Chapter 6), as it not only reduces the wasteful generic propagation patterns but it directs beams towards the intended receiver upon a designed pattern. This function also helps in reducing interfering energy.

For the communication protocols, some well-designed lightweight algorithms governing communication can also minimize energy consumption without much compromise on other performance aspects of the WSN such as speed, error rates, and risk of missing critical sensing. The protocols themselves require energy of a different form; computing. This can be also simplified through unnecessary generic routines and also taking on techniques such as cross-layer techniques and network coding. The error correction coding techniques, now mature and perfected can be easily adopted to enhance the performance for weaker signals.

It is also known that using low energy components through integrated design approaches can reduce energy wastes that can in turn result in gains by cutting down the device's energy consumption. Dense integration being limited by Moore's law and other practical problems has been progressing continually for the low cost components of COTS systems due to their greater global market worth considered for new designs [15].

On the other hand, we should consider better use of passive and batteryless sensors in the design. These sensors principally work similar to RFID devices, but instead of sending the identification code in a regulated sequence, they measure and send a set of sensing data in a group [16].

Another line of interest for energy aspects of development to address a sensor's energy problems is to supply the required energy through wireless energy transmission from one device to another. Direct transfer or using feeding points are both currently available and practical. The only remaining problem for these methods is the efficiency of the transferring of energy. Then, as another option we have the harvesting and scavenging energy methods to feed the sensors independently using some form of abundantly available sources in the media.

In general energy has been and will remain to be increasing as an important aspect of the industrial process. There are wide ranges of particular cases and each needs to be designed in its own right. An interesting source of design is Ouellette's thesis [4], where a systematic approach for examining so-called energy harvesting paradigms. Figure 11.8 shows a generic model and configuration of the three stages of the energy harvesting process within a smart sensor.

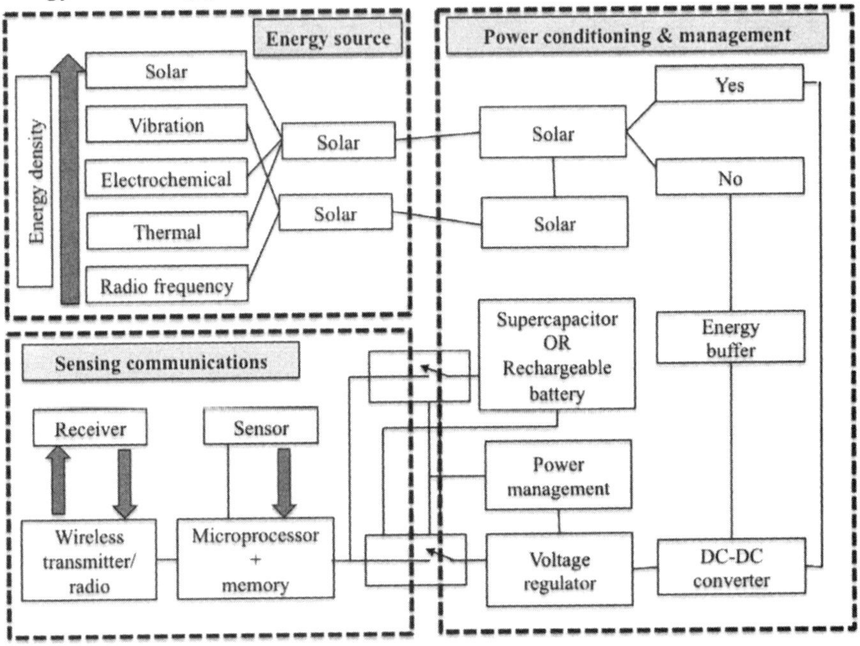

Figure 11.8 A generic model of the energy harvesting process within a smart sensor, drawn based on [4].

In Figure 11.8 we have three sections of the energy (a) source, (b) management, and (c) consumption. The energy source here consists of five options of solar, vibration, electrochemical, thermal, and radio frequency in the order of their energy density. They normally convert the source into either DC or AC based on the existing device technologies. In the energy management section of the smart sensors, we need to check if the amount of energy supplied by the source is sufficient for the existing consumption to supplement as from an internal source or another energy buffer. The use of supercapacitors and rechargeable batteries often becomes essential if the source is not always available or varies significantly throughout time [4].

11.4.2 Platform Deployment

The provision of a platform in conjunction with the WSN technologies can benefit both significantly. The platform could be tiny, small, large, or very large. The smaller ones are normally used as a base or gateway for monitoring, surveillance, and investigating units where, WSN services becomes an integral part of the operation. Due to its domination of the WSN services we call it an *enhanced smart sensor platform* (ESSP). The role of WSN services in large platform applications compromises the three distinct services of (local) housekeeping, *platform information and automation services* (PIAS), *centralized sensing information services* (CSIS) and *global sensing information services* (GSIS).

For a large platform deployment we can discuss the design in a generic format only where, the configuration of the WSN and volume of information being handled can vary extensively from one platform to another. That is, the design optimization process should divide the tasks in specific proportions for PIAS, CSIS, and GSIS as three interrelated projects prior to any detailed design. Figure 11.9 shows a generic process for the integrated WSN design.

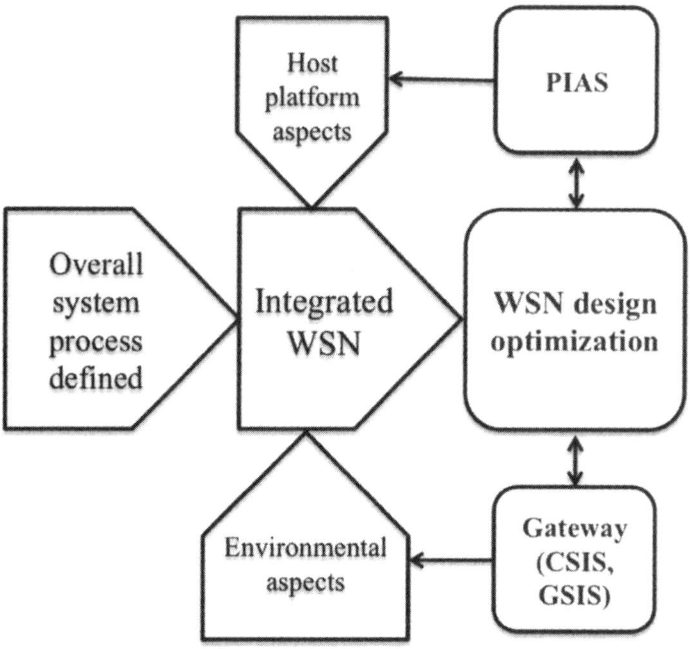

Figure 11.9 Integrated WSN design process for large platforms.

For example in monitoring the railway system various sections of the system can be regarded as a large distributed platform. Then, the information rich (ITS) monitoring system can be deployed to help with maintaining operation of the train service. However, from a WSN adaptation point of view this process can be considered as a distributed gateway system for optimizing the interactions and interworking with other networks and databases. Also as a railway system operating the trains requires a significant level of safety and a significant number and variety of sensors and reliable actuators to ensure its safe future operations with potential benefits including fault detection, smooth running, and building user confidence [17].

11.4.3 Networking Aspects

The networking systems and the complicated standards aspects of the WSN require thorough understanding due to global market obligations and their critical impacts on their success as an industrial process.

This is mainly due to the way our future markets will be operating. However, from the smart sensor design process every WSN deployment requires its own unique scenario dedicated design objectives. We therefore consider the following networking standards for compatibility and associated COTS benefits could be quite limited for most of the extreme environment WSN applications; extra complexity that does not help with the progress. The IoT, however, is moving the complex communication standards towards our WSN style market developments but under its own generalization it seems to be adding more obstacles to the path than helping WSN applications. The industrial aspects of the IoT (discussed in Section 11.3.3) can be helpful at the production level but for designing small and moderate stand alone WSN applications cannot be much help.

11.5 EVALUATION OF INNOVATIVE SOLUTIONS

Our generic proposed solutions for the industrial and transportation aspects of the smart sensors (IWSN), mainly listed in this chapter, are here to guide the engineers and industries to see the bigger picture behind our marvelous technology to flourish with a smooth adoption of the WSN into their own unique technological arena. In order to clarify our WSN development route for further incremental innovations we provide two inter-related tables highlighting the potentials and gaps of WSN developments under the test of extreme environments.

Table 11.2 shows the various innovative aspects of WSN product innovation in known extreme environments. All grading in the columns of Table 11.2 are subjective but can be quantized in four levels of a to d, 'a' for the best and 'd' for the worst. Column 1 indicates the progress (state of the art): a. mature; b. progressive; c. lagging; and d. very low. Column 2 indicates the complexity: a. acceptable; b. moderate; c. high; and d. very high. Column 3 indicates the innovation side effects: a. insignificant; b. minor; c. moderate; and d. high. Column 4 indicates the energy: a. insignificant; b. minimal; c. considerable; and d. demanding.

As shown in Table 11.2 the only mature technology is space. This is mainly due to the fact that space industries operate in a very confined and closed working environment. Their design, development, tests, and deployments are free from any common interference. The most important and specific fact of their success is related to the safety aspects of the flight missions. That is the cost and complexity of adopting wireless smart sensors in comparison to the large and hugely expensive vehicles allows any financial margins to ensure a vehicle's operation using relatively low cost sensors and associated WSN systems. This is confirmed by the second column's high complexity. Then Table 11.2 indicates the lagging progress for underground industries, which is mainly due to common negligence in assessing the hardness of the ground[33]. On the innovation side effects column the grades are too high to be overlooked. Then the last column shows poor grading indicating that we should follow the motto of save energy save the sensors.

Table 11.3 shows the various innovative aspects of the WSN production process in known extreme environments. All grading in the columns of Table 11.3 are also subjective quantized in four levels of a to d, 'a' for the best and 'd' for the worst. Column 1 indicates 'quality of production': a. as required; b. acceptable; c. low; and d. poor. Column 2 indicates 'speed of production': a. good; b. require improvements; c. lagging; and d. unacceptable. Column 3 indicates innovation side effects: a. insignificant; b. minor; c. moderate; and d. high. Column 4 indicates 'compatibility': a. fully compatible; b. mostly compatible; c. some compatibility; and d. not compatible.

---

[33] The thriving large agricultural developments, empowered mostly by chemicals and heavy machinery, cannot be counted for an appropriate use of smart sensors.

As shown in Table 11.3 the space industry stays in good condition due to its full industrial production. Other industries should make an effort to take advantage of unique timely WSN prime business paradigms before it is too late.

Table 11.2

Innovative Aspects of WSN Services for Known Extreme Environments

Environment	State of art	Complexity	Side effects	Energy issue
Underwater	Progressive	High	High	Demanding
Underground	Lagging	Very high	Moderate	Demanding
Industrial	Progressive	Moderate	Acceptable	Minimal
Space	Mature	High	Acceptable	Considerable
Other EEs	Lagging	Very high	Minimal	Depends upon application

Table 11.3

Innovative Aspects of WSN Enabled Production for Some Extreme Environments

Environment	Quality of production	Speed of production	Side effects	Compatibility
Underwater	Low	Require improvements	High	Mostly compatible
Underground	Poor	Unacceptable	High	Some compatibility
Industrial	Acceptable	Acceptable	Moderate	Mostly compatible
Space	As required	Good	Insignificant	Highly compatible
Other EEs	Poor	Unacceptable	High	Some compatibility

11.6 CONCLUDING REMARKS

This chapter discusses practical aspects of innovative WSN-enabled production, products, and service provision for the extreme environments such as underwater, underground, manufacturing, space, and transport systems. In order to optimize the production systems and enable sensor rich service deployment for the extreme environments we analyze interaction of industrial IoT and developments of the IWSN for those environments with some samples of progress to demonstrate each environment's specific demand, progress, and production of WSN solutions.

References

[1] Braun, E., and D. Elliott, "T302 Technology: Innovation: Design Environment and Strategy - Block 6, Global Patterns In Technological Innovation," Open University, 1999 (ISBN 10: 0749274506).

[2] Pant, R. B., "Wireless Sensor Networking with Lab- Scale Intermediate Measurement Node for Extension of WSN Coverage," Master's Thesis 2011, Telemark University College, Norway.

[3] Earthworks,"Troubled Waters: How mine waste dumping is poisoning our oceans, rivers, and lakes?" Earthworks and mining watch Canada, February 2012, https://miningwatch.ca/sites/default/files/Troubled-Waters_Full.pdf, https://earthworks.org/.

[4] Ouellette, S. A., "Energy Harvesting Paradigms for Autonomously-Powered Sensor Networks," Electronic Theses and Dissertations, UC San Diego, 2015.

[5] Sahoo, R., "Application of Sensor in Mining Machinery to Recognise Rock Surfaces," *International Journal of Computer & Communication Technology (IJCCT),* Vol. 2, Is.VI, 2011.

[6] Kanellakis, C., and G. Nikolakopoulos, "Evaluation of Visual Localization Systems in Underground Mining," *IEEE, 24th Mediterranean Conference on Control and Automation (MED),* June 21–24, 2016, Athens, Greece.

[7] Sell, R., and M. Tamre, "Design templates for mobile robot conceptual design," *2007 IEEE/ASME international conference on advanced intelligent mechatronics*, Zurich, 2007, pp. 1–6.

[8] Pippin II, C. E., "Trust and reputation for formation and evolution of multi-robot teams," PhD thesis, Georgia Institute of Technology, December 2013.

[9] Internet of things – Wikipedia (https://en.wikipedia.org/wiki/Internet_of_things).

[10] Vermesan, O. (SINTEF, Norway), and P. Friess (EU, Belgium), "Internet of Things: Converging Technologies for Smart Environments and Integrated Ecosystems," River Publishers, 2013.

[11] Intelligent transportation system – Wikipedia.

[12] Hautala, R. et al., "Smart sustainable mobility," VTT Technical Research Centre of Finland, 2014.

[13] Scotland's Trunk Road and Motorway Network "Future Intelligent Transport Systems Strategy", Transport Scotland, 2017 (https://www.transport.gov.scot/media/40406/its-strategy-2017-final.pdf).

[14] Turnbull, K. F., "Towards Road Transport Automation: Opportunities in Public-Private Collaboration," The National Academies Press, 2015.

[15] Mammela, A., and A. Anttonen, "Why Will Computing Power Need Particular Attention in Future Wireless Devices?" in *IEEE Circuits and Systems Magazine*, Vol.17, No. 1, pp. 12–26, First Quarter 2017.

[16] Abedi, A., 'Wireless sensors without batteries,' *High Frequency Electronics Magazine*, Nov, 22–26, 2012.

[17] Hodge, V. J., S. O'Keefe, M. Weeks, and A. Moulds, "Wireless Sensor Networks for Condition Monitoring in the Railway Industry: A Survey," in *IEEE Transactions on Intelligent Transportation Systems*, Vol. 16, No. 3, pp. 1088–1106, June 2015.

Chapter 12

Space Applications of Wireless Sensor Systems

This chapter presents a comprehensive view of wireless sensor systems applications in space explorations. The NASA Technology Roadmap (shown in Figure 12.1) [1] presents various technology areas to guide technology development for space exploration. Most of these technical areas present opportunities for wireless sensor systems. Utilizing wireless sensor systems not only reduces the design and future modifications costs, but also renders lower weight systems.

12.1 INTRODUCTION

Several space application case scenarios introduced in [2] that may benefit from wireless technology are discussed in more depth in the sequel. We start with transportation means to space (i.e., space vehicles) and consider opportunities for utilizing wireless sensor systems in both manned and unmanned vehicles. We continue with the study of satellites and payloads, which poses a different set of boundary conditions and design requirements. Once a vehicle reaches a new planet, different environmental conditions pose new challenges for surface exploration designs, which are discussed next.

All of the above systems undergo a rigorous set of tests to ensure their reliability. These ground systems may be the low hanging fruits in proving the advantages of wireless sensor systems. Last but not least, deep space exploration and missions with extended durations require design and maintenance of habitat systems, which are also very good candidates for deploying wireless sensor systems. Each use case scenario and its specific boundary conditions relevant to wireless sensor system design are presented next.

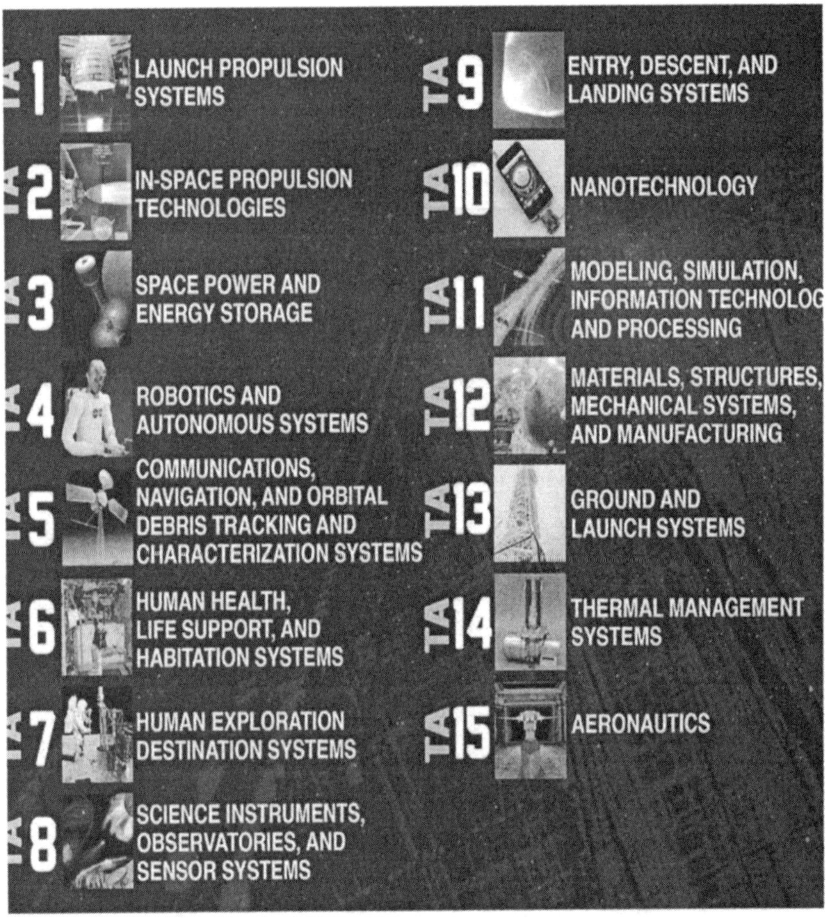

Figure 12.1 NASA Technology Roadmap [1].

12.2 WIRELESS SENSOR SYSTEMS USE CASE SCENARIOS

12.2.1 Space Vehicles

Space vehicles such as the Space Launch System (SLS) shown in Figure 12.2 need to be designed to withstand the harsh space environment. Monitoring critical systems such as thermal and pressure systems, cryogenic fluid management, heating ventilation and air conditioning (HVAC), environmental control and life support system (ECLSS), lighting monitoring, docking, and rendezvous systems requires specific considerations as discussed in this section.

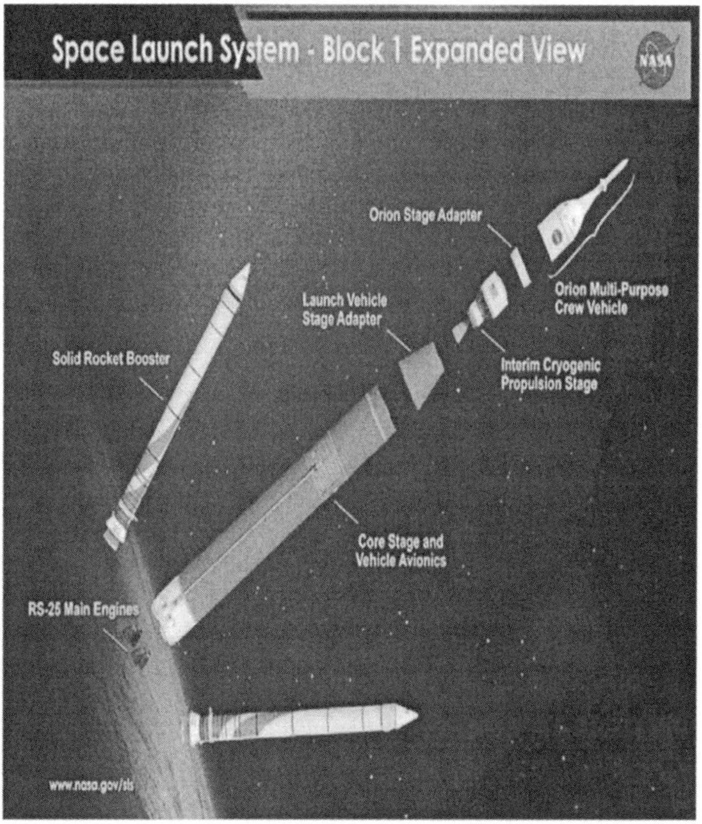

Figure 12.2 Space Launch System (SLS); Next generation of deep space launch vehicles [2].

Most of these systems are in hard to reach areas that are not easily accessible. The wireless system design needs to accommodate operations in confined spaces or enclosed metallic chambers, or inside fluid or gaseous environments at extremely high or low temperatures. Monitoring the engine and heat shield require sensors that can operate in high temperature environments with harsh chemical vapors present. Real time monitoring of pressure and temperature at various locations of the engine can assist in accurate performance measurement and dynamic performance control using sensors and actuator networks.

Although wireless sensor systems present a high potential for benefits, their usage in close proximity of fuel lines and tanks in launch vehicles should be reviewed from a safety point of view to avoid any potential case of radio waves heating up fuel or causing sparks or premature ignitions.

The following challenges are identified when wireless sensors are used in space vehicle applications:

- Wide range of temperature variations;
- Vibration tolerance;
- Accessibility in confined environments;
- Signal propagation in metallic enclosures.

The benefits of wireless sensing systems may overcome these challenges if the system is designed properly. Here are some highlighted benefits in space vehicles:

- Acquiring more data from supporting structures and engine itself;
- Reducing weight due to cable elimination;
- Dynamic performance control with wireless sensing and actuation.

12.2.2 Satellites and Payloads

Satellites and payloads continuously operate in harsh environments unattended with low or no chance of access for tuning or repairs. This calls for constant monitoring and protection against extreme temperature changes, radiation, and even micrometeoroid impacts. At a first glance, it may not seem logical to replace short wires inside a small satellite with wireless radios; however, in some applications, where volume and interior space is limited or when signals are required to get in and out of an enclosed heated or cooled box, wireless sensor systems may become desirable.

Eliminating the need to drill a hole in the payload or satellite's exterior body can save heat and energy to say the least. Transferring power between two disjointed sections may also be another opportunity for using wireless systems. Short-range magnetic coupling is a good candidate for this application.

Figure 12.3 Satellite rotating around earth [3].

The following challenges are identified when wireless sensor systems are used in satellites (e.g., see Figure 12.3) and payloads:

- Wide range of temperature and radiation variations;
- Size and weight limitations;
- Signal propagation in metallic enclosures;
- Power constraints.

Similar to space vehicles, the benefits of wireless sensor systems often overweigh the shortcomings, making them an attractive alternative to wired systems that are currently in use within satellite and payloads. Some of these benefits include the following:

- Wireless connection between two disjoint sections;
- Reducing heat loss by avoiding drilling holes;

- More efficient use of harvested power.

12.2.3 Surface Explorations

Surface exploration may require a plethora of technologies ranging from multi surface traction systems, energy production and storage, computer vision, and robotics. These applications often require navigating, collecting and analyzing samples, and returning preserved samples to base for further analysis.

All these applications can benefit from wireless sensors. For instance, infrared sensors can be used alongside visible light cameras for object detection and classification and assisting the robotics arms to maneuver accurately. Humidity and temperature sensors can be used during ground drilling, while vibration sensors can monitor drill operation. Navigation without global positioning system (GPS) on planetary surfaces requires dedicated active wireless links with precise time of arrival measurements. One such an example is the ultra wide band (UWB) radios developed at NASA JSC to precisely measure location [4].

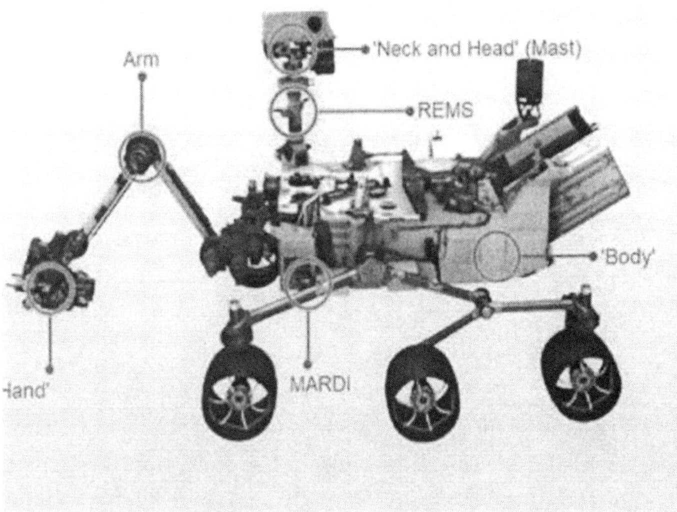

Figure 12.4 Mars rover/science laboratory designed by NASA JPL [5].

Similar to other use case scenarios discussed before, there are challenges that need to be addressed before wireless systems are used in surface explorations (e.g., rovers as shown in Figure 12.4). Here are some of those challenges:

- Capability to operate and survive dust or radiation storm;
- Mobile chemical and biological sensor units;
- Long range reliable link back to base with navigation capabilities.

The added advantages of using wireless sensors in surface explorations are the following:

- In situ testing of samples;
- Navigation without need for GPS;
- Dynamic control of robotic arms using wireless sensors and actuators.

12.2.4 Ground Systems

Ground testing is necessary for verifying the design and ensuring the reliability of space systems. Test setups often require structural sensors such as strain gauges, accelerometers, and deflection sensors as well as leak detection sensors, hydrogen (or other gas) sensors, humidity, and temperature sensors. These tests are usually conducted in a controlled environment with thermal cycles that mimic space conditions. Therefore, all these instrumentation, although used on earth, need this to be capable of operating in harsh environments.

Wiring and cabling is the most costly and time consuming part of these tests. Wireless sensing in ground systems can open up many opportunities to gather more critical data at a much lower cost and project duration since wire diagram design and implementation time is significantly reduced. Figure 12.5 depicts a view of the wiring and cabling for a small portion of the next generation space launch system at the NASA MSFC system integration lab to put the complexity of this process into perspective.

Some challenges in utilizing wireless sensors in ground testing are as follows:

- High precision in sampling and data transfer;
- Interference management for a large number of sensors working within the limitations of the test setup at specific distances.

Figure 12.5 View of cabling and wiring complexity at system integration lab, NASA MSFC.

The following benefits of using wireless sensors in lieu of wired testing systems in ground tests are identified:

- Acquiring more data for structural analysis;
- Reducing the cost of tests due to cabling elimination;
- Flexibility of tests for adding more sensors later without redesigning the whole wiring plan;
- Versatility in programming test beds for future tests.

12.2.5 Habitats

Habitats need to be designed in such a way that they are low in weight, but highly reliable and durable. Impacts include large temperature variations in space, harmful radiation, and even micrometeoroids. Monitoring the structural integrity of the habitats is of prime concern. Various techniques may be used depending on the habitat type. Some designs rely on using lunar or martian soil to build the habitat and only require transportation of supporting structures, while other designs follow inflatable multilayer fabric structures, which are low weight and easy to set up and maintain. An example of inflatable lunar habitat is illustrated in Figure 12.6.

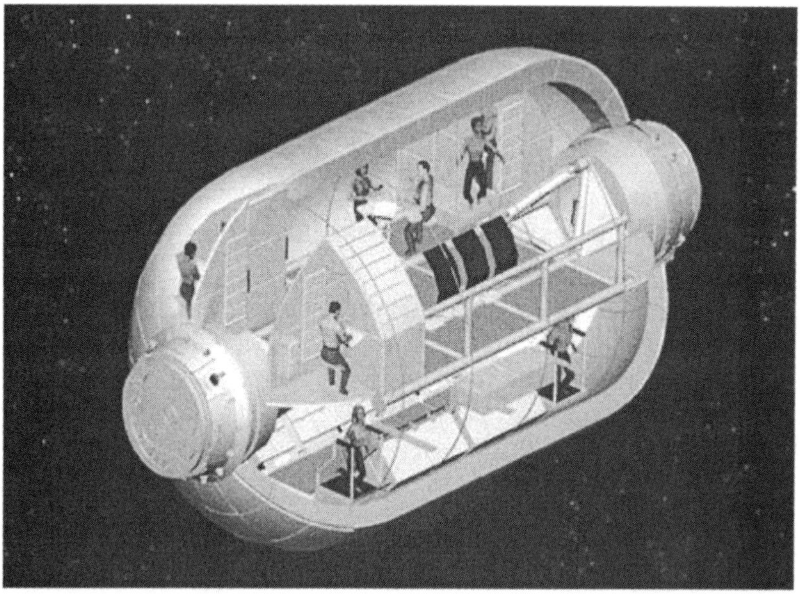

Figure 12.6 Bigelow inflatable space habitat [6].

Monitoring the living conditions inside habitats is another potential application area for using wireless sensor systems. The oxygen level, the presence of harmful chemicals and gases such as CO and CO_2, and the detection of mold and mildew on interior layers with poor air circulation are just a few important monitoring tasks that wireless sensors can easily tackle.

Tracking and finding inventories of goods and items inside a habitat is another cumbersome task if traditional barcode based technology is used. RFID based systems may be a more efficient alternative to the optical systems when it comes to tracking inventory. RFID based inventory-tracking methods for autonomous logistical management (ALM) is being developed at JSC and can tie into habitat monitoring [7].

Integrating sensors as loads on RFID devices and reading the changes in the response in addition to ID numbers is a promising approach in this direction. Monitoring living conditions including physical (temperature, humidity, and radiation), chemical (air and water quality) and biological (mold and mildew or other airborne bacteria) are some critical applications that benefit from wireless sensors. Another important aspect of monitoring habitat systems is evaluating cognitive changes of its inhabitants (i.e., crew health monitoring). Real time vital signs tracking and wireless sensors for sleep behavior monitoring are essential for ensuring mission success.

Some of the major challenges in using wireless sensing systems including RFIDs in habitats are the following:

- Wide range of temperature and radiation variations;
- Aggregation challenges in multimodal sensor data with different sampling rates and precisions;
- Signal propagation in metallic enclosures.

The benefits of wireless sensing in habitats are summarized as follows:

- Acquiring more data from habitat structure;
- Reducing weight due to cable elimination;
- Flexibility of change in design and sensor location after deployment.

12.3 APPLICATION SPECIFIC REQUIREMENTS

In this section, several examples from the NASA technology roadmap with a high potential for deploying wireless sensor systems are presented. We specifically focus on the following technology areas: 2, 3, 5, 6, 7, and 10. Each area contains several subtopics. The examples list the relevant subtopics that can benefit from the use of wireless sensor systems.

12.3.1 Technical Area 2 Subtopic 4.1: Engine Health Monitoring

Definition: "Use of simulation and data processing to determine and mitigate operational, safety, and reliability risks and issues in the propulsion system. In general, the key metrics for health monitoring for in-space propulsion are reliability, weight, and cost."

This technology area is closely connected to the space vehicle use case scenario. A combination of wireless solutions such as RF, optical, and acoustic may be needed to cover all subsystems of a space vehicle in a reliable manner. Most sensors used in these applications such as temperature, pressure, fuel level, and gas sensors produce low data rates with the exception of engine health monitoring, which requires high speed sampling of accelerometers or load sensors. Operation in extremely low or high temperatures may require specially coated materials to withstand those temperatures. An example of an engine with sensors is shown in Figure 12.7.

Video monitoring of environment for short duration events or monitoring vehicles for performance verification is also of interest in space applications. For example, a short video of landers thrusters sent to control station for verification purposes can help in improving design and real time performance adjustments. High data rate WPAN standards such as IEEE 802.15.3 are good standards suitable for these kinds of applications.

12.3.2 Technical Area 3 Subtopic 3.4: Wireless Power Transmission

Definition: "This area describes needed enhancements in short-range, low power wireless power transmission for battery charging and instrumentation and in longer range, high power surface element applications."

Figure 12.7 RS-25 engine powering SLS [8].

Wireless sensor systems in space applications face a major challenge from a power consumption perspective. They have to be either powered with a battery or operate using passive reflective technologies. The former requires frequent battery replacement, while adding to the cost and weight. The latter suffers from the range problem due to the passive nature of responding sensors. Passive sensors also lack memory and processing capabilities of active battery power sensors.

Energy transfer seems like a middle ground, providing remote energy to active sensors with processing power address both range and power consumption issue as well as onboard processing capability.

Design challenges that need to be considered in these systems include the following trade-offs:

- *Sensor sampling frequency*; A low sampling rate conserves power and requires less frequent charging, while a high sampling rate adds more data points and further accuracy.
- *Transmission frequency*; When a wireless channel is in poor condition (low SNR) or the battery power is low, it may be more power efficient to store packets and wait until the next time slot to transmit. Periodic transmission wastes energy if the transmitter has any indication that the packet may arrive at the receiver corrupted (due to low SNR).
- *Charging frequency*; Periodic charging may seem like a simple method with low complexity. However, if the channel conditions are not ideal or the sensor already has enough energy, a more efficient decision-making algorithm may need to be implemented to save the overall network energy and avoid unnecessary transmissions.

One point that requires clarification is that the notion of frequency is used to refer to how often energy or data is transmitted. Another meaning of the frequency is the radio wave frequency measured in hertz, which is discussed in the radio frequency section.

One recent example of wireless energy transfer system optimization is presented in [9]. In this work, NASA MSFC and University of Maine researchers worked on an optimum scheduling method that models data and energy channels separately, but optimizes the transmission or wait decisions jointly based on current channel conditions for both data and energy as well as current battery level.

In this work, a closed form equation for outage probability is calculated for various types of data and energy channel models. Additive white Gaussian noise (AWGN), Rayleigh, and Rician channels models were used for both data and energy channels to cover a variety of applications with a static or mobile charger and/or sensors (a total of 9 cases). Figure 12.8 presents outage probability versus threshold. The threshold effect indicates that as long as we limit transmission in a low quality channel or low battery states, the outage can be controlled under a certain desirable limit. Otherwise, the outage may be too large to allow for a reliable sensing.

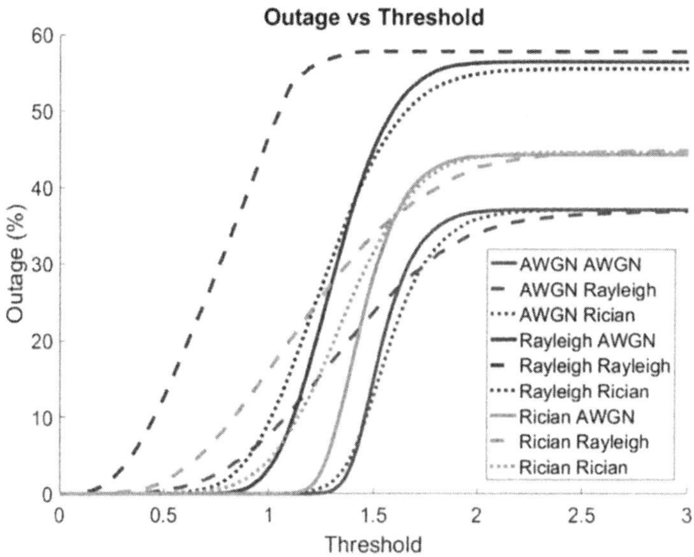

Figure 12.8 Outage probability versus threshold for various wireless channels [9].

12.3.3 Technical Area 5 Subtopic 2: Radio Frequency Communications

Definition: "Radio Frequency (RF) Communications strives to dramatically accelerate techniques in use today for NASA's missions. RF technology development concentrates on getting more productivity out of the constrained spectrum bands that are allocated to space users. Although it is quite a bit more mature than optical communications, there is still a great deal of promise for technology breakthroughs in the RF domain. The focus of RF technology development will be on the RF spectrum allocated and needed for space use by the International Telecommunication Union

(ITU), where adequate bandwidth would provide a useful service, or where the application is beyond the near-Earth environment."

Spectrum is a scarce resource in radio frequency based communications. With the exponential increase in the number of users on earth and demand for higher speeds on their wireless devices, the need for more spectrum efficient communication technologies is felt more than ever. In space applications, the large number of sensors used for monitoring various subsystems of a spacecraft or habitat, while maintaining highly reliable and accurate data links, poses a similar challenge in large scale wireless sensor system deployment. Spectrum efficiency if optimized jointly with power consumption can deliver even more benefits in space applications where power is equally as scarce.

Traditional power and spectrum allocation methods work independently of each other and although each method may be optimum in their domain the overall solution may not be globally optimum. Joint power and spectrum allocation for a large number of wireless sensors with link reliability constraints including specific bit error rate or acceptable throughput level is a complex optimization problem.

A distributed method for rate and power control in CDMA systems was proposed in [10]. This method maximizes the system throughput while keeping the interference to the minimum. Centralized approaches may seem more predictable, but they require a substantial amount of computing power at a central location with many communication links between all sensors and the controller, making distributed systems look more attractive.

Joint power allocation and spectrum sharing in femtocell networks may be achieved using a bi-level game theoretical framework [11]. The data rates in these applications are fixed, while the interference is minimized.

Most current wireless systems operate in half-duplex, meaning that each radio can only receive (T1<t<T2) or transmit (T2<t<T3) at a given time. Recent developments in self-interference cancellation (SIC) methods paved the way for new full-duplex systems where transmission and reception can happen simultaneously (T1<t<T2). This method is almost twice as spectrum efficient as half-duplex methods. However, the major challenge is self-interference caused by high power transmitter signals on the receiver, which is designed for low power signal reception. Interference is avoided in half-duplex systems using a circulator and

dividing the time between transmitter and receiver operations (see Figure 12.9).

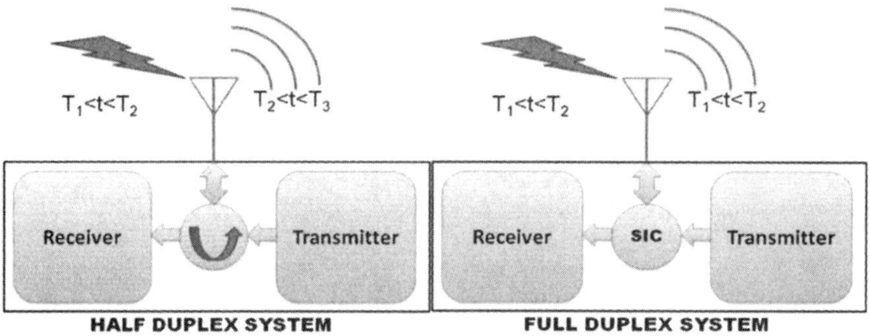

Figure 12.9 Half-duplex with circulator versus full-duplex with self-interference canceller (SIC).

However, in full-duplex systems since both transmitter and receiver operate at the same time, either analog or digital or both methods for self-interference cancellation need to be used. Resource allocation in these systems is a mixed combinatorial and nonconvex optimization problem with high computation complexity that is studied in [12]. Discrete stochastic optimization methods are used to address the problem in full-duplex methods considering the residual self-interference in each cell.

12.3.4 Technical Area 5 Subtopic 5.3: Cognitive Networks

Definition: "Cognitive Networks: Communications system in which each communication node on the network is dynamically aware of the state and configuration of the other nodes to autonomously optimize their operational parameters in response to changes in user needs or environmental conditions."

Space applications of wireless sensors may require two types of critical and noncritical sensors. Allocating dedicated spectrum to critical sensors ensures their operations carried out with desired high reliability. On the other hand, noncritical sensors can use the same spectrum provided when it is not in use by critical sensors. This is similar to the concept of cognitive radio networks, where primary owners of spectrum allow secondary users to transmit when primaries are idle [13]. Power and spectrum allocation discussed in the previous example is even more complicated in cognitive networks. Game theoretical approaches offer a tractable solution to analyze such systems [14].

Assuming two different groups of sensors with critical and noncritical functions, called CS and NS, we can formulate the problem as a time division problem, where a time slot is divided into three partitions called CST, NSR, and NST, where the transmission is managed as shown in Table 12.1. Either a central coordinator or critical sensors may determine the length of CST and NSR and leave the remaining time to NST.

Table 12.1

Cooperation Schedule in Cognitive Sensor Networks

Sensor type / Time slot	**CST**	**NSR**	**NST**
Critical Sensors (CS)	Transmit	Receive	Idle
Noncritical Sensors (NS)	Idle	Relay	Transmit

This combined cooperative and cognitive radio approach may be the key for future implementation of cognitive networks in space. Noting that complete interference alleviation is unavoidable, if high priority traffic from critical users is always prioritized over less sensitive traffic, low priority users can never get any data across. Cooperative relaying is the price that low priority users pay to gain access to the spectrum during a predetermined, but dynamic portion of each time slots.

12.3.5 Technical Area 6 Subtopic 4.1: Environmental Monitoring and Sensors

Definition: "Sensors: Air, Water, Microbial, and Acoustic: The objective of this area is to provide future spacecraft with advanced, networks of integrated sensors to monitor environmental health and accurately determine and control the physical, chemical, and biological environments of crew living areas and their environmental control systems."

Sensors and wireless systems required for environmental monitoring inside the vehicle or habitats can be designed with less stringent requirements as opposed to sensors deployed outside habitat or space vehicles. The interior environment is mainly shielded from extreme thermal fluctuations and radiations, but pose other challenges such as multipath reflections and scattering of wireless signals no matter if they are RF, acoustic, or even optical.

RF signals reflect from metallic interior surfaces of objects creating a highly scattered environment. Appropriate wireless communication techniques such as space-time coding [15–17] and MIMO [18] need to be used to address this challenge. These techniques developed in the past decade are capable of using the multipath signal to their advantage as opposed to traditional methods that treat them as interference.

Acoustic communication can be conducted in two ways: either through solid body material or through air. The former provides means for data or power transfer through metals and is prone to less interference [19]. The latter method needs to deal with environmental scattering and reflections using beam-forming techniques [20].

Optical wireless communications is another means of communicating between sensors and a data processing center if line of sight exists. Note that free space wireless optical communication is different from fiber optics wired systems. High data rates have been achieved in free space optical communications using recent technologies such as LiFi [21–22].

No matter which mode of communication (RF MHz, Acoustic KHz, or optical > THz) is used, a large number of environmental sensors is required to ensure that physical, chemical, and biological parameters are monitored with high accuracy at all times. Air quality sensors such as CO and CO2 as well as water quality sensors such as dissolved oxygen, PH, conductivity, and temperature are few examples.

Biological sensors capable of detecting mold and mildew due to the condensation effect on the interior part of the habitats are also of interest. Biofouling starts with an irreversible attachment of bacteria to the inner surface of the habitat and if not treated properly may lead to a dispersion state causing airborne bacteria to spread inside the habitat or space vehicles.

To avoid this problem, either the interior surface of the pressure-tolerating bladder should be coated to minimize possibility of initial attachment, or proper air circulation system should be designed to keep the surface dry.

12.3.6 Technical Area 7 Subtopic 4.1: Habitat Systems

Definition: "Integrated Habitat Systems: Addresses acoustical treatments and noise reduction; solar optic lighting and heating; low-toxicity, fire-retardant textiles; antimicrobial and surface coatings; and embedded sensors that monitor system performance. Additional dependency technologies that support Integrated Habitat Systems capabilities are being developed under other technology areas, such as TA 4 Robotics and Autonomous Systems, TA 6 Human Health, Life Support, and Habitation Systems, TA 8 Science Instruments, Observatories, and Sensor Systems, TA 11 Modeling, Simulation, Information Technology and Processing, and TA 12 Materials, Structures, Mechanical Systems and Manufacturing."

Habitats may be built from metal, wood, fabric, or other additive manufacturing materials. Each construction method may require different modes of communication, being RF or acoustic or even optical depending on the use case. The interior size and shape of a metal cavity will determine the best frequency to use and what frequencies to avoid.

The inflatable example (built by NASA JSC and instrumented by UMaine) is a versatile test-bed for technology demonstration as well as outfitting including a human factors perspective (see Figure 12.10). It is important that both wireless technology developers and human factor researchers work together when creating monitoring systems for habitats.

Human factor research at NASA entails a variety of studies including psychological and cognitive challenges of working in small confined spaces. These issues need to be studied in an interdisciplinary team of medical doctors, researchers, and engineers to understand problems, collect data, and address the challenges. Wireless sensors that can monitor our daily activities such as movement, sleep, and vital signs can be very useful in these studies.

12.3.7 Technical Area 10 Subtopic 4.1: Sensors and Actuators

Definition: "Sensors and Actuators: Nanotechnology-based sensors include systems for the detection of chemical and biological species to support planetary exploration and astronaut health, in addition to state (temperature, pressure, strain, damage) sensors for use in vehicle health management. Nanotechnology can lead to low-volume, less invasive sensors and actuators with better performance and lower power demand for new designs of morphing vehicle control surfaces, rovers, and robotic systems."

Eliminating wires from sensors' and actuators' networks will certainly present significant savings in weight and cost. However, new challenges such as delay and noise in the actuator control network need to be addressed first. Several applications of sensors and actuators require a feedback control system to act on sensed data in a timely fashion and still remain stable. Wireless feedback control systems with noise and delay in feedback links have been studied recently.

Figure 12.10 NASA's inflatable lunar habitat hosted at UMaine Wireless Sensing Laboratory and instrumented with 172 wireless sensors.

The effect of noise in the wireless channel is modeled using AWGN noise, while delay is assumed to be constant (a function of end to end wireless system delay). Results indicate that new controllers can be designed to operate satisfactorily in these environments as long as the noise and delay are below specific thresholds.

Although the focus of this technical area TA 10.4 is on nanosensors, it is noteworthy to mention that microsensors capable of detecting biochemical species are also a viable candidate to consider for sensor actuator networks. Technologies such as surface acoustic waves with thin film deposited on their exterior surface can be designed to be sensitive to a variety of physical, chemical, or biological substances.

12.4 CONCLUDING REMARKS

Wireless sensing for space applications projects a significant saving in weight and cost of future space missions. The efficiencies are not limited to cost and weight of wires, but rather include cabling, bundling, fixtures, and routing costs as well. Another advantage of wireless sensors is in their added flexibility for future changes or additions. Wired systems require redesigning the routes, new study of electromagnetic compatibility, and implementation cost and time.

One of the common challenges in the design and implementation of sensing systems for field operation (e.g., inside space vehicles or ground test facilities) is the lack of sufficient and ongoing coordination with the crew or actual users of the system. It is very difficult to predict which other tasks are going to require access to various parts of the vehicle or habitat systems interfering with the current wiring or sensor location. With wireless sensors, these challenges are all alleviated since sensor location can be easily changed without the need for rewiring the whole system.

The first step in the design and implementation of a wireless sensor system for space applications is the requirement analysis. Exterior-sensing hardware prone to radiation and high energy particles damage may require special considerations such as use of radiation hardened components and circuit boards. Interior systems, on the other hand, may be designed with lower grade materials without the need for thermal or radiation protection. However, the interior spectrum should be analyzed carefully to avoid unintentional interference. Modern techniques such as cooperative communication and cognitive networking may be used to enable future additions of new sensor systems without the need to change the current critical wireless systems design. The most important consideration before replacing wired sensor systems with wireless sensor networks is the reliability of these systems and selecting applications where reliability, speed, and latency render acceptable performance. Energy cost per transmitted bit is the next notable metric to be considered when it comes to developing new systems for space applications. This chapter attempts to summarize the benefits and challenges of using wireless sensors in space and extreme environments.

References

[1] NASA Technology Roadmaps, www.nasa.gov/offices/oct/home/roadmaps/.

[2] A. Abedi, D. Wilkerson, "Applications of Wireless Technology in Space," NASA Technical Report, August 2016.

[3] Webster dictionary, www.merriam-webster.com/dictionary/satellite/.

[4] J. Ni, et. al., "UWB Tracking System Design for Lunar/Mars Exploration," *IEEE International Conference on Wireless*, March 13–16, 2006, Sydney, Australia.

[5] Mars Science Laboratory, https://mars.nasa.gov/msl/.

[6] Bigelow Inflatable Space Habitat, rocketry.files.wordpress.com/2012/05/bigelow.jpg.

[7] Fink, P., RFID Smart Storage Enclosure, ntrs.nasa.gov/archive/nasa/casi.ntrs.nasa.gov/20160009157.pdf.

[8] RS25 Engine, Picture courtesy of Aerojet-Rocketdyne, https://www.nasa.gov/exploration/systems/sls/rs25-engine-powers-sls.html.

[9] Veilleux, S., A. Almaghasilah, A. Abedi, and D. Wilkerson, "Stochastic modelling of wireless energy transfer," *2017 IEEE International Conference on Wireless for Space and Extreme Environments (WiSEE)*, Montreal, QC, 2017, pp. 153–155.

[10] Berggren, F., and Seong-Lyun Kim, "Energy-efficient control of rate and power in DS-CDMA systems," in *IEEE Transactions on Wireless Communications*, Vol. 3, No. 3, pp. 725–733, May 2004.

[11] Ahmad, I. et al., "Game theoretic approach for joint resource allocation in spectrum sharing femtocell networks," in *Journal of Communications and Networks*, Vol. 16, No. 6, pp. 627–638, December 2014.

[12] Liu, G., F. R. Yu, H. Ji, and V. C. M. Leung, "Energy-Efficient Resource Allocation in Cellular Networks With Shared Full-Duplex Relaying," in *IEEE Transactions on Vehicular Technology*, Vol. 64, No. 8, pp. 3711–3724, August 2015.

[13] Setoodeh, P., and S. Haykin, *Fundamentals of Cognitive Radio*, Wiley, Hoboken, NJ, 2017.

[14] Afghah, F., et al., "A Reputation-based Stackelberg Game Approach for Spectrum Sharing with Cognitive Cooperation," in *Proceedings of 52nd IEEE Conference on Decision and Control*, December 2013, Florence, Italy.

[15] Tarokh, V., N. Seshadri, and A. R. Calderbank, "Space-time codes for high data rate wireless communication: performance criterion and code construction," in *IEEE Transactions on Information Theory*, Vol. 44, No. 2, pp. 744–765, March 1998.

[16] Tarokh, V., H. Jafarkhani, and A. R. Calderbank, "Space-time block codes from orthogonal designs," in *IEEE Transactions on Information Theory*, Vol. 45, No. 5, pp.1456-1467, July 1999.

[17] Tarokh, V., H. Jafarkhani, and A. R. Calderbank, "Space-time block coding for wireless communications: performance results," in *IEEE Journal on Selected Areas in Communications*, Vol. 17, No. 3, pp. 451–460, March 1999.

[18] Zheng, L., and D. N. C. Tse, "Diversity and multiplexing: a fundamental tradeoff in multiple-antenna channels," in *IEEE Transactions on Information Theory*, Vol. 49, No. 5, pp. 1073–1096, May 2003.

[19] Lawry, T. J., et al, "A high-performance ultrasonic system for the simultaneous transmission of data and power through solid metal barriers," in *IEEE Transactions on Ultrasonics, Ferroelectrics, and Frequency Control*, Vol. 60, No. 1, pp. 194–203, January 2013.

[20] Roufarshbaf, H., J. Castro, F. Schwaner, and A. Abedi, "Sub-optimum fast Bayesian techniques for joint leak detection and localisation," in *IET Wireless Sensor Systems*, Vol. 3, No. 3, pp. 239–246, September 2013.

[21] Deng, P., and M. Kavehrad, "Software defined adaptive MIMO visible light communications after an obstruction," *2017 Optical Fiber Communications Conference and Exhibition (OFC)*, Los Angeles, CA, 2017, pp. 1–3.

[22] Haas, H., L. Yin, Y. Wang, and C. Chen, "What is LiFi?," in *Journal of Lightwave Technology*, Vol. 34, No. 6, pp. 1533–1544, March 15, 2016

Chapter 13

WSN Devices and Systems for Extreme Environments

WSN can be regarded as a special distributed system, or a compilation of networkable smart sensors, with some common features and some specialized features integrated as the nodes of the network. Although their application scenarios assume different configurations sensors are usually manufactured in great numbers to facilitate cost effective solutions.

Despite their commonality, due to their wide range of interactive functions (media properties, strengths, and sizes), we have hundreds of sensors that have been made to suit the vast variety of WSN systems working in various *extreme environments* (EEs). That is, for each design we may only need a few types, usually in larger numbers with the right combination to optimize a particular solution.

Therefore, the main objectives of this chapter are: (a) facilitating the understanding of the device end of the solutions provided in the design chapters; (b) reducing duplications and repeated definitions throughout the book; and (c) classifying extreme environment WSN devices and systems in new set of categories to help designers with the commercial off-the-shelf (COTS) parts and standardization.

Due to the wide range, many types, and variety of features of such devices our coverage is by no means exhaustive, and will require continual updating.

13.1 INTRODUCTION

To develop a new classification of sensors we should set out some simple rules and begin by dividing them into groups, with minimum crossover ambiguities. Therefore, we need to separate sensors into smart sensing devices that can configure some basic networking. That is, we assume that swarm style networking and optimization features of large networks do not require the node's autonomy functions to remain within the cluster heads, or above parts of the WSN's hierarchy. This assumption is fair, as smart sensor nodes need to be agile and lightweight for sensing, making basic decisions and performing simple maneuvering actions as required.

This is shown in Figure 13.1 as sensors, the first group representing smart sensor nodes of a WSN solution, in an application scenario. Sensors are discussed in Sections 13.2–13.6. Then we have other devices in Section 13.7 of the chapter followed by the WSN systems, with some typical features to complement those discussed in the other chapters in Section 13.8.

The first part in Figure 13.1, justifies the provision of a dedicated chapter for device classification, which is partially explained earlier. To do this we use a new simple five-digit code for the items appearing in various sections of this chapter to mark the product. For example, 13300 is allocated to underwater WSN applications and the 13500 series is saved for over the ground (avionics, satellite, and outer space) WSN applications. The prefix 13 indicates our coding system and reduces the actual coding space to three digits, sufficient to accommodate the required specifications of the working environment for sensors' and associated devices' environment. [34]

In order to ease the coding classification we start each section of this chapter with the following set of brief data:

- *Group*: Title, appears only if different from the section's subtitle;
- *Code*: Five-digit number as mentioned above;
- *Expected functionality*: Any known specific features;
- *Typical items*: A compact list to give sufficient indication of the items.

[34] Based on their definition most sensors and their associated systems are required to be specifically designed and calibrated to perform work in an extreme environment. Therefore, allocation of a coding system for such devices is regarded fundamental.

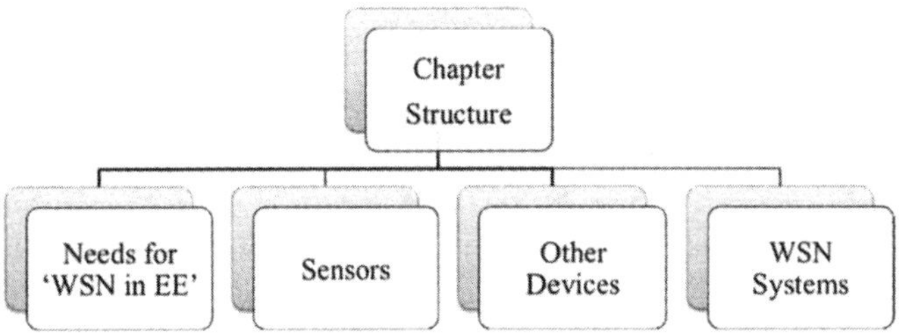

Figure 13.1 Structure of this chapter with respect to classification of sensors and associated devices and systems.

13.2 SENSORS

Generic Sensors (Intended to Perform in EE Smart Sensor Applications)

Code: 13100

Expected functionality: All common sensors that can be used in the extreme environments.

Typical items: List includes temperature, pressure, humidity, gravity, and directional sensors. Most specialized applications being physical, chemical, and biological sensors are not extremely environment-specific though these sensors should work under the particular extreme environment they are designed to work for.

EE Generic Sensors

Code: 13200

Expected functionality: Two groups: (1) enhanced common sensors to function in the extreme environment and (2) those specifically designed for the working extreme environments.

Typical items: See the relevant section for their own specific sensors.

Basically, smart sensors used these days for extreme environment applications carry more functionality weights than their earlier parents, the transducers converting one form of energy to another. For saving energy and cost, sensors usually appear in designs in the form of integrated and multiple sensing, which simplifies the design process. We, therefore, do not provide a *class code* for common sensors such as temperature unless they have been modified or tested to work for the extreme environments.

Figure 13.2 shows our family tree of extreme environment sensors. To help with the familiarization of our classification, the family tree in Figure 13.2 matches this chapter's sectional structure.

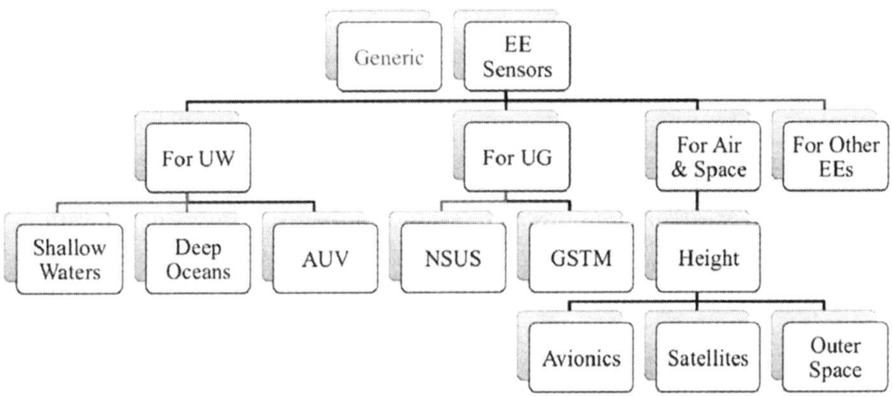

Figure 13.2 Family tree for the extreme environments smart sensors. See previous chapters for EEs.

Generic sensors appearing in the form of smart sensors can be accountable as the nodes of a WSN application, as long as they can perform under the harsh specialized UW working conditions.

Pressure Sensors

There are a wide range of sensors for measuring the pressure apprearing in forms of static and dynamic forces caused due to some differential pressure. Converting seismic waves or acoustic signals into electrical signals, for example, is commonly make use of lead (Pb) zirconate titanate (PZT) in form of flexible ceramic material for its high sensitivity. A pressure sensor unit shown in Figure 13.3 for an easy to produce short cylinder (also called ring) that brings into the vectorial sum of the pressure apprearing on the surface of the cylinder. Figure 13.4 shows an array of ten units for directional sensitivity while Figure 13.5 enhances the preformance for a rapidly changing system used in real-time pressure controlled applications [1, 2].

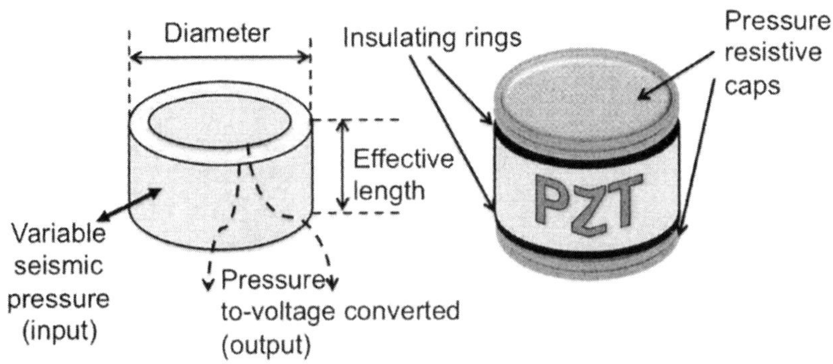

Figure 13.3 Basic PZT pressure sensor produced in the form of ring. Due to the flexibility of their production PZT materials can be formed into any shape to suit the application. In many application cases the cylinder types suit the product best.

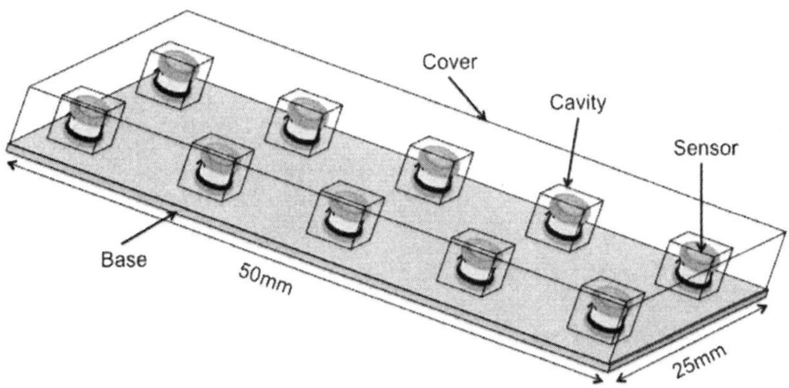

Figure 13.4 A typical medium-size PZT pressure sensor array product, drawn based on [1].

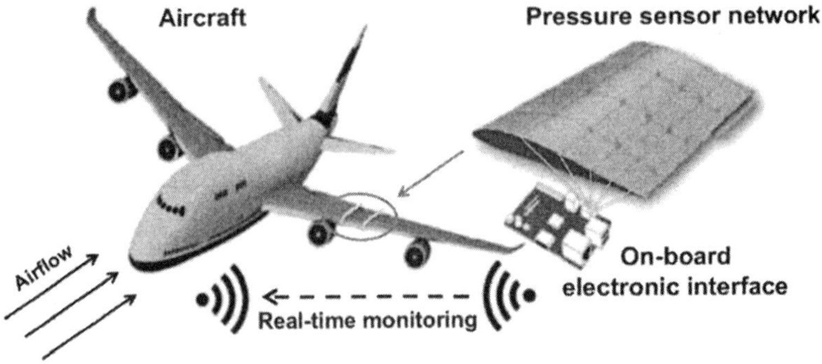

Figure 13.5 The micro electromechanical system (MEMS) technology PZT pressure sensors have been used in various industries, such as aerospace, automotive, and healthcare. The absolute pressure sensing system has the advantages of low cost, compatibility, and flexibility, and can be a core component of sensing in avionics applications, with permission [2].

The process of making such devices, however, is quite complicated for which one needs an instrumental setup, as shown in Figure 13.6 for the silicon and Figure 13.7 for the array.

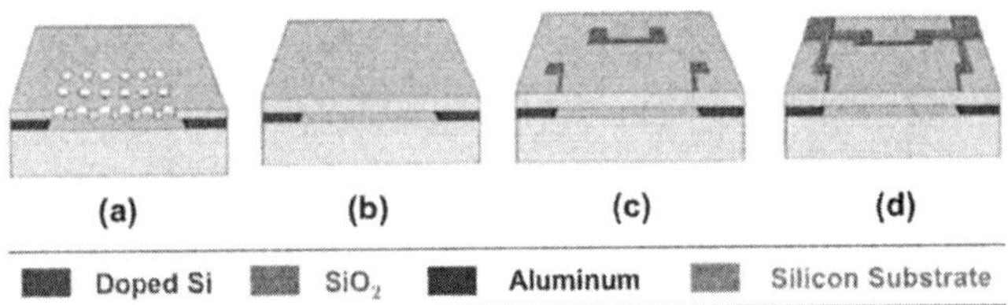

Figure 13.6 The process of preparing right structure for the base. [2].

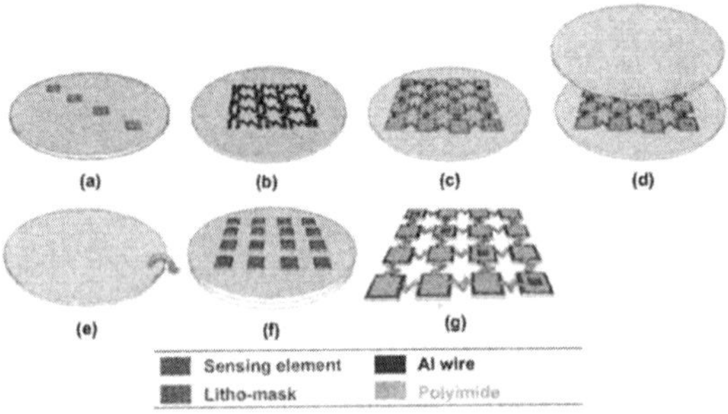

Figure 13.7 The process of constructing the 2D array of pressure sensors. [2].

In practice building models and prototypes to test and measure the device's functionality and performance requires professional designers equipped with the proper tools, often in a laboratory. Figure 13.8 shows a prototype to investigate the device prior to production.

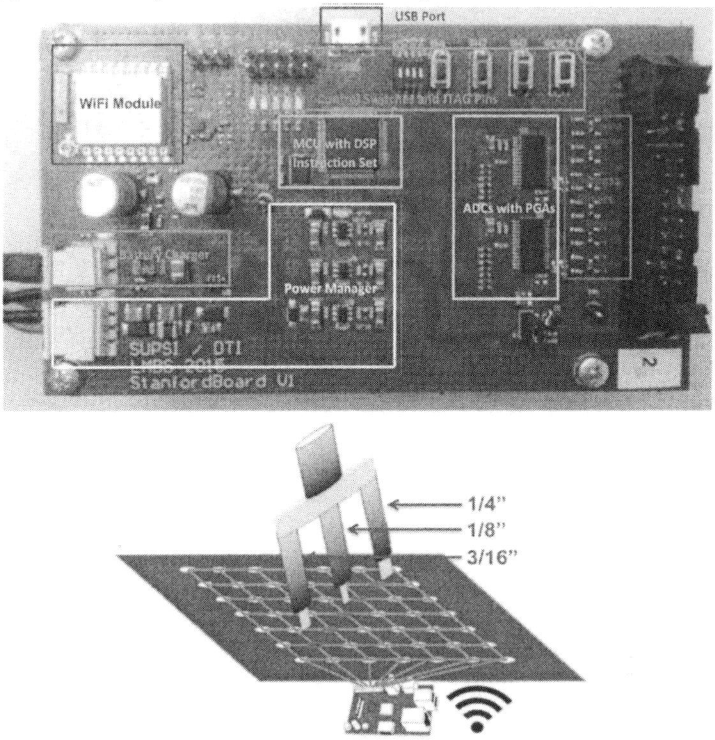

Figure 13.8 A printed circuit style design and the proposed technique used for the measurement [2].

Oil Spill Sensors

With many industries rushing into making the maximum use of low cost natural oil opportunities, we need to stop oil spillage incidents from destroying the environment. The pollution of water is a common case. Some products use xenon beam technology to detect oil.

Figure 13.9 shows a simplified mechanism of filtered beams being detected when positive. Figures 13.10 and 13.11 show industrial and marine applications, respectively. Figures 13.12 shows an oil-loading incident, where a sophisticated alarming system enables further steps for removing the oil.

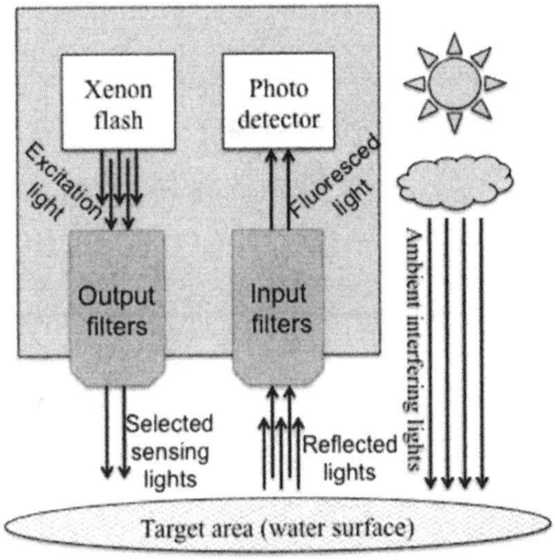

Figure 13.9 Principle structure of xenon light beam sensing method, where the source light beam is filtered first selecting useable wavelength components. Then the reflected beam collected by the receiving filter clears the unwanted spectrums of the reflected beam to the photo detector part to analyze for the existence oil properties, drawn based on [3].

Figure 13.10 An installation of the 'oil detector' in the drainage and outlets of a factory. Courtesy of Slick Sleuth™ [3].

Figure 13.11 Application of above 'oil detector' technology installed on a fuel pier and a marine terminal. Alarm output of the sensor upon monitoring positive sends a SMS messaging to key personnel. Photo courtesy Royal Australian Navy and Courtesy of Slick Sleuth™ [3].

Figure 13.12 Offshore application of 'oil detector' installed on a SBM-type 10m loading buoy. In case of a positive detection spill alarms will be transmitted to shore from six points on series of three buoys. Photo courtesy Chinese Petroleum Corp. and Slick Sleuth™ [3].

13.3 UNDERWATER

Underwater Specialized Sensors

Code: 13300

Expected functionality: Two groups: (1) enhanced common sensors used in the UW environment, which should be protected against the water and its pressure, to work as expected from other sensors or better, and (2) those sensors specifically designed for working underwater as per the required specifications.

Typical items: High pressure, salinity, flow, acoustic sensing modems (acoustic vector sensors and hydrophones).

Hydrophone

A hydrophone is regarded as the receiver of an acoustic propagation; originally they were very large mechanical devices but with piezoelectric they are now much smaller, usually in the form of an array, they convert any change of pressure in the water into useable information processed for acoustic communications. A hydrophone in operation is similar to geophones but is designed to work underwater. These days remote sensing and advanced maintenance for extreme environments such as underwater sensors could be developed with in built communication capabilities to ease up the cost and complexity. For example, the HART-enabled velocity sensor shown in Figure 13.10 is not only immune to the hazards of the working environment but can also be used proactively as part of UWSN systems [4].

Figure 13.13 Samples of specially designed sensors that can be used in hydrophones in the sea, courtesy of Wilcoxon Sensing Technologies [4Wilc].

Vector Sensors

A vector sensor normally is an array of PZT sensors that combines persistent multipath signals to detect and locate the origin of the source. Due to the pressure feature of acoustic signals it is commonly used in underwater situations as shown in Figures 13.14–13.16 [5, 6, 7].

Figures 13.14 shows a typical multipath scenario of three dominating paths for the underwater cases, where with mature beamforming mathematics used for vector processing the receiver can adapt and produce the required data for practical distances. In case of line-of-sight path unavailability, normally replaced by a few other dominating paths.

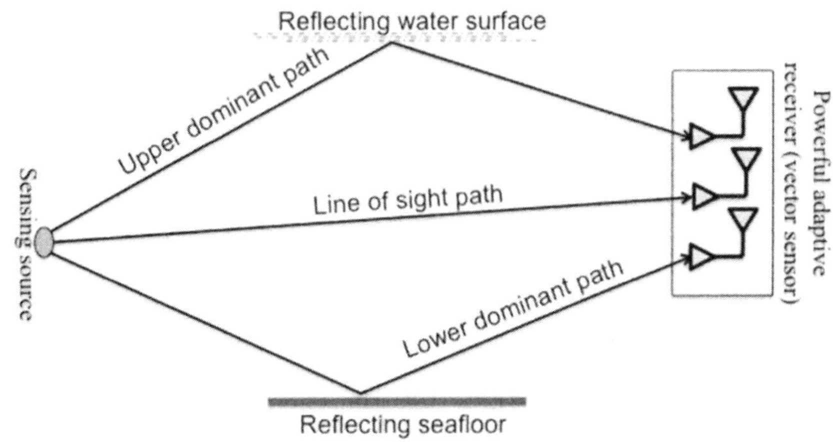

Figure 13.14 Principle of using vector sensor for underwater applications. Here main design task falls within signal processing and effective tuning, which could be helped remotely minimizing the required complexity and associated energy, service, and management requirements.

Due to the length of the acoustic wavelength the size of the classic vector sensors require special design structure. Figure 13.15 shows the schematic of a multi-ring sensor product for higher performance and sensitivity. They are also called circular.

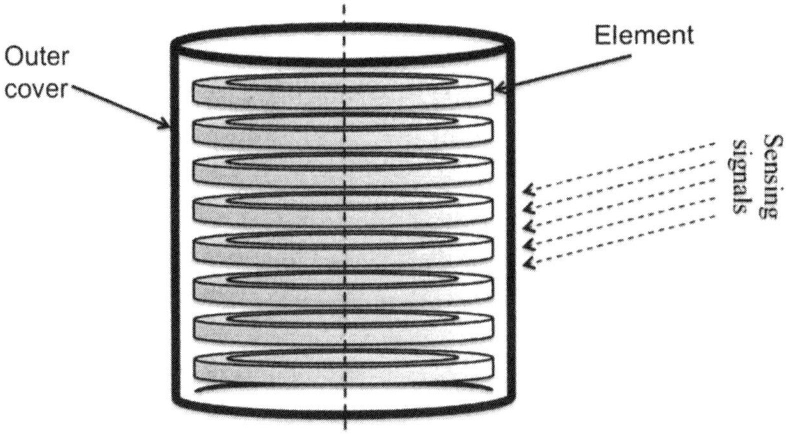

Figure 13.15 A circular acoustic vector sensor for underwater applications, drawn based on [6].

Figure 13.16 A sample of low frequency vector sensor from Wilcoxon Sensing Technologies [4].

13.3.1 Shallow Waters

The use of smart sensors for areas such as shallow waters, coastal, and offshore installations has been increasing for decades.

Buoy as a Sensor Platform

Buoys have been used for safeguarding boats and ships for many years. Nowadays with new low cost sensors they can be used for much greater tasks. They have shown to be very practical and cost effective for connecting UWSNs to terrestrial networks (see Figures 13.17).

Figure 13.17 A buoy installed in the Indian Ocean in cooperation with NOAA. Sensors attached to these buoys are for disaster warnings and climate changes such as temperature and CO_2. But they can be designed for any underwater specific industrial applications [7].

13.3.2 Deep Ocean

The deep ocean and its floor provide the best opportunity for installing and using UG sensors and measurement devices away from possible tempering and interference.

13.3.3 AUV

Autonomous underwater vehicles (AUVs) have been evolving by making better use of smart sensor and array technologies. Using these services they can integrate many new sensing features with mobility as a perfect exploring machine to bring visibility and control to the vast UW environment. Typical features include accelerometers, current profiling, acoustic modem, navigation control, cameras, fluxgate compass, GPS, gyroscopes, magnetic sensors, radio beacon depth meter, sonic ranging and positioning, speedometer, ultrasonic pressure, temperature, sonar and other measuring capabilities.

13.4 UNDERGROUND

Underground Specialized Sensors

Code: 13400

Expected functionality: Two groups: (1) enhanced common sensors used in the UG environment to work for their expected life without any service faults, or operational issues and (2) specialized UG sensors designed to perform under UG conditions.

Typical items: Seismic and vibration sensors, geophones, magnetic and acoustic gradiometers, magnetic (MI) smart sensors, array seismic sensors, power transfer and harvesting devices.

Due to a lack of suitable wideband radio frequency propagation through the ground, various applications of WUSNs such as agriculture, mine industry, and archaeological explorations tend to use new MI devices, which are expected to compete with high frequency seismic technologies. Therefore, identifying the *magnetic* properties of the ground is becoming important.

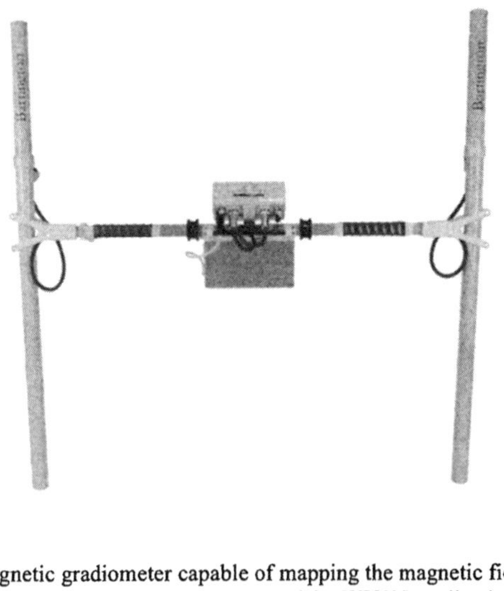

Figure 13.18 A typical magnetic gradiometer capable of mapping the magnetic field, detecting metals, and check on interfering magnetic field for MI devices used for WUSN application (See Chapter 10), "Image © Bartington Instruments Limited, used under licence" [8].

Geophone as a Sensor

Seismic waves carry earthly information through the layers of earth. Their energy is propagated through the ground and water. Geophones were invented to detect these waves with their pressure properties using various techniques. With the advent of PZT now all new products of this kind use this popular technology. The earlier versions made use of a magnet being moved by the seismic or acoustic waves. See Figures 13.19–13.21, [9].

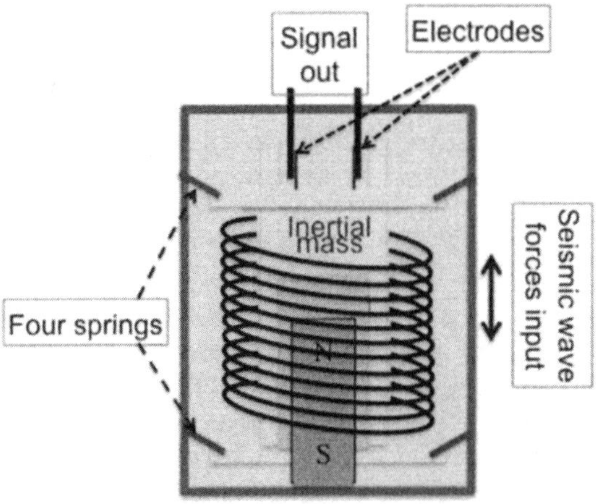

Figure 13.19 Structure of a magnetic geophone, where a spring stabilized magnet moves with the arrival of seismic energy generating electrical signal in proportion to the vibration being inducted to the sensor. To maximize the detection in UG application (e.g., earthquakes) the geophone is rigidly attached to the ground. However, MEMS technology such as PZT is becoming an important part of the industry, drawn based on [9].

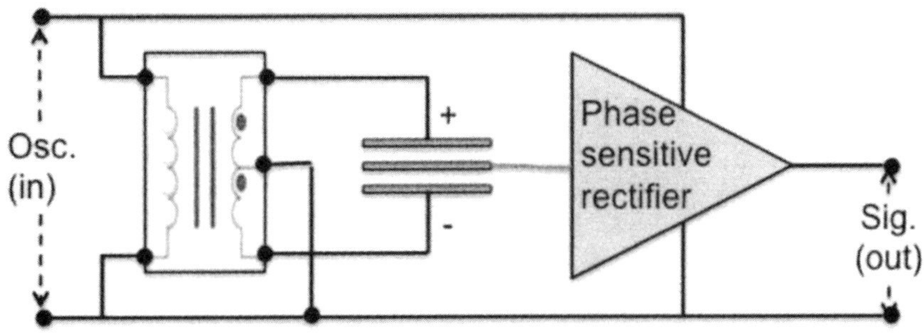

Figure 13.20 Amplification capability of small signal detection is a critical part of sensing seismic waves' strengths. Using a specially designed capacitor can improve geophone sensitivity, drawn based on [9].

13.4.1 NSUS

The use of near surface UG services growing everyday and many are used for profiling the ground for a wide range of applications including mining and agricultural. These properties may save considerable costs if the ground is profiled before any major projects.

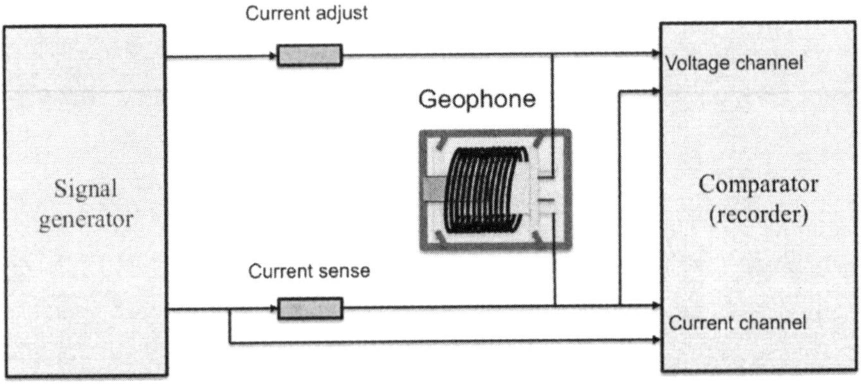

Figure 13.21 Calibrating configuration of a seismic geophone, drawn based on [9].

The use of geophones and wide frequency range accelerometers take their great share of automating WUSN applications throughout the ground and underground industries. Figure 13.22 shows two interesting sensors to be used for new applications and 13.23 showing their potential for a safe mining industry [4, 10].

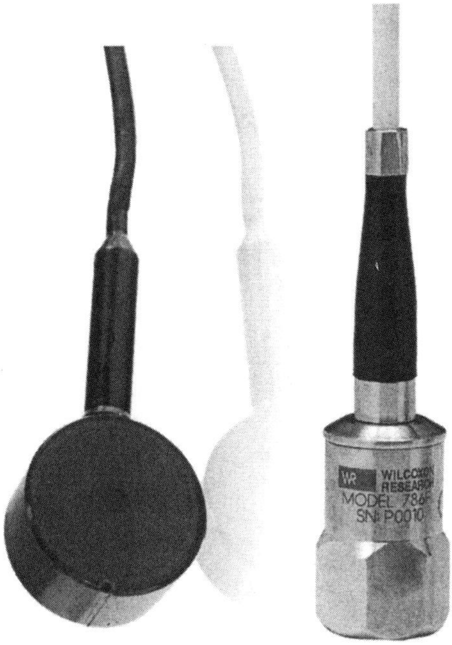

Figure 13.22 Two generic sensors for wide range of underground applications [4].

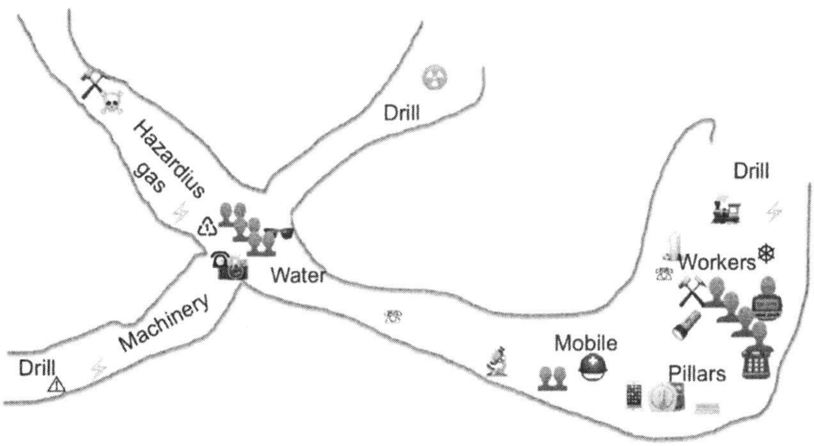

Figure 13.23 Mine as an EE working place–limited movement, hazardous, and dangerous, drawn based on [10].

13.4.2 GSTM

We discuss in Chapter 10 that the source and nature of landslides, earthquakes and associated tsunamis that need to be investigated. There are special sensors to use (see Figures 13.24 and 13.25).

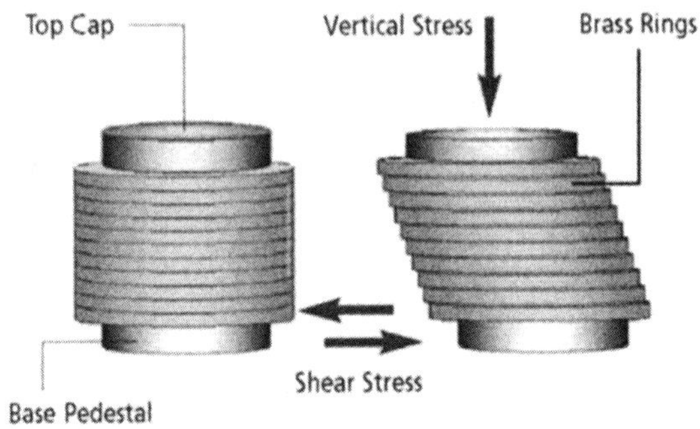

Figure 13.24 Model demonstrating how landslides could happen as a result of earth moving plates in large scale (Pangaea theory) or just soft soil liquefaction in smaller scales, drawn based on [11].

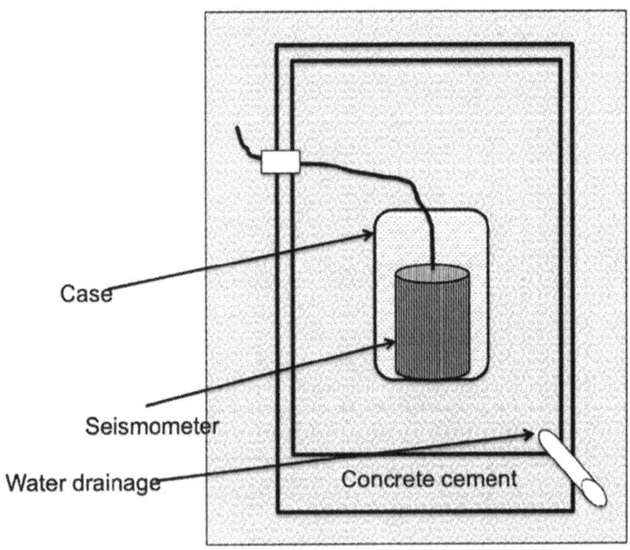

Figure 13.25 Installation of UG seismic sensors, drawn based on [12].

13.5 SPACE

Space Specialized Sensors

Code: 13500

Expected functionality: Two groups: (1) enhanced common sensors useable in the space environment to work as required and (2) space specialized sensors designed to perform under the required working conditions.

Typical items: Lightweight radio frequency sensors and many specially designed high performance TWSN devices with the required capabilities (e.g., battery less sensors).

Although this section is entitled with the word space, we know very well that we also have a growing need for air transportation and satellite technologies providing global communications.

13.5.1 Avionics

This section of the aerospace industry is expected to grow much faster then other industries, and with it there will be a much greater demand for smart sensors for their products (unmanned vehicles, airplanes), and production process. Figure 13.26 shows the predicted trends for various sections of airplanes.

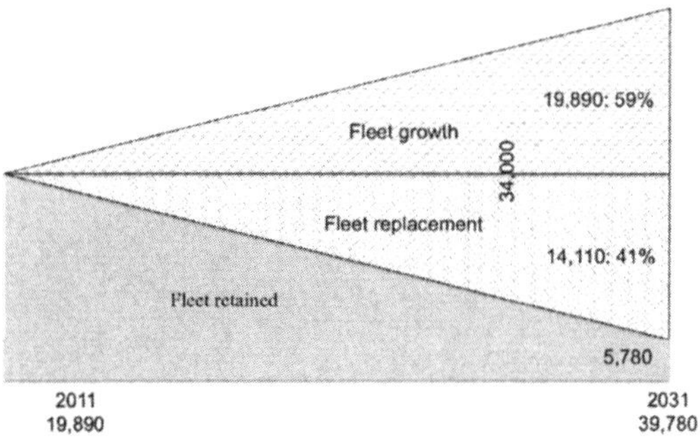

Figure 13.26 Boeing outlook market for 2011-2031 indicates its three parts expected to be 50% new products, 35% enhanced replacement, and 15% of the existing products. That is, 85% innovative production requires extensive use of smart sensors and WSN for their products and production lines. With this comes associated WSN applications including: fire fighting, disaster relief; agricultural services; power and pipeline monitoring, pollution monitoring, atmospheric and oceanographic monitoring, wildlife monitoring and management, maritime and shipping monitoring, traffic monitoring; law enforcement and investigation, surveillance, search and rescue; imagery and mapping, advertising and sky banners, entertainment and news media, science and research, drawn based on [13].

13.5.2 Satellites

Using satellite radars, accountable as sensors, can help us to monitor and act upon problems on land and water. One great benefit comes from locating the ever-increasing oil spills in the seas and oceans, where reflection sensitive radar can spot an incident and inform the right authorities to act (see Figure 13.27) [14].

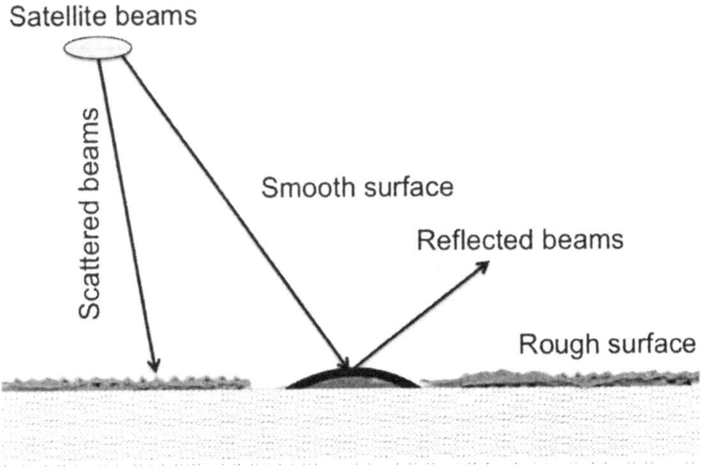

Figure 13.27 The black sea has been counted as one of the seas with very high crude oil traffic, with black spots of oil spillage as shown on the right. Radar sensing method is now available to help with this ever-growing problem, drawn based on [14].

Lower earth orbit (LEO) satellites have been particularly useful for many industrial and exploration services small satellites. Their extremely low cost production and low maintenance make them the most effective members of the satellite systems. They, however, suffer from a lack of a basic source of energy, as they cannot draw this from the sun due to using a lower orbit. Therefore for a full coverage of the earth they need to work in large numbers, or groups, to pass on information and sensor signals from one corner of the earth to another. For implementing this service they normally communicate through ground base stations and make use of DTN protocols.

Historically, the deployment of LEO satellites progressed through eight stages of development, namely: OSCAR-1 (1961), TELESTAR-1 (1962), MAGSAT (1979), UoSAT-OSCAR-9 (1981), UoSAT-3 (1990), Iridium (1998), Earth Observation satellites (2003), and finally Mega-constellations (2017). More specialized brief given in Figure 13.28.

13.5.3 Outer Space

NASA as the leading space agency in the world has been engaged in outer space discoveries, as well as helping innovative technological developments in various aspects of air transportation, health, safety, and humanities (see Figures 13.29–13.32) [15].

LEO Mega and Sparse Constellations

By embracing hundreds and even thousands of satellites, mega-constellations (i.e., O3B, OneWeb, SpaceX's Starlink) are promised to provide highly dynamic and synchronous communications [1]. In other words, an end-to-end connection is possible at all times. As a result, mega-constellations comply with the synchronous principle and provide real-time Internet, voice and video services. However, mega-constellations demand several resourceful satellites with enough power (size, and mass) to sustain several continuous data links simultaneously. Such infrastructure requires tremendous investment only affordable by few institutions in the world, which have been a source of controversy in the past (i.e., Iridium and Globalstar bankruptcy) [5]. Furthermore, mega-constellations are based on closed and proprietary technology, and are still a matter of debate in the community because of space debris [4], frequency spectrum allocation [3], and rocket launchers pollution.

A more democratic type of space-terrestrial networks is emerging by means of the so-called sparse-constellations. Relaxing the synchronous constraint forbids real-time end-to-end communications, but also removes most of the controversies behind mega-constellations. Recent innovative missions based on Commercial off The Shelf (COTS) components on-board of inexpensive cubesat platforms are enabling new sparse-constellation services opportunities (i.e., GomX-4, NASA's EDSN, S-Net, SAS) [2]. Because of their low-cost, easy procuration of components, and absence of continuous coverage requirements, sparse-constellations are allowing more countries, institutions and private companies to access space, so as to develop Earth optical and radar observation, aircraft tracking, remote sensor data collection, among others [5]. Nonetheless, as a few and power-constrained orbiting satellites need to establish opportunistic contacts with neighboring satellites and with ground stations disruption forbids the utilization of Internet protocols and favors asynchronous data transmission approaches, also known as Delay-Tolerant Networking (DTN) [6].

Figure 13.28a The 'LEO..' main text box.

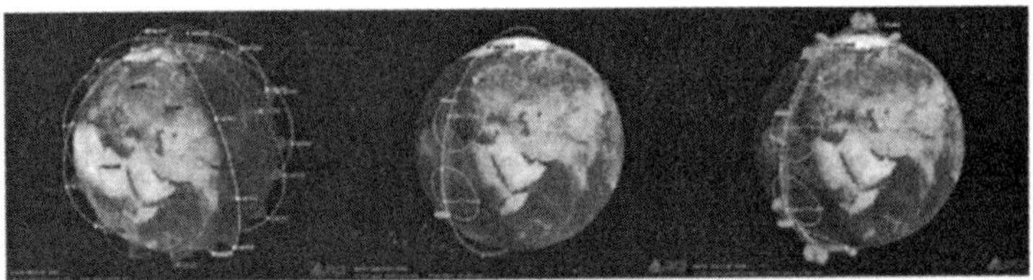

Iridium mega-constellation (left), GOMX-4 cubesats (center), and Ulloriaq sparse-constellation (right)

[1] O. Kodheli, A. Guidotti, and A. Vanelli-Coralli. Integration of satellites in 5g through leo constellations. In *GLOBECOM 2017 - 2017 IEEE Global Communications Conference*, pages 1–6, Dec 2017.

[2] Paul Muri and Janise McNair. A survey of communication sub-systems for intersatellite linked systems and cubesat missions. *Journal of Communications*, 7(4), 2012.

[3] Lai Shun-xiang. Study on the frequency spectrum management model of satellite radio communication systems. In *2010 Asia-Pacific International Symposium on Electromagnetic Compatibility*, pages 76–78, April 2010.

[4] A. H. SÃ¡nchez, T. Soares, and A. Wolahan. Reliability aspects of mega-constellation satellites and their impact on the space debris environment. In *2017 Annual Reliability and Maintainability Symposium (RAMS)*, pages 1–5, Jan 2017.

[5] M. N. Sweeting. Modern small satellites-changing the economics of space. *Proceedings of the IEEE*, 106(3):343–361, March 2018.

[6] J. A. Fraire, S. Burleigh, and J. M. Finochietto. Disruption-tolerant satellite networks. ArtechHouse.

Figure 13.28b The 'LEO..' the references box.

Figure 13.29 Development, design, and launching of shuttles and spacecraft require precision and entirely new and advanced skills in virtually all aspects of engineering innovation. Such significant level of skill has been leading the nation and the world for over fifty decades, courtesy of NASA [15].

Figure 13.30 Triggered by NASA engineering a new family of turbo fan engines have been developed that revolutionized commercial aircraft engine design featuring light weight, quiet, and more fuel efficient, courtesy of NASA [15].

Figure 13.31 Development of safety of long traveling spacecraft passengers has triggered safety and new procedures to save life of the pilots and the passengers in small and large aircrafts. Such a reliable system only works well due to highly reliable smart sensors, courtesy of NASA [15].

Figure 13.32 Development of specially air-conditioned clothing for worker in the EEs has been advancing technology for better health under harsh working conditions, courtesy of NASA [15].

13.6 OTHER EXTREME ENVIRONMENT SENSORS

Other Specialized Sensors

Code: 13600

Expected functionality: All three main groups of industrial, transportation, and medical applications of WSN need specially featured sensors.

Typical items: All three main groups of industrial, transportation, and medical applications have special needs. A wide range of solutions for IWSN and other environments is becoming available.

Industrial sensors and other specialized devices are changing rapidly enabling new industries to thrive once again. Figure 13.33 shows the common trends for the growth of sensors. Wireless sensors are catching up and are expected to exceed their wired counterparts [16].

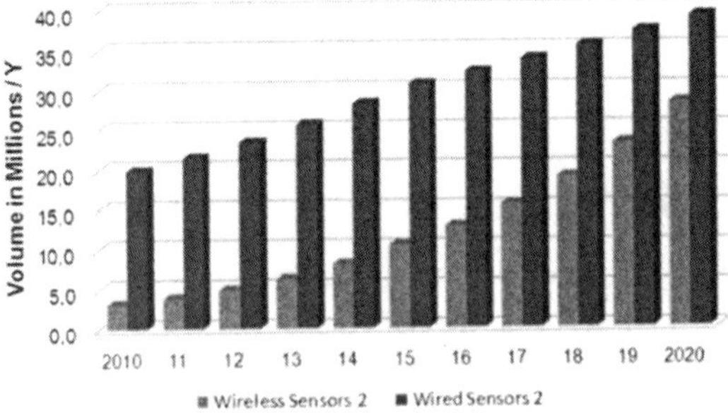

Figure 13.33 Trends for wired and wireless sensor markets for industrial applications [16].

13.7 OTHER DEVICES

Other Devices for EE

Code: 13700

Expected functionality: Although they sound complementary, in general WSN applications cannot work with sensors only; they need many other devices and equipment to enable the deployment of WSN in the extreme environments.

Typical items: A wide range of solutions for IWSN and other environments is becoming available. Examples include signal generators, beacons, accelerometers, actuators, RFID, microcontrollers, and industrial IoT devices.

Pipelines have always been of great help to safety and hygiene in both domestic and industrial environments. Despite their progressive improvements and resistance against aging and corrosions their bursts or leaks cause serious problems. Using robotic style devices running up and down the pipes sounds like a great idea. Figure 13.34 shows three common conventional methods. EEs are especially eager to use these for their own convenience.

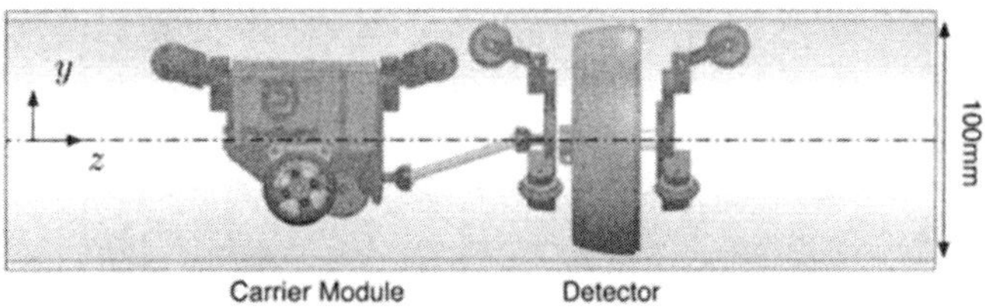

[Top]

[Bottom]

Figure 13.34 A robotic pipeline cleaner [17].

Energy has been and will continue to be the critical success or failure factor for industries. Energy harvesting is far from being matured especially for the EEs using WSN and I^{2}oT technologies. Figure 13.35 shows three common conventional methods. EEs are especially eager to use for their own convenience.

Figure 13.35 Three progressive energy-harvesting methods common to EEs [16].

On the contrary, industries are prone to pollution; air pollution is particularly annoying and often dangerous. Industrial sensors and other specialized devices are changing rapidly and are enabling new industries to thrive once again. Figure 13.36 shows three main sources of industrial pollution.

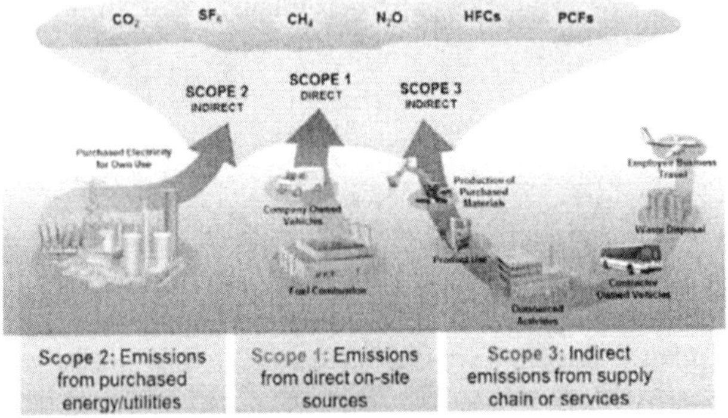

Figure 13.36 Three main sources of industrial pollution [16].

Transportation has been a source of great debate with many proposed solutions. Transport industries are prone to pollution, and air pollution is particularly annoying and often dangerous. New ideas for enabling people to travel could help; e.g., combined transportation. Figure 13.37 shows the ticketing part of the process.

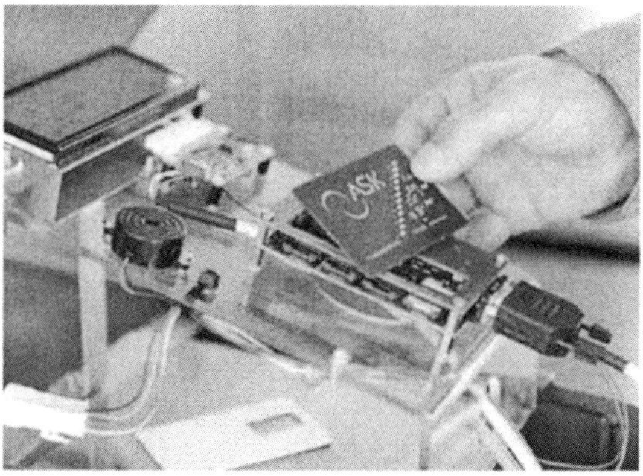

Figure 13.37 Ticketing for combined routes may solve the problem [16].

13.8 WSN SYSTEMS

WSN Systems for EE

Code: 13800

Expected functionality: As we have discussed in various parts of this book we tend to use the term 'system' instead of 'network' (WSS) due to the practical difficulties and excessive cost. EE applications cannot grow too large and are often made simple using agile and lightweight algorithms. Therefore niche markets and specialized products and applications are expected to capture their views before any large-scale global application paradigms. This lengthy slope of development will bring two important design processes to the top of new projects: (1) application specific developments to create a global public view of what new opportunities extreme environments WSN products can initiate, followed by (2) software enabled modular design processes to integrate the growing scattered sporadic developments together into valuable service provisions. In order to speed up the process a new family of processor systems based on the mote idea, but specialized for extreme environments, should be addressed.

Typical items: A wide range of applications and service solutions from small-scale IoT enabled scanning to tomography and monitoring applications growing into global-scale services to control global warming, earth crust level modeling (GSTM) and global industrial cooperation under the IWSN.

Modular Design as Design System

The modular design concept has its own unique design features of flexibility and rapid prototyping, which suits many products with long development periods. This concept can be used for the integration of two different systems as it happens frequently in industry. However, it cannot survive a competitive environment if the process becomes very slow and resource hungry, especially if we take it through the various stages of the intermediary processes (see Figure 13.38).

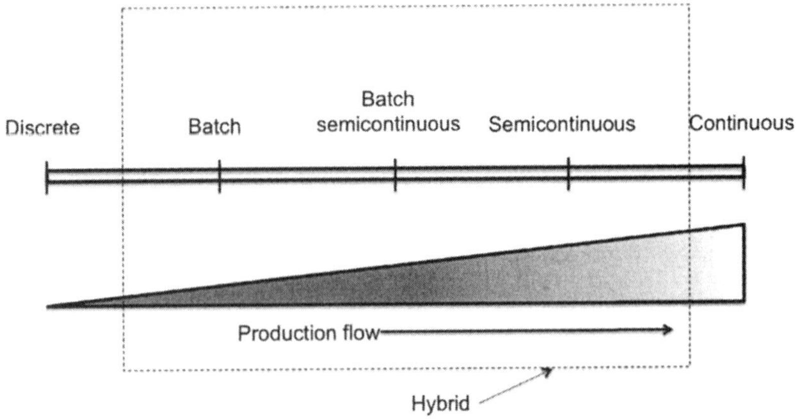

Figure 13.38 Stages of integration required for a hybrid system to take through different continuity flows, drawn based on [18].

13.9 CONCLUDING REMARKS

Scanning through sample sensors, products, applications, and services of networkable smart sensors systems (WSNs) for extreme environments we have shown four essential design aspects: (a) grouping and coding the devices that will help with new standards toward the generation of extreme environments sensors and devices; such a grouping would help industries to develop functional sensors and devices at a far lower cost and complexity; (b) due to little commonality in WSN solutions for various extreme environments the effort of development would face a heavy start emerging only through dedicated projects; (c) the existing standard processing systems (e.g., mote family) could only help with initial extreme environment services but due to their TWSN design nature cannot be regarded as optimized for all extreme environment applications and so there is an urgent requirement for a new family of agile and lightweight core processors designed to integrate the old industrial standards with the new functional requirements of the extreme environment (e.g., a core lightweight low energy base processor adoptable to the extreme environment family); (d) with the modular system design in mind the required software and middleware algorithms can produce flexible, programmable, and optimized solutions for all upcoming WSN-EE systems and services.

References

[1] Asadnia, M., et al., "Flexible and Surface-Mountable Piezoelectric Sensor Arrays for Underwater Sensing in Marine Vehicles," " in *IEEE Sensors Journal*, Vol. 13, No. 10, pp. 3918–3925, October 2013.

[2] Guo, Y., et al., ."Stand-Alone Stretchable Absolute Pressure Sensing System for Industrial Applications," in *IEEE Transactions on Industrial Electronics*, Vol. 64, No.11, pp. 8739–8746, November 2017.

[3] Chase, C.R., and L. G. Roberts, "Advanced Detection Technology For Early Warning – The Key To Oil Spill Prevention," *International Oil Spill Conference*, May 2008 – Savannah, Georgia, USA, www.SlickSleuth.com.

[4] Wilcoxon Research, www.wilcoxon.com.

[5] He, J., "Joint Space-Time Parameter Estimation for Underwater Communication Channels with Velocity Vector Sensor Arrays," in *IEEE Transactions on Wireless Communications*, Vol. 11, No. 11, pp. 3869–3877, November 2012.

[6] Zou, N., and A. Nehorai, "Circular Acoustic Vector-Sensor Array for Mode Beamforming," in *IEEE Transactions on Signal Processing*, Vol. 57, No. 8, pp. 3041–3052, August 2009.

[7] NOAA Research News, "NOAA teams up with India to strengthen ocean observations," Tuesday, June 12, 2018, (https://research.noaa.gov/article/ArtMID/587/ArticleID/2363/NOAA-teams-up-with-India-to-streng.

[8] Grad601 Gradiometer System, Bartington™ (www.bartington.com).

[9] Discovery of Sound in the Sea, https://dosits.org.

[10] Yarkan, S., et al., "Underground Mine Communications: A Survey," in *IEEE Communications Surveys & Tutorials*, Vol. 11, No. 3, pp. 125–142, 3rd Quarter 2009.

[11] Committee on State of the Art and Practice in Earthquake Induced Soil Liquefaction Assessment, NAP Report, "State of the Art and Practice in the Assessment of Earthquake-Induced Soil Liquefaction and Its Consequences," http://nap.edu/23474.

[12] Bormann, P. (Ed), "New Manual of Seismological Observatory Practice (2002)," revised version, electronically published 2009, ftp://ftp.ingv.it/pub/elisabetta.danastasio/grot/manual_seismological_observatory-2002.pdf.

[13] Materna, R., and F. W. Deck, "Aerospace Industry Report," Third Edition, Facts, Figures & Outlook for the Aviation and Aerospace Manufacturing Industry, Aerospace Industries Association of America, Inc. 2013.

[14] Guido Ferraro and MEASURE team, "Monitoring sea-based oil pollution in the Black Sea: JRC activities," Joint Research Centre (JRC), https://www.osce.org/eea/32917?download=true.

[15] NASA SPINOFF, "Technology Transfer Program", 2015 https://www.nasa.gov.

[16] IEC, Internet of Things: Wireless Sensor Networks, white paper, www.iec.ch/whitepaper /pdf/iecWP-internetofthings-LR-en.pdf.

[17] Chatzigeorgiou, D., et al., "Design of a Novel In-Pipe Reliable Leak Detector," *IEEE/ASME Transactions on Mechatronics*, Vol. 20, No. 2, April 2015, pp. 824–833.

[18] Yang, C.H., and V., Vyatkin, "Design and validation of distributed control with decentralized intelligence in process industries: A survey," 2008 *6th IEEE International Conference on Industrial Informatics*, Daejeon, 2008, pp. 1395–1400.

About the Authors

Habib F. Rashvand received his PhD (Distinction) in electronics from University of Kent at Canterbury in 1981. He joined the University of Zambia as head of communications, School of Engineering in 1981. He joined Racal Data Groups as head of Advanced Developments, Reading in 1983. He joined Cable & Wireless Plc. for a joint program with Coventry University as an academic advisor, London in 1990. He joined University of Magdeburg as European Professor of Data Communication in 2001 and moved to University of Warwick in 2004. Professor Rashvand extracurricula activities include teaching security techniques at Portsmouth University, research association with Southampton University, teaching innovation design at Open University, and chief editor and advisor for IET research Journals (COM, IFS, NET, WSS). His keynotes and invited speeches include: "Telecentres & Telemedicine;" "Internet Traffic Engineering;" "WiMax, Cybercity & Next Generation Networks;" "Innovation for Engineers;" "Network Security;" "Wireless Communications;" "Wireless Sensor Systems;" "ICT, Innovation, & Technological Revolutions;" "Distributed Intelligence;" "Ubiquitous Overlay Technologies;" "DSS & IoT;" "Multi Mobile Robotic Platforms;" "Smartphone Intelligent Applications;" "Ad Hoc Communication for High Speed Trains;" "Distributed Intelligence & Agent Technology;" "Mobile Technology for Monitoring Health;" "Distributed Intelligence & MAS;" "Overlay Communication Infrastructure for Heterogeneous Multi Agents;" "Role of Computer in Third Technological Revolution;" "Wireless Distributed Sensing Development and Prospects;" "Promises of Wireless Communication Networking;" and "Wireless Communication Technology Research Prospective." His technical books are: *Distributed Sensor Systems* (Wiley 2012), *Cross-Layer Communications* (IGI Global 2012), *Dynamic Ad Hoc Networks* (IET 2013), *Wireless Sensor Systems for*

Extreme Environments (Wiley 2017). As Director of Advanced Communication Systems involving global academics and innovative industries, he helps to build a sustainable global village.

Ali Abedi received his PhD in electrical and computer engineering from the University of Waterloo in 2004. He joined the University of Maine, Orono, in 2005, where he is currently Professor of Electrical and Computer Engineering, Cooperating Professor of Computing and Information Sciences, and Assistant Vice President for Research. He was an adjunct professor at Queens University, Kingston, Ontario (2004), a visiting associate professor at the University of Maryland, College Park, MD (2012), guest researcher at NIST (2012), and faculty fellow at NASA (2016). Dr. Abedi served as principal investigator on several NASA, NSF, and NIH funded projects, including wireless sensing of lunar habitats and leak detection for the International Space Station. Dr. Abedi has received a number of awards and recognition from the Natural Sciences and Engineering Research Council of Canada (NSERC), Japan Society for the Promotion of Science (JSPS), Canadian Space Agency (CSA), NASA, and IEEE. He has published over 100 articles in IEEE journals and conferences, including several books and patents. Dr. Abedi is a senior member of IEEE and has served on several IEEE committees at local, regional, national, and international levels as well as on organizing committees of several IEEE International conferences and editorial boards of IEEE, KICS, and IET journals. He is cofounder of two startup companies and coinventor of Wireless Sensors for Brain Injury Detection.

Index

Recent Titles in the Artech House Integrated Microsystems Series

Acoustic Wave and Electromechanical Resonators: Concept to Key Applications, Humberto Campanella

Adaptive Cooling of Integrated Circuits Using Digital Microfluidics, Philip Y. Paik, Krishnendu Chakrabarty, and Vamsee K. Pamula

Applications of Energy Harvesting Technologies in Buildings, Joseph W. Matiko and Stephen P. Beeby

Cost-Driven Design of Smart Microsystems, Michael Niedermayer

Design Solutions for Wireless Sensor Networks in Extreme Environments, Habib F. Rashvand and Ali Abedi

Fundamentals and Applications of Microfluidics, Second Edition, Nam-Trung Nguyen and Steven T. Wereley

Hermeticity Testing of MEMS and Microelectronic Packages, Suzanne Costello and Marc P. Y. Desmulliez

Highly Integrated Microfluidics Design, Dan E. Angelescu

Integrated Interconnect Technologies for 3D Nanoelectronic Systems, Muhannad S. Bakir and James D. Meindl, editors

Introduction to Microelectromechanical (MEM) Microwave Systems, Héctor J. De Los Santos

An Introduction to Microelectromechanical Systems Engineering, Nadim Maluf

Lab-on-a-Chip, Techniques, Circuits, and Biomedical Applications, Yehya H. Ghallab and Wael Badawy

MEMS Mechanical Sensors, Stephen Beeby et al.

Micro and Nano Manipulations for Biomedical Applications, Tachung C. Yihllie Talpasanu

Microfabrication for Microfluidics, Sang-Joon John Lee and Narayan Sundararajan

Microfluidics for Biotechnology, Second Edition, Jean Berthier and Pascal Silberzan

Organic and Inorganic Nanostructures, Alexei Nabok

Post-Processing Techniques for Integrated MEMS, Sherif Sedky

Pressure-Driven Microfluidics, Václav Tesař

RFID Design Principles, Second Edition, Harvey Lehpamer

RFID-Enabled Sensor Design and Applicatons, Amin Rida, Li Yang, and Manos Tentzeris

RF MEMS Circuit Design for Wireless Communications, Héctor J. De Los Santos

Understanding Smart Sensors, Third Edition, Randy Frank

Wafer-Level Testing and Test During Burn-in for Integrated Circuits, Sudarshan Bahukudumbi Krishnendu Chakrabarty

A Whole-System Approach to High-Performance Green Buildings, David Strong and Victoria Burrows

Wireless Sensor Network, Nirupama Bulusu and Sanjay Jha

For further information on these and other Artech House titles, including previously considered out-of-print books now available through our In-Print-Forever® (IPF®) program, contact:

Artech House
685 Canton Street
Norwood, MA 02062
Phone: 781-769-9750
Fax: 781-769-6334
e-mail: artech@artechhouse.com

Artech House
16 Sussex Street
London SW1V 4RW UK
Phone: +44 (0)20 7596-8750
Fax: +44 (0)20 7630-0166
e-mail: artech-uk@artechhouse.com

Find us on the World Wide Web at: www.artechhouse.com